Interstellar Flow and Star Formation

Online at: https://doi.org/10.1088/2514-3433/adcb28

AMERICAN ASTRONOMICAL SOCIETY **IOP** | **ebooks**™

About the program:

AAS-IOP Astronomy ebooks is the official book program of the American Astronomical Society (AAS) and aims to share in depth the most fascinating areas of astronomy, astrophysics, solar physics and planetary science. The program includes publications in the following topics:

- GALAXIES AND COSMOLOGY
- INTERSTELLAR MATTER AND THE LOCAL UNIVERSE
- STARS AND STELLAR PHYSICS
- EDUCATION, OUTREACH, AND HERITAGE
- HIGH-ENERGY PHENOMENA AND FUNDAMENTAL PHYSICS
- THE SUN AND THE HELIOSPHERE
- THE SOLAR SYSTEM, EXOPLANETS, AND ASTROBIOLOGY
- LABORATORY ASTROPHYSICS, INSTRUMENTATION, SOFTWARE, AND DATA

Books in the program range in level from short introductory texts on fast-moving areas, graduate and upper-level undergraduate textbooks, research monographs, and practical handbooks.

For a complete list of published and forthcoming titles, please visit iopscience.org/books/aas.

About the American Astronomical Society

The American Astronomical Society (aas.org), established 1899, is the major organization of professional astronomers in North America. The membership (~7,000) also includes physicists, mathematicians, geologists, engineers, and others whose research interests lie within the broad spectrum of subjects now comprising the contemporary astronomical sciences. The mission of the Society is to enhance and share humanity's scientific understanding of the universe.

Interstellar Flow and Star Formation

Enrique Vázquez-Semadeni
Instituto de Radioastronomía y Astrofísica,
Universidad Nacional Autónoma de México, Campus Morelia, México City, México

IOP Publishing, Bristol, UK

ISBN 978-0-7503-1290-5 (ebook)
ISBN 978-0-7503-1291-2 (print)
ISBN 978-0-7503-1834-1 (myPrint)
ISBN 978-0-7503-1292-9 (mobi)

DOI 10.1088/2514-3433/adcb28

Multimedia content is available for this book from https://doi.org/10.1088/2514-3433/adcb28.

Version: 20250901

AAS–IOP Astronomy
ISSN 2514-3433 (online)
ISSN 2515-141X (print)

British Library Cataloguing-in-Publication Data: A catalogue record for this book is available from the British Library.

Published by IOP Publishing, wholly owned by The Institute of Physics, London

IOP Publishing, No.2 The Distillery, Glassfields, Avon Street, Bristol, BS2 0GR, UK

US Office: IOP Publishing, Inc., 190 North Independence Mall West, Suite 601, Philadelphia, PA 19106, USA

To my wife Elizabeth, my son Luis Enrique, and the memory of my parents, Enrique and María Eugenia.

Contents

Preface

One of the fundamental properties of scientific research is constant progress, and one of my main motivations for writing this book, which probably shows throughout it, is my continuous amazement at how much our understanding of the interstellar medium (ISM) and star formation (SF) has changed through my own professional lifespan, while at the same time, some key ideas, which were developed many decades ago by some of the pioneers in the field, ended up forgotten, missed, or even rediscovered among the avalanche of literature produced in modern days. Thus, this book attempts to present a coherent view of the ISM and SF that combines classical theoretical concepts and solutions to the fluid dynamics equations with a modern interpretation based on the lessons learned from numerical simulations and key observational findings. Part I first presents some of the classical theoretical foundations with this "expanded" view, and then Part II applies them to our present understanding of the ISM and SF.

The book's title reflects its continuous emphasis on the *fluid* nature of the ISM; that is, that the medium flows from one place to another, is often out of thermal and dynamical equilibrium, and develops streams and currents that form clouds, which in turn evolve in time. This view is closer to a meteorological description of atmospheric and oceanic flow on Earth than to classical object-like descriptions of the ISM based on simplifying assumptions such as hydrostatic equilibrium, pressure confinement, and ballistically-moving, isolated clouds with well-defined sizes and masses. Much emphasis is also made on the physical interpretation of the mathematical results, since I am fully convinced that this is an essential requirement for their correct application.

Since the turn of the 21st century, the role and interplay between turbulence and gravity in the dynamics of the ISM and the SF process have become increasingly recognized, although often simplifications such as modeling turbulence as just "additional heat", and gravitational collapse as a homologous and spherically symmetric process, have continued to guide theoretical models and the interpretation of numerical simulations and observations. Thus, the chapters on turbulence and gravitational contraction aim to provide a view that goes beyond those simplifications and presents them as the highly complex, chaotic, and multi-scale phenomena that they are.

Another crucial player in ISM (thermo)dynamics is the medium's ability to cool upon compression. Consequently, the chapter on thermodynamics and compressibility, after presenting the parts of the classical theory relevant for the warm and cold ISM and SF, goes into more modern topics such as the nonlinear triggering of thermal instability in the linearly stable warm atomic ISM and the flow among the classical phases of the atomic ISM. Similarly, the chapter on the virial theorem, after covering the standard applications, goes into less idealized cases such as non-equilibrium configurations, flow across boundaries—including magnetic support in the presence of accretion—, and the balance between turbulent energy injection and dissipation.

Armed with this theoretical toolbox, Part II of the book then presents a dynamical and evolutionary view of clouds in the ISM, from their initial stages as thin, veil-like clouds of cold atomic gas, through their growth, conversion to a mostly molecular composition, and gravitational contraction, to their final stages of star formation and dispersal. Where possible, this view is contrasted to the more stationary, quasi-hydrostatic view of turbulence-supported clouds that arose at the turn of the century, and is still considered as a dominant paradigm for molecular clouds and the control of their rate of star formation.

The book is thus aimed equally to the graduate student approaching the ISM and SF for the first time and to the experienced researcher who seeks to delve into a more fluid and dynamic view of the ISM and its consequences on SF. Given the heavy reliance of Part II on the material developed in Part I, there is constant cross-referencing among sections and equations. Taking advantage of this being an e-book, my suggested form to approach it is with two copies of it open simultaneously, so that the reader can go over the astrophysical applications on one copy, while having the equations and theoretical discussions handy on the other. At least, that's the way I wrote it.

Undoubtedly, this book reflects my own perspective, to which I arrived after having devoted many years to the study of turbulence in molecular clouds, and then concluding that gravity is at least as important in their dynamics. Thus, this book should be regarded mostly as a thorough reference, from the theory to the application, of the evolutionary scenario for star-forming clouds and their star formation activity. Nevertheless, I have attempted to present the material in a coherent and self-contained manner, that hopefully provides the reader with a full perspective of the scenario, and how it compares to its alternative on the description of ISM dynamics, cloud evolution, and star formation.

There are certainly major topics that have been left out of this book, most prominently magneto-hydrodynamic (MHD) turbulent flow in Part I, and the origin of the stellar initial mass function (IMF) in Part II. This is due in part to space and time conservation, while excellent references already exist concerning them, and in part to my lesser expertise on these subjects.

Acknowledgments

My deepest appreciation goes to my many friends, colleagues, and family who have continuously provided the scientific and moral support that have made this several-year project a reality. First and foremost, I wish to thank my friend and colleague Gilberto Gómez, who not only generously agreed to read the whole manuscript and provide commentary, but has also been my closest collaborator for about one and a half decades, with whom many of the ideas in this book were developed and turned into analytic calculations and numerical simulations. Enormous gratitude also goes to John Dickey, who generously also offered to read the book, and provided important comments in record time. I also deeply thank my first graduate student ever, and now decades-old good friend and colleague, Javier Ballesteros-Paredes, who, through our many discussions, enlightened me about some of the key concepts outlined here, and my colleague, Mordecai-Mark Mac Low, for incepting in my mind the possibility of large-scale gravity domination during our writing of a chapter for *Protostars and Planets V*. Warmest thanks also go to my graduate students over the years, Manuel Zamora-Avilés, Raúl Naranjo-Romero, Vianey Camacho, Guido Granda-Muñoz, Griselda Arroyo-Chávez, and Rubén Guerrero-Gamboa, alongside whom I learned a great deal of what is described here. I also express my deepest gratitude to my departed parents, Enrique Vázquez D. and María Eugenia Semadeni, for making me what I am; to my sister, María Eugenia Vázquez, for her continual assistance, company, and comradeship; and to my PhD supervisor and friend, John Scalo, who not only had me learn about turbulence long before it entered the ISM field and taught me to have an always inquiring and digging mind, but also shared his passion for jazz guitar with me. Lastly, but most dearly, I thank my son Luis Enrique, for being the energy-injection mechanism of my life, and my wife, Elizabeth, for her love, partnership, and never-ending cheering throughout our journey together.

Author Biography

Enrique Vázquez-Semadeni

Enrique Vázquez-Semadeni obtained his BSc in Physics at the National Autonomous University of México (UNAM) and his PhD in Astronomy at the University of Texas at Austin. He is a Professor and former Director of the Institute of Radioastronomy and Astrophysics (IRyA) at UNAM's Morelia Campus. He has served at UNAM's Computing Advisory Council and Supercomputing Committee, and has received the Mexican Physical Society Research Award in 2019 and the National University Award in Exact Sciences in 2021, among others.

The Drunkard and the Policeman

A policeman doing his midnight shift sees a drunkard searching for something near a lamp post, and asks him "What are you looking for?"

"My house keys," the drunkard replies.

"Did you lose them here?"

"No, I dropped them over there," says the drunkard, pointing afar.

"So, why are you searching here?"

"Because there's more light…"

Part I

Theoretical Background

Enrique Vázquez-Semadeni

Chapter 1

The ISM and Star Formation

1.1 The ISM as a Flowing Continuum

The interstellar medium (ISM) of galaxies, as its name indicates, is the gaseous and dusty medium permeating the space between stars in galaxies. It is made up of gas in various states (ionized, atomic or molecular) and dust (solid particles of sizes $\sim$0.1 μm). Sharing the same volume as the ISM, there are magnetic fields, radiation, and cosmic rays (energetic light atom nuclei), that interact with it, and energize it. The gaseous and dusty components of the ISM are furthermore the raw material from which stars and planetary systems form.

Traditional approaches to the ISM have focused strongly on its microphysical (atomic and molecular) properties, and treated interstellar features, such as its various *phases* (regions of very different densities and temperatures but similar pressures) and dense clouds and their substructures, as isolated structures, often in dynamical equilibrium (e.g., Spitzer 1978). In this book, instead, we will be concerned primarily with the *fluid* properties of the ISM; that is, in the manner it *flows* within galaxies, in much the same way as the atmosphere and oceans in our planet do. In particular, we will show that many structures within the ISM should be considered as *flow features*, like streams, currents, stagnation points, etc. Such features may be quasi-stationary in the flow, in spite of not being made of the same set of atoms or molecules over time, and are thus analogous to a river on Earth. A river is a slowly-evolving, long-lived structure, despite the fact that the water molecules passing through are never the same. The river is effectively the *locus* of the water molecules flowing from higher to lower altitudes, and is permanently connected to its sources (the water sliding down the mountain slopes) and its final destination (a lake or the ocean).

This view is also in contrast with the classical approach of treating many of the interstellar features essentially as well-defined *objects* that consist of a fixed set of fluid particles. Such an approach is applicable to isolated and highly differentiated objects, like stars and planets. However, it does not apply to blurry entities such as

doi:10.1088/2514-3433/adcb28ch1

spiral arms and atomic and molecular clouds in galaxies, as well as their substructures (e.g., filaments, clumps and cores, which we will define below). Also, in the cases where a "flow" has been truly considered, it often is calculated under stationary conditions. So, in general, relatively little discussion has been devoted to the *evolution* of the interstellar features. Adopting the fluid, evolutionary vantage point in relation to the problem of interstellar clouds and their role in the formation of stars is one of the main goals of this book.

In the first part of this book we will cover the theoretical background underlying the formation and evolution of atomic and molecular clouds in the ISM, and the formation of stars within them, like the gas dynamics equations, ISM thermodynamics, turbulence, the virial theorem, hydrostatic equilibrium, and gravitational collapse. Although these theoretical concepts are well known and have been discussed in many books and reviews, here we will address them from the continuum, dynamic, and out-of-equilibrium vantage point mentioned above. In the second part we will then apply this theoretical background to describe the formation and evolution of clouds, and the resulting evolution of the star formation process, viewed from a collective perspective (the formation of star clusters). Also, wherever appropriate, we will adopt a viewpoint consistent with the novel perspective of nonlinear, complex systems.

1.2 The Components of the ISM

Excellent monographs exist covering the physical properties of the gaseous and dusty components of the ISM, as well as the radiative processes that allow it to heat and cool (e.g., Spitzer 1978; Dyson & Williams 1997; Lequeux 2005; see also the review by Ferrière 2001), and so here we just provide a brief introduction to its components and to the various types of clouds it contains, in order to provide the context for the dynamic and thermodynamic discussions in Part I of this book.

The ISM consists primarily of a gaseous component, which is made up of hydrogen (~70.5% by mass, ~91% by number), helium (~28% by mass, ~9% by number) and heavier elements (~1.5% by mass, 0.1% by number). It spans a wide range of densities and temperatures, and components with different combinations of density and temperature are known as *phases*, and receive different names according to their physicochemical state. The various phases are listed in Table 1.1. The so-called *hot*, *warm*, and *cold* phases are true thermodynamic phases, in the sense that they have similar values of the thermal pressures $P = nkT$, where n is the particle *number* density, with units of inverse volume (e.g., cm^{-3}), k is the Boltzmann constant, and T is the temperature. The origin of these different phases is the dependence of the radiative heating and cooling processes acting on the interstellar gas on density and temperature, and will be discussed in Chapter 4. A fourth component is the molecular gas, which is not really a different thermodynamic phase, as it is generally at a higher thermal pressure than the other phases, and the pressure increases with density within this component. Its origin is most likely the gravitational compression of these regions, together with dynamical compressions from expanding ionized H II regions and supernova (SN) explosions.

Table 1.1. ISM States (*Phases*; Ferrière 2001; Heiles & Troland 2003; Kalberla & Kerp 2009)

Name	T^a [K]	n^b [cm^{-3})	h^c [pc]	M^d [$M_\odot$]	L_J^e [pc]	M_J^f [$M_\odot$]	Main Component
Hot (coronal)	$\sim 10^6$	$\sim 6 \times 10^{-3}$	$\gtrsim$1000	$\sim 10^8$ (?)			H^+
Warm ionized (WIM)	$\sim 10^4$	~0.1	~1000	$\sim 10^9$	~5000	$\sim 2 \times 10^8$	H^+
Warm atomic (WNM)	~5000	~0.5	~1000	$\sim 2 \times 10^9$	~1600	$\sim 3.3 \times 10^7$	H^0
Cold atomic (CNM)	~50	~50	~100		~16	~3300	H^0
Cold molecular	$\gtrsim$10	$\gtrsim$300	~100	$\sim 1.5 \times 10^9$	~1.4	~27	H_2

[a] Temperature.
[b] Density.
[c] Scale height above midplane.
[d] Total mass.
[e] Jeans length (Equation (6.14)).
[f] Jeans mass (Equation (6.16)).

In addition, connected[1] regions of high density in the cold atomic phase and the molecular gas are known as *clouds*, and their main types are listed in Table 1.2, together with the classification of the molecular cloud substructures, known as "clumps", "filaments", and "cores". It is important to note that cold atomic clouds can appear either in isolation (i.e., just as coherent regions populated only by cold atomic gas), or as the *envelopes* of molecular clouds. In the first case, they have been determined to have extremely flattened, veil-like shapes, having extensions in the plane of the sky (POS) up to $\sim$300 times larger than their thicknesses (Heiles & Troland 2003).

Denser clouds can already be partially or completely molecular, when the gas becomes sufficiently opaque to photodissociating radiation. The gas is then said to *self-shield* from this radiation, and this condition typically requires a certain column density threshold. The largest and lowest-density molecular structures, giant molecular clouds (GMCs), probably still contain a significant fraction of their volume occupied by "pockets" of atomic gas (Goldsmith & Li 2005). In regions of higher density, such as clumps and cores, however, the hydrogen is almost exclusively in molecular form. Stars then form in the densest regions (the *dense cores*) of molecular clouds, and, after some time, they then destroy their parent cloud, through the various forms of *stellar feedback* they produce, such as bipolar jets, photoionizing radiation, stellar winds, radiation pressure, and supernova explosions.

One of the main goals of this book is to describe the entire evolutionary process of clouds, from their formation to their destruction, and how one type transforms, or fragments, into another, thus providing a coherent, unified picture in both space and time of the cloud and star formation process.

1.2.1 What Is a Cloud? The Types and Arbitrariness of Cloud Boundaries

It is of great importance for the context of this book to note that "clouds" and their substructures are not necessarily well-defined, isolated objects that could be made up of the same mass over time. This is because, over scales much larger than the mean free path between particles, the ISM can be considered as a *continuum*, and therefore all regions within it are in contact with the rest of the fluid. Therefore, clouds are closer to being *flow features*; that is, particular regions in the flow through which the gas *passes*, and acquires certain values of its physical variables, such as its density, temperature, velocity, ionization fraction, etc.

The most extreme analogy to this view of clouds is that of a *flame* in a candle.[2] Although a flame appears to us as a long-lived, stationary object, no single gas molecule is static within it. Rather, a flame is the *locus* (the region along the gas flow) where combustion is taking place. Air enters from the side, mixes with the vaporized

[1] "Connected" here is used in the mathematical sense, meaning that all member points (pixels in two-dimensional [2D] or three-dimensional [3D] observational or numerical data) have neighbors that are also members of the data set.

[2] I am indebted to my father, Enrique Vázquez Domínguez, a chemical engineer, for sharing his intrigued thoughts with me about the nature of a flame early in my undergraduate years.

Table 1.2. ISM Clouds

Name	T^a [K]	n^b [cm^{-3}]	L^c [pc]	M^d [$M_\odot$]	Line Width [km s^{-3}]	Notes
Diffuse atomic clouds	~50–100	~30–100	~ 0.1 × 30	~ 10^3–10^5 (?)		Often sheetlike[e]
Giant molecular clouds (GMCs)	~7–15	~100–500	~10–60	~ 10^4–10^6	~5–15	Probably contain HI pockets[f]
Individual molecular clouds	~10–30	~ 10^2–10^4	~2–20	~ 10^2–10^4	~1–10	
Clumps	~10–30	~ 10^3–10^5	~0.1–2	~10–10^3	~0.3–3	
Dense cores	~7–15	$\gtrsim 10^5$	$\lesssim$0.1	~0.1–10	~0.1–0.7	

[a] Temperature.
[b] Density.
[c] Size.
[d] Mass.
[e] Heiles & Troland (2003).
[f] Li & Goldsmith (2003).

wax from the candle around the wick, and undergoes an exothermic chemical reaction with it (combusts), releasing heat that maintains the area hot. The hot gases released from the combustion rise, leaving behind a low-pressure region, so that more air is pushed by the surroundings into the flame region. So, a flame is a *self-sustaining process*, not an object, and does not consist of the same material over time. Terrestrial examples of particular regions of flows where specific processes are occurring are clouds, storms and hurricanes in the atmosphere, currents and tsunamis in the oceans, or rivers on the land.

In astronomy, the concept of *cloud* in the ISM is a fundamental milestone in our understanding of the star formation process. However, it is very often neglected that it is a highly elusive concept. Clouds are generally defined *operationally* as high-density connected regions in the ISM; that is, as connected regions emitting above the noise level in observations of a certain tracer, or above some density threshold in numerical simulations. This threshold then constitutes the cloud's *boundary*.

However, as should be obvious from the definition, the term "high density" is relative, and can mean "high" with respect to the mean density of the whole ISM, or to the mean density of a region already denser than the whole ISM. That is, a hierarchy of overdensities exists. This is even more so, since the ISM is a gaseous medium that can be compressed by a variety of mechanisms (e.g., supersonic turbulence, thermal instability, self-gravity, etc.) that will be discussed in the following chapters. So, a new class of overdensities can arise if, within the overdensities previously produced by one of those processes, another process is triggered by the enhanced density there. This leads to the notions of "atomic" and "molecular" clouds, and to "clumps", "cores", "fragments", within molecular clouds as successive density enhancements over the previous class, as classified in Table 1.2. Moreover, depending on the nature of the density-enhancing process, the subsequent levels of the hierarchy may be mediated by smooth or abrupt changes of the density. In the remainder of this book, we will use the terms "cloud" and "clump" quite generically, but they can refer to any of these categories.

The other crucial consideration is that "clouds" in the ISM, together with their substructures, cannot be unambiguously defined, since their boundaries are a matter of convention, rather than physical and truly *bounding*, in the sense of confining or isolating the clouds. Specifically, the clouds and clumps cannot in general be considered in isolation from their surroundings, since they can freely exchange mass and energy with their surroundings. After all, in order to become denser than its surroundings, the cloud must have received material from its environment, concentrating it in its own volume. For example, if the gas is flowing from an atomic envelope into a molecular cloud, then the "boundary" of the molecular cloud can be defined as the limit of the molecular region, but this may correspond to the *front* where the gas is changing from atomic to molecular as it enters the molecular region. In this sense, the atomic envelope does not "confine" the molecular cloud, but rather "feeds" it.

In practice, cloud boundaries can be of one of the following categories:

- *Eulerian:* These are static boundaries defined in some frame of reference external to the flow; i.e., by means of a fixed contour in space, invariant in time. In numerical simulations, generally they are defined in the coordinates of the numerical domain, while in observations they can be defined in the corresponding coordinates. Since these boundaries are fixed in an external reference frame, in which the fluid flows, in general there is a mass, momentum and energy flux across them.

- *Lagrangian:* These are boundaries fixed on (or "following") the flowing fluid. That is, once initially defined, a Lagrangian boundary subsequently moves and becomes distorted, according to how each of its constituent points follows the fluid's motion. Since these boundaries are fixed on the flow, there is no flux across them. However, a cloud defined with this type of boundaries is expected to continuously change shape, being stretched, compressed or distorted by the flow. Lagrangian boundaries can be approximately defined in particle-based numerical simulations (e.g., using the smoothed-particle hydrodynamics, or SPH, prescription; Monaghan, 1985), by tagging and subsequently tracking the particles in the boundary of a region. However, they are difficult or impossible to implement in grid-based codes and in observations. In the latter, the precise flow regime of the fluid is in fact not known.
- *Feature-tracking:* These are boundaries defined by means of some feature in the flow, such as a front of some kind (a shock, a phase transition, a chemical transition, etc.), or a threshold in some physical or chemical variable, or in emission or absorption intensity, in observations. When only one temporal snapshot is available, as in observations, these boundaries can be considered Eulerian. However, in numerical simulations, where the temporal evolution is available, it is clear that these features in general do not remain fixed in the simulation's reference frame, but rather travel *through* the gas, although not *with* it. For example, the *locus* where the density of a cloud reaches a certain prescribed value can move, but in general does not follow the flow (see, e.g., Figure 12.2). Therefore, boundaries of this type are neither Eulerian nor Lagrangian. However, they are the most common way of defining cloud boundaries in observations and simulations of the ISM. In any case, like Eulerian boundaries, feature-tracking boundaries do not follow the flow, and therefore there is also mass, momentum and energy flux across them.

In conclusion, the "boundaries" of clouds are in most cases a matter of convention, and, even if they correspond to sharp transitions in the gas properties, in general they do not imply that the cloud is "confined" in any way. It will be important to keep this in mind in the chapter on the virial theorem (VT, Chapter 5), where we will see that the exchange with the environment can be described by the terms evaluated on the surface of the cloud. In that chapter, we will also stress its application for time-dependent, evolving clouds, besides the traditional applications to equilibrium situations.

Armed with this very brief introduction to the ISM components and structures, as well as to its inherent fluid nature, in the following chapters we examine the fundamental physical processes determining their properties and evolution.

References

Dyson, J. E., & Williams, D. A. 1997, The Physics of the Interstellar Medium (Boca Raton, FL: CRC Press)
Ferrière, K. M. 2001, RvMP, 73, 1031
Goldsmith, P. F., & Li, D. 2005, ApJ, 622, 938
Heiles, C., & Troland, T. H. 2003, ApJ, 586, 1067
Kalberla, P. M. W., & Kerp, J. 2009, ARA&A, 47, 27
Lequeux, J. 2005, The Interstellar Medium (Berlin: Springer)
Li, D., & Goldsmith, P. F. 2003, ApJ, 585, 823
Monaghan, J. J. 1985, CoPhR, 3, 71
Spitzer, L. 1978, Physical Processes in the Interstellar Medium (New York: Wiley)

Chapter 2

The Hydrodynamic Equations

2.1 The Fluid Approach

All matter is composed of discrete particles, such as atoms and molecules (at least from the point of view of classical mechanics). Yet, under certain circumstances, it is adequate to neglect the discrete nature of matter, and treat it as a continuum. Fluid dynamics is the treatment of fluids (liquids, gases, and plasmas) as continua. The basic condition for this approach to be valid is that the number of particles in a region of space much smaller than the dimensions of the system under study is still so large that they appear as a uniform mass distribution. This condition is satisfied if the *collisional mean free path* of the particles, ℓ, satisfies $\ell \ll L$, where L is the size of the system. In this case, it is possible to define a *volume ("fluid") element* (or *fluid parcel*) ΔV in space such that $\Delta V \ll L^3$, yet it contains a sufficiently large number of discrete particles (atoms, molecules) that spatial and velocity averages within it can be performed. We refer to these as *local* averages.

In particular, the *fluid* (or *bulk*) velocity of a given volume element can be defined as

$$\boldsymbol{u} \equiv \frac{1}{N_{\Delta V}} \sum_{n=1}^{N_{\Delta V}} \boldsymbol{v}_n \equiv \langle \boldsymbol{v}_n \rangle_{\Delta V}, \tag{2.1}$$

where $N_{\Delta V}$ is the total number of particles in the volume element ΔV, and $\boldsymbol{v}_n$ is the total velocity of the n-th particle. Moreover, the total velocity of a particle, $\boldsymbol{v}_n$, can then be decomposed into the fluid velocity plus the particle's *peculiar* velocity, $\boldsymbol{w}_n$; that is,

$$\boldsymbol{v}_n = \boldsymbol{u} + \boldsymbol{w}_n. \tag{2.2}$$

As we will see later, the typical magnitude of the peculiar velocities, $w = |\boldsymbol{w}|$, is related to the speed of sound in the medium.

doi:10.1088/2514-3433/adcb28ch2

Correspondingly, the *mass density* ρ of the fluid element ΔV is defined as

$$\rho \equiv \frac{1}{\Delta V}\sum_{n=1}^{N_{\Delta V}} m_n, \tag{2.3}$$

where m_n is the mass of the n-th particle, while the *temperature* T can be defined in terms of the *root mean square speed*, $v_{\rm rms}$, according to the well-known relation from the kinetic theory of gases,

$$\frac{1}{2} m v_{\rm rms}^2 = \frac{3}{2} k_{\rm B} T, \tag{2.4}$$

where m is the mass of the gas particles, $k_{\rm B}$ is Boltzmann's constant, and

$$v_{\rm rms} \equiv \left(\frac{1}{\Delta V}\sum_{n=1}^{N_{\Delta V}} v_n^2\right)^{1/2}, \tag{2.5}$$

with $v_n^2 \equiv \boldsymbol{v}_n \cdot \boldsymbol{v}_n$.

If ΔV is still very small with respect to the size of the system, then a position $\boldsymbol{x}$ can be ascribed to it, so that the bulk velocity and the density can be considered functions of position, $\boldsymbol{u} = \boldsymbol{u}(\boldsymbol{x})$ and $\rho = \rho(\boldsymbol{x})$, noting that $\boldsymbol{x}$ cannot be located in space with an accuracy greater than $\sim(\Delta V)^{1/3}$.

2.2 The Conservation Equations

Reference: "Fundamental Mechanics of Fluids", Currie, 1974, Chapter 1.

2.2.1 Eulerian and Lagrangian Descriptions

The fundamental equations of fluid mechanics are simply the expressions of the conservation of mass, momentum, and energy. These equations are generally written following the convention that their left-hand side indicates the rate of change of the variable in question, while the right-hand side groups all sources and sinks for this variable. In the case when there are no sources nor sinks, the variable is absolutely conserved.

Now, the conservation laws are easily postulated in a reference frame that "follows the fluid". For example, it is evident that, if we consider a certain subset of the molecules of a fluid, and track them in space as they move (i.e., as they "flow"), the mass of this subset will be conserved because by definition it consists always of the same set of particles. However, mathematically it is easier to deal with reference frames that are "fixed" in space, with different molecules passing through a certain location as time proceeds. The first kind of frame, "following the flow", is called *Lagrangian*, while the second kind, "fixed in space", is known as *Eulerian*.[1]

In the Eulerian description, the fluid particles move through a fixed grid of coordinates. In this case, the independent variables are the spatial coordinates x, y, and z, and the time t. Because the Eulerian coordinate frame is assumed to remain fixed in time, the spatial coordinates, x, y, and z, are considered *absolute*, and not depending on each other, nor on time. Instead, in the Lagrangian description, it is the spatial locations themselves that "pass" through each fluid particle at different times. In this case, the independent variables are x_0, y_0, and z_0, the initial coordinates of a given fluid particle, which identify it, and the time t. The subsequent coordinates that "pass" through the fluid element at later times are functions of time, so that the instantaneous position vector of the particle is a function of time, $\boldsymbol{x} = \boldsymbol{x}(t)$. Before deriving the conservation equations themselves, then, we need to derive the necessary transformation laws between Eulerian and Lagrangian frames.

2.2.2 The Transformation between Lagrangian and Eulerian Frames

2.2.2.1 The Material Derivative

In order to calculate the amount by which a certain quantity varies as the fluid flows through space and time, it is necessary to consider the amount of change brought in by the displacement of the fluid parcel through space and time, as well as any contribution from sinks and sources of the quantity. This is easiest to set up in a Lagrangian description, but if we want to have an Eulerian description as well, it is necessary to know how to convert between the two descriptions. A first conversion rule is provided by the *material (or Lagrangian, or total) derivative*, which expresses the temporal rate of change of a physical property of a fluid element.

Let α be any property per unit volume of the fluid—such as its mass density, momentum density in some direction, or energy density—defined as a function of position and time in an Eulerian frame, so that

$$\alpha = \alpha(\boldsymbol{x}, t), \tag{2.6}$$

where $\boldsymbol{x}$ is the position vector. In a Lagrangian description, however, the independent variables are $\boldsymbol{x}_0$, the initial position of the fluid parcel, and t, the time, while the subsequent positions of the parcel are a function of time; that is, $\boldsymbol{x} = \boldsymbol{x}(t)$, and

$$\alpha = \alpha(\boldsymbol{x}(t), t). \tag{2.7}$$

Therefore, applying the chain rule of calculus, the rate of change of α in the Lagrangian frame (i.e., following the fluid parcel) in terms of the Eulerian

[1] As discussed in Section 1.2.1, in practice a third type of reference frame is commonly used in astronomy, which we referred to as *feature-tracking*, with boundaries that are precisely associated to some feature of the flow, such as a shock front, a phase transition, a threshold in density or a chemical tracer, etc. These boundaries do not remain fixed in space, but do not follow the flow, either. Unfortunately, even though this type of boundary is the most commonly used one in observations, and often emulated in numerical simulations, no well-defined mathematical treatment has been developed for them.

derivatives (i.e., the partial derivatives, which consider the rate of change of the physical variable only with respect to each coordinate) is

$$\frac{d\alpha}{dt} = \frac{\partial \alpha}{\partial t} + \frac{\partial \alpha}{\partial x_i}\frac{dx_i}{dt}, \tag{2.8}$$

where, for convenience, we have switched to tensor notation (see Appendix A), expressing any vector in terms of its components as $\boldsymbol{x} = x_i, i = 1, 2, 3$, and applying the convention that repeated indices denote a summation, so that

$$\frac{\partial \alpha}{\partial x_i}\frac{dx_i}{dt} \equiv \sum_i \frac{\partial \alpha}{\partial x_i}\frac{dx_i}{dt}, \tag{2.9}$$

and therefore denotes a *scalar* term, while free indices denote *vector* components. Now, noting that the time derivative of the position of a fluid parcel is its velocity, i.e., $dx_i/dt = u_i$, we can rewrite Equation (2.8) as

$$\frac{d\alpha}{dt} = \frac{\partial \alpha}{\partial t} + u_i\frac{\partial \alpha}{\partial x_i}, \tag{2.10}$$

or, back to vector notation,

$$\frac{d\alpha}{dt} = \frac{\partial \alpha}{\partial t} + \boldsymbol{u} \cdot \nabla \alpha. \tag{2.11}$$

More generally, one can think of d/dt as an *operator*, defined as

$$\frac{d}{dt} = \frac{\partial}{\partial t} + \boldsymbol{u} \cdot \nabla . \tag{2.12}$$

The physical meaning of Equation (2.11) is that the rate of change of a physical quantity α, *following the flow*, is given by the intrinsic temporal rate of change of the quantity at a given location in space ($\partial\alpha/\partial t$), plus the *transport* of the spatial change of α (represented by $\nabla\alpha$) by the fluid velocity $\boldsymbol{u}$. Indeed, we notice that $\boldsymbol{u} \cdot \nabla \alpha$ is the projection of the gradient of α along the velocity field $\boldsymbol{u}$. That is, part of the total change in α is brought about because the fluid motion (represented by $\boldsymbol{u}$), brings in a different value of α that was previously at a different location. For this reason, the term $\boldsymbol{u} \cdot \nabla \alpha$ is known as the *advective* (or transport) term in a conservation equation, and $\boldsymbol{u} \cdot \nabla$ is known as the *advection operator*.

2.2.2.2 Reynolds' Transport Theorem

We will now derive another very useful result, known as Reynolds' transport theorem. It allows us to express the temporal rate of change of a fluid property within a *finite* (rather than infinitesimal) volume in terms of the local and instantaneous rates of change of the fluid properties inside the volume. That is, it is the integral version of the material derivative.

Consider a certain mass of the fluid, contained in Lagrangian volume V, at a certain time, and follow it for a short period of time δt as it flows. Note that the volume itself is a function of time, because it is following the flow. Let α be any property of the fluid, as in the previous section. The volume-integrated value[2] of α over V at time t is

$$A(t) \equiv \int_{V(t)} \alpha(t) dV. \tag{2.13}$$

We are interested in describing the evolution of the physical properties of finite-volume[3] regions within the flow.[4] Thus, we are interested in the temporal rate of change of the physical quantity within the volume. Applying the definition of the time derivative, this can be written as

$$\frac{d}{dt}\int_{V(t)} \alpha(t) dV = \lim_{\delta t \to 0} \left\{ \frac{1}{\delta t} \left[\int_{V(t+\delta t)} \alpha(t+\delta t) dV - \int_{V(t)} \alpha(t) dV \right] \right\}, \tag{2.14}$$

Now, adding and subtracting $\int_{V(t)} \alpha(t+\delta t) dV$ inside the square brackets, we obtain

$$\begin{aligned} \frac{d}{dt}\int_{V(t)} \alpha(t) dV &= \lim_{\delta t \to 0} \left\{ \frac{1}{\delta t} \left[\int_{V(t+\delta t)} \alpha(t+\delta t) dV - \int_{V(t)} \alpha(t+\delta t) dV \right] \right\} \\ &+ \lim_{\delta t \to 0} \left\{ \frac{1}{\delta t} \left[\int_{V(t)} \alpha(t+\delta t) dV - \int_{V(t)} \alpha(t) dV \right] \right\}. \end{aligned}$$

The first limit in this expression corresponds to taking the time derivative keeping the integrand fixed and allowing the volume V to vary in time, and the subtraction implies that the integration will only be performed over the difference volume $\Delta V = V(t+\delta t) - V(t)$. On the other hand, the second limit corresponds to holding V fixed and permitting the integrand to vary, which therefore corresponds to $\int_{V(t)} \partial\alpha/\partial t dV$. So, the above expression can be rewritten as

$$\frac{d}{dt}\int_{V(t)} \alpha(t) dV = \lim_{\delta t \to 0} \left\{ \frac{1}{\delta t} \left[\int_{\Delta V} \alpha(t+\delta t) dV \right] \right\} + \int_{V(t)} \frac{\partial \alpha}{\partial t} dV.$$

Let us now evaluate the first term on the right-hand side of the last equation using geometric considerations. Let S be the surface bounding the volume V. Since S is fixed on the fluid, each point on this surface moves with the fluid velocity $\boldsymbol{u}$, but only the component of $\boldsymbol{u}$ parallel to the local normal to the surface, $\boldsymbol{n}$, brings along a change in the volume V, since a motion tangent to the surface leaves the surface unchanged. Thus, a small element of the difference volume ΔV of area dS and depth $\boldsymbol{u} \cdot \boldsymbol{n}\, \delta t$ (see Figure 2.1) is given by

$$\delta \Delta V = \boldsymbol{u} \cdot \boldsymbol{n}\, dS\, \delta t.$$

[2] For example, if $\alpha = \rho$, then its volume integral is the mass contained in volume V.

[3] That is, as opposed to infinitesimal regions, such as the ones considered in Equations (2.1) and (2.3).

[4] These can constitute, for example, clouds in the interstellar medium.

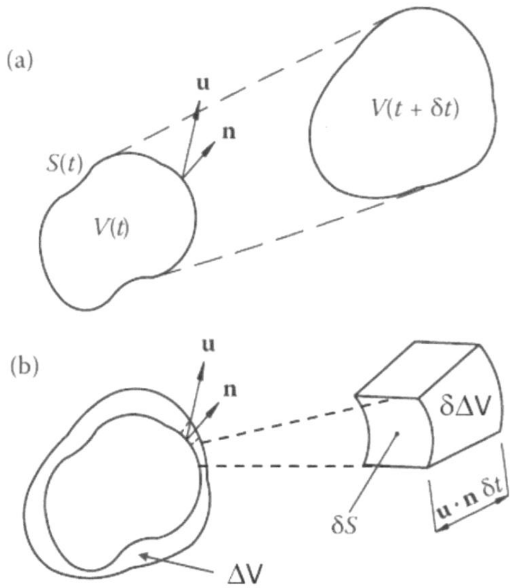

Figure 2.1. *Top*: Evolution of the volume V over a time δt, including both a translation and a deformation. The surface S bounds the volume. *Bottom left*: The initial and evolved volumes after removal of the translational part of the velocity (the mean velocity over the volume). The difference volume contained between the two surfaces is ΔV. *Bottom right*: Volume element bounded by a surface element dS and the displacement $\boldsymbol{u} \cdot \boldsymbol{n}\ \delta t$. From Currie (2012), reprinted by permission of the publisher (Taylor & Francis Ltd, http://www.tandfonline.com.)

Therefore, the integral over ΔV can be computed as an integral over $dS\delta t$, as follows

$$\begin{aligned}\frac{d}{dt}\int_{V(t)} \alpha(t) dV &= \lim_{\delta t\to 0}\left\{\frac{1}{\delta t}\left[\int_{S(t)} \alpha(t+\delta t)\ \boldsymbol{u}\cdot\boldsymbol{n}\ \delta t\ dS\right]\right\} + \int_{V(t)}\frac{\partial\alpha}{\partial t}dV \\ &= \lim_{\delta t\to 0}\int_{S(t)}\frac{1}{\delta t}[\alpha(t+\delta t)\ \boldsymbol{u}\cdot\boldsymbol{n}\ dS]\delta t + \int_{V(t)}\frac{\partial\alpha}{\partial t}dV \\ &= \int_{S(t)}\alpha(t)\ \boldsymbol{u}\cdot\boldsymbol{n}\ dS + \int_{V(t)}\frac{\partial\alpha}{\partial t}dV,\end{aligned}$$

where the first term after the third equality follows from the cancellation of δt inside the integral, and we note that $\lim_{\delta t\to 0}\alpha(t+\delta t) = \alpha(t)$. Finally, applying Gauss' Theorem to the first term on the right-hand side, we can finally write Reynolds' Transport Theorem as

$$\frac{d}{dt}\int_V \alpha dV = \int_V\left[\frac{\partial\alpha}{\partial t} + \nabla\cdot(\alpha\boldsymbol{u})\right]dV. \tag{2.15}$$

Or, in tensor notation,

$$\frac{d}{dt}\int_V \alpha dV = \int_V\left[\frac{\partial\alpha}{\partial t} + \frac{\partial}{\partial x_i}(\alpha u_i)\right]dV. \tag{2.16}$$

2.2.2.3 Discussion

It is interesting to compare the transformation between Lagrangian and Eulerian coordinates in its differential (the material derivative, Equation (2.11)) and integral (Reynolds' Transport Theorem, Equation (2.15)) forms. Comparing these two equations, we note that the transformations are similar, but not identical. While the material derivative contains the term $\boldsymbol{u} \cdot \nabla \alpha$, Reynolds' Theorem contains $\nabla \cdot (\alpha \boldsymbol{u}) = \boldsymbol{u} \cdot \nabla \alpha + \alpha \nabla \cdot \boldsymbol{u}$. The second term in this last expression, which appears only in the integral form, corresponds to the change in the integral of α over volume V due to the change in V itself. This contribution does not appear in the differential form (the material derivative) because the volume considered in that case is infinitesimal. We will encounter this contribution explicitly in the equation for mass conservation in the next section, and also in the derivation of the Virial Theorem, in Section 5.1.

2.2.3 Mass Conservation (Continuity Equation)

We now start the derivation of the conservation equations, considering first the conservation of mass, so that we replace our generic variable α by the mass density ρ. Because in our classical approach there are no sources or sinks of mass, our conservation principle is simply that the total mass in a Lagrangian volume V is conserved. That is,

$$\frac{d}{dt} \int_V \rho dV = 0.$$

According to Reynolds' Transport Theorem, Equation (2.15), this is equivalent to

$$\int_V \left[\frac{\partial \rho}{\partial t} + \nabla \cdot (\alpha \boldsymbol{u}) \right] dV = 0.$$

Now, because the volume V is arbitrary, it is necessary that the integrand in the above equation vanishes identically, so that the differential expression of mass conservation, known as the *equation of continuity*, is

$$\frac{\partial \rho}{\partial t} + \nabla \cdot (\rho \boldsymbol{u}) = 0, \tag{2.17}$$

or, in tensor notation,

$$\frac{\partial \rho}{\partial t} + \frac{\partial}{\partial x_j}(\rho u_j) = 0. \tag{2.18}$$

The continuity equation can be rewritten as

$$\frac{\partial \rho}{\partial t} + \boldsymbol{u} \cdot \nabla \rho = -\rho \nabla \cdot \boldsymbol{u}, \tag{2.19}$$

where we recognize the left-hand side of this equation as the material derivative of the density, so that this equation can also be written as

$$\frac{d\rho}{dt} = -\rho \, \nabla \cdot \boldsymbol{u}. \tag{2.20}$$

This equation indicates that the only way to induce an increase of the density of a Lagrangian fluid parcel is for the flow there to have a negative divergence (i.e., a *convergence*) of the velocity field. Physically, this means that the volume of the parcel is shrinking and, since its mass is constant, its density increases. Another way to see this is to consider the *volume per unit mass*, or *specific volume*, $\upsilon = 1/\rho$. In terms of this variable, Equation (2.20) becomes

$$\frac{d\upsilon}{dt} = \upsilon \, \nabla \cdot \boldsymbol{u}, \tag{2.21}$$

indicating that the local rate of change of the volume per unit mass is proportional to the velocity divergence. That is, a negative velocity divergence causes the volume per unit mass to decrease, or the density to increase.

2.2.3.1 Incompressible Flow

An important particular case is that of an *incompressible* fluid, which is applicable in a wide variety of terrestrial applications. In this case, the density of each fluid parcel remains constant, so that

$$\frac{d\rho}{dt} = 0 \quad \text{(Incompressible flow).}$$

Thus, from Equation (2.20) it follows that

$$\nabla \cdot \boldsymbol{u} = 0 \quad \text{(Incompressible flow).} \tag{2.22}$$

2.2.4 Conservation of Momentum

The equation of momentum conservation we will derive next is, effectively, an expression of Newton's second law of motion to a fluid. It states that the rate of change of the momentum of a fluid parcel equals the sum of the forces applied on it. These can be either *body* forces, applied to the entire volume of the fluid parcel, such as gravity (and electromagnetic forces in the case of a plasma) or *surface* forces, such as pressure or viscous forces, which are applied only on the bounding surface of the volume.

We will consider the rate of change of the momentum *per unit volume*, $\rho\boldsymbol{u}$, of a finite fluid mass contained in a volume V. Through Reynolds' Theorem, we know how to compute its global time derivative. Also, denoting the body force per unit volume applied to this mass by $\boldsymbol{f}$, the total force applied onto the entire volume V is simply $\int_V \boldsymbol{f} \, dV$. However, the treatment of the surface forces requires some further discussion.

The surface forces are the consequence of the interactions between neighboring microscopic particles. Thus, if one considers a certain surface in the medium, defined by its normal $\boldsymbol{n}$, the particles in its immediate vicinity will exert forces on it through collisions. However, the force exerted on this surface can have any relative orientation with respect to $\boldsymbol{n}$, and so the effect of the surface forces needs to be described by a quantity that accounts for both the direction of the force and for the orientation of the surface (the direction of its normal). Such a quantity is a *tensor*, and the tensor describing the *force per unit area* in direction i acting on a surface whose normal is in direction j is called the *stress tensor*, denoted σ_{ij} or $\overleftrightarrow{\boldsymbol{\sigma}}$. In turn, the resulting vector force acting on the surface element $n_j dS$ is the projection of the stress tensor along it,

$$dF_i = \sigma_{ij} n_j dS. \tag{2.23}$$

or, in vector notation,

$$d\boldsymbol{F} = \overleftrightarrow{\boldsymbol{\sigma}}\cdot\boldsymbol{n}\ dS, \tag{2.24}$$

The total force exerted on the surface bounding volume V is then $\int_S \sigma_{ij} n_j\ dS$. We can thus write the conservation of momentum for this volume, including surface and body forces, as

$$\frac{d}{dt}\int_V \rho\boldsymbol{u}\ dV = \int_S \overleftrightarrow{\boldsymbol{\sigma}}\cdot\boldsymbol{n}\ dS + \int_V \boldsymbol{f}\ dV.$$

Applying Reynolds' Theorem to the left-hand side, setting $\alpha = \rho\boldsymbol{u}$, and applying Gauss' Theorem to the surface integral, the above equation can be written as

$$\int_V \left[\frac{\partial}{\partial t}(\rho\boldsymbol{u}) + \nabla\cdot(\rho\boldsymbol{u}\boldsymbol{u})\right] dV = \int_V \nabla\cdot\overleftrightarrow{\boldsymbol{\sigma}} dV + \int_V \boldsymbol{f}\ dV.$$

Note that the quantity $\boldsymbol{uu}$ is the tensor product of the velocity field with itself, denoted in tensor notation by $u_i u_j$. Finally, because the volume V is arbitrary, the equality must hold at the level of the integrands, and so we obtain the expression for conservation of momentum *per unit volume*

$$\frac{\partial}{\partial t}(\rho\boldsymbol{u}) + \nabla\cdot(\rho\boldsymbol{u}\boldsymbol{u}) = \nabla\cdot\overleftrightarrow{\boldsymbol{\sigma}} + \boldsymbol{f}, \tag{2.25}$$

or, in tensor notation,

$$\frac{\partial}{\partial t}(\rho u_i) + \frac{\partial}{\partial x_j}(\rho u_i u_j) = \frac{\partial\sigma_{ij}}{\partial x_j} + f_i. \tag{2.26}$$

The law of conservation of momentum can also be written *per unit mass* by expanding and regrouping the derivatives on the left-hand side, separating the product $\rho u_i u_j$ as $(\rho u_j)u_i$, to obtain

$$u_i\left[\frac{\partial\rho}{\partial t} + \frac{\partial}{\partial x_j}(\rho u_j)\right] + \rho\left[\frac{\partial u_i}{\partial t} + u_j\frac{\partial u_i}{\partial x_j}\right] = \frac{\partial\sigma_{ij}}{\partial x_j} + f_i.$$

The first term in the brackets is seen to be the left-hand side of the equation of continuity, Equation (2.18), which is identically equal to zero. So, the *momentum equation per unit mass* is finally

$$\frac{\partial u_i}{\partial t} + u_j \frac{\partial u_i}{\partial x_j} = \frac{1}{\rho} \frac{\partial \sigma_{ij}}{\partial x_j} + a_i, \tag{2.27}$$

where $a_i = f_i / \rho$ is the local acceleration. In vector notation,

$$\frac{\partial \boldsymbol{u}}{\partial t} + \boldsymbol{u} \cdot \nabla \boldsymbol{u} = \frac{1}{\rho} \nabla \cdot \overleftrightarrow{\boldsymbol{\sigma}} + \mathbf{a}. \tag{2.28}$$

2.2.5 Energy Conservation

The equation of energy conservation for a fluid volume V will eventually describe the application of the first law of thermodynamics to it. However, to derive it, it is necessary to consider the conservation of *total* energy, because the exertion of a force onto the volume will have an effect on both its internal and its kinetic energy. The conservation of kinetic energy can then be removed using the momentum equation, to finally obtain the conservation of internal energy alone.

We thus start by writing down the rate of change of the total energy contained in volume V, and equate it to the rate of application of mechanical work and heat exchange. The latter will be described by a vector $\boldsymbol{q}$, which represents the conductive or radiative heat flux *leaving* the volume. The net heat lost by the volume is then obtained by integration of the dot product of $\boldsymbol{q}$ with the surface normal over the surface of the volume. Note that we do not include *convective* flux, as this form of heat conduction involves the motion of the fluid, and therefore is represented by the hydrodynamic equations themselves—specifically, by the advective term.

To describe the application of mechanical work to our control volume, we consider the same two kinds of forces as we did in Section 2.2.4; that is, body and surface forces. As is well known, the rate of application of work (i.e., the *power*) exerted by a force is obtained by dotting it with the velocity–in this case, the flow velocity $\boldsymbol{u}$. Also, we denote the *internal energy per unit mass* of the flow by e, and the *kinetic energy per unit mass* as $\boldsymbol{u} \cdot \boldsymbol{u} = u_i u_i = u^2$. Thus, we write the energy conservation condition per unit volume as

$$\frac{d}{dt} \int_V \left(\rho e + \frac{1}{2} \rho u^2 \right) dV = \int_S \boldsymbol{u} \cdot \overleftrightarrow{\boldsymbol{\sigma}} \cdot \boldsymbol{n} \, dS + \int_V \boldsymbol{u} \cdot \boldsymbol{f} \, dV - \int_S \boldsymbol{q} \cdot \boldsymbol{n} \, dS, \tag{2.29}$$

where the heat flux term takes a minus sign because of the convention that $\boldsymbol{q}$ represents heat leaving the volume.

At this point, it is convenient to switch to tensor notation, so we write

$$\frac{d}{dt} \int_V \left(\rho e + \frac{1}{2} \rho u_i u_i \right) dV = \int_S u_i \sigma_{ij} n_j \, dS + \int_V u_i f_i \, dV - \int_S q_i n_i \, dS. \tag{2.30}$$

Applying Reynolds' theorem to the left-hand side and Gauss' Theorem to the surface integrals on the right-hand side, we obtain

$$\int_V \left\{ \frac{\partial}{\partial t}\left(\rho e + \frac{1}{2}\rho u_i u_i\right) + \frac{\partial}{\partial x_j}\left[\left(\rho e + \frac{1}{2}\rho u_i u_i\right)u_j\right]\right\} dV$$
$$= \int_V \left[\frac{\partial}{\partial x_j}(u_i\sigma_{ij}) + u_i f_i - \frac{\partial q_i}{\partial x_i}\right] dV.$$

As it is by now a familiar procedure to us, we now require that, since the volume V is arbitrary, the equality must hold at the level of the integrands, and so we write

$$\frac{\partial}{\partial t}\left(\rho e + \frac{1}{2}\rho u_i u_i\right) + \frac{\partial}{\partial x_j}\left[\left(\rho e + \frac{1}{2}\rho u_i u_i\right)u_j\right]$$
$$= \frac{\partial}{\partial x_j}(u_i\sigma_{ij}) + u_i f_i - \frac{\partial q_i}{\partial x_i}. \tag{2.31}$$

This expression can be simplified by suitably expanding the derivatives. In the first term of the left-hand side, we split the product $\frac{1}{2}\rho u_i u_i$ as $(\rho)(\frac{1}{2}u_i u_i)$ so that, after some regrouping,

$$\frac{\partial}{\partial t}\left(\rho e + \frac{1}{2}\rho u_i u_i\right) = \rho\frac{\partial}{\partial t}\left(e + \frac{1}{2}u_i u_i\right) + \left(e + \frac{1}{2}u_i u_i\right)\frac{\partial \rho}{\partial t}. \tag{2.32}$$

Similarly, the second term on the left-hand side of Equation (2.31) can be expanded by separating the products $\rho e u_j$ and $\frac{1}{2}\rho u_i u_i u_j$, respectively, as $(e)(\rho u_j)$ and $(\rho u_j)(\frac{1}{2}u_i u_i)$, so that

$$\frac{\partial}{\partial x_j}\left[\left(\rho e + \frac{1}{2}\rho u_i u_i\right)u_j\right] = e\frac{\partial}{\partial x_j}(\rho u_j) + \rho u_j\frac{\partial e}{\partial x_j}$$
$$+ \frac{1}{2}u_i u_i\frac{\partial}{\partial x_j}(\rho u_j) + \rho u_j\frac{\partial}{\partial x_j}\left(\frac{1}{2}u_i u_i\right). \tag{2.33}$$

Now, according to the continuity equation, Equation (2.18), the term $(\partial/\partial x_j)(\rho u_j)$ can be replaced by $-\partial\rho/\partial t$. After doing this and regrouping the terms, we obtain

$$\frac{\partial}{\partial x_j}\left[\left(\rho e + \frac{1}{2}\rho u_i u_i\right)u_j\right] =$$
$$-\frac{\partial \rho}{\partial t}\left(e + \frac{1}{2}u_i u_i\right) + \rho u_j\left[\frac{\partial}{\partial x_j}\left(e + \frac{1}{2}u_i u_i\right)\right]. \tag{2.34}$$

Therefore, the summation indicated on the left-hand side of Equation (2.31), amounting to the sum of Equations (2.32) and (2.34), becomes, after canceling similar terms on the right-hand sides of the two equations and replacing $1/2(\partial u_i u_i/\partial t)$ by $u_i\partial u_i/\partial t$,

$$\frac{\partial}{\partial t}\left(\rho e+\frac{1}{2}\rho u_i u_i\right)+\frac{\partial}{\partial x_j}\left[\left(\rho e+\frac{1}{2}\rho u_i u_i\right)u_j\right]$$
$$=\rho\frac{\partial}{\partial t}\left(e+\frac{1}{2}u_i u_i\right)+\rho u_j\left[\frac{\partial}{\partial x_j}\left(e+\frac{1}{2}u_i u_i\right)\right]$$
$$=\rho\frac{\partial e}{\partial t}+\rho u_i\frac{\partial u_i}{\partial t}+\rho u_j\frac{\partial e}{\partial x_j}+\rho u_i u_j\frac{\partial u_i}{\partial x_j}.$$

After some regrouping, we can then write the full Equation (2.31) as

$$\rho\left(\frac{\partial e}{\partial t}+u_j\frac{\partial e}{\partial x_j}\right)+\rho u_i\left(\frac{\partial u_i}{\partial t}+u_j\frac{\partial u_i}{\partial x_j}\right)$$
$$=u_i\frac{\partial \sigma_{ij}}{\partial x_j}+\sigma_{ij}\frac{\partial u_i}{\partial x_j}+u_i f_i-\frac{\partial q_i}{\partial x_i}.$$

We now note that, in the last equation, the second term on the left-hand side equals the first and third terms on the right-hand side by virtue of the momentum equation, Equation (2.27), so they drop from the equation, and we finally obtain

$$\frac{\partial e}{\partial t}+u_j\frac{\partial e}{\partial x_j}=\frac{\sigma_{ij}}{\rho}\frac{\partial u_i}{\partial x_j}-\frac{1}{\rho}\frac{\partial q_i}{\partial x_i}, \tag{2.35}$$

which is the equation of *internal energy conservation per unit mass*. Written in vector notation, it reads

$$\frac{\partial e}{\partial t}+\boldsymbol{u}\cdot\nabla e=\frac{1}{\rho}(\overleftrightarrow{\boldsymbol{\sigma}}\cdot\nabla)\cdot\boldsymbol{u}-\frac{1}{\rho}\nabla\cdot\boldsymbol{q}. \tag{2.36}$$

Per unit volume, the equation reads

$$\rho\left(\frac{\partial e}{\partial t}+u_j\frac{\partial e}{\partial x_j}\right)=\sigma_{ij}\frac{\partial u_i}{\partial x_j}-\frac{\partial q_i}{\partial x_i}, \tag{2.37}$$

and

$$\rho\left(\frac{\partial e}{\partial t}+\boldsymbol{u}\cdot\nabla e\right)=(\overleftrightarrow{\boldsymbol{\sigma}}\cdot\nabla)\cdot\boldsymbol{u}-\nabla\cdot\boldsymbol{q}, \tag{2.38}$$

in tensor and vector notation, respectively. This equation expresses the fact that the internal energy of a fluid parcel can be varied either by the work performed by the mechanical stresses applied onto it, or by a net heat flux into/away from it.

2.3 Constitutive Relations

We now turn to the task of determining the stress tensor σ_{ij}. The equations involving σ_{ij} relate the intensity of an external force to its effect on the fluid itself (stretching, rotation, compression, etc.). Relations of this kind are called *constitutive equations*.

In turn, the quantity that describes the response of the fluid to these forces is the so-called *deformation-rate tensor*, given by

$$d_{ij} \equiv \frac{\partial u_i}{\partial x_j}. \tag{2.39}$$

Note that this is nothing other than the velocity gradient tensor $\nabla \boldsymbol{u}$, although its units are inverse time, and therefore it is indeed a "rate".

2.3.1 Shearing, Rotation, and Compression Rates

The deformation-rate tensor can be written as the sum of a symmetric and an antisymmetric part, as follows:

$$d_{ij} = s_{ij} + r_{ij}, \tag{2.40}$$

where the symmetric part, corresponding to the *shearing rate*, is given by

$$s_{ij} \equiv \frac{1}{2}\left(\frac{\partial u_i}{\partial x_j} + \frac{\partial u_j}{\partial x_i}\right), \tag{2.41}$$

and the antisymmetric part, which is necessarily traceless and corresponds to the *rotation rate*, is given by

$$r_{ij} \equiv \frac{1}{2}\left(\frac{\partial u_i}{\partial x_j} - \frac{\partial u_j}{\partial x_i}\right). \tag{2.42}$$

While we do not give a formal derivation of these rates here,[5] it is useful to see a graphical representation of how these quantities are indeed related to shear and rotation, as illustrated in Figure 2.2. In this figure, the left panel shows a simple two-dimensional flow in which the velocity is given by $u_x = y/2$ and $u_y = x/2$. It is easy to check that, in this case,

$$s_{ij} = \frac{1}{2}\begin{pmatrix} 0 & 1 \\ 1 & 0 \end{pmatrix}.$$

while

$$r_{ij} = \begin{pmatrix} 0 & 0 \\ 0 & 0 \end{pmatrix},$$

Graphically, from the figure it is apparent that, if we consider a region of the flow whose sides lie parallel to the axes, as illustrated by the solid-line square, these will evolve as indicated by the velocity arrows, so that the vertical lines will be rotated clockwise, while the horizontal lines will be rotated counterclockwise, producing the diamond-shaped structure depicted with dotted lines. It can be seen that, during this process, the opposite sides of the initial square have shifted (sheared) away from each

[5] For a derivation, see, for example, Currie (2012).

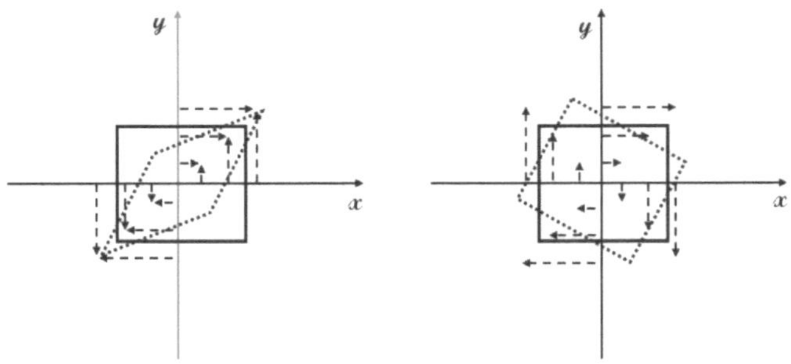

Figure 2.2. *Left*: An example of a simple flow in two dimensions for which the rotation rate is zero while the shearing rate nonzero and positive. The dashed arrows depict the flow velocity. The opposite sides of the initial square (black solid line) are seen to have shifted (sheared) with respect to each other after some time δt to form a diamond-shaped figure (dotted lines). *Right*: A similar example, for which the shearing rate is zero while the rotation rate is nonzero. In this case, the initial square is seen to have only rotated around its center, as indicated by the dotted-line square.

other. We also note that this flow *stretches* the region, making it more elongated, while preserving its area, as implied by the fact that $\nabla \cdot \boldsymbol{u} = \partial u_x/\partial x + \partial u_y/\partial y = 0$ for this flow. On the other hand, this region has no tendency to rotate as a solid body, in agreement with its rotation rate being identically zero.

Conversely, the right panel shows a flow defined by $u_x = y/2$ and $u_y = -x/2$. In this case, we find that

$$r_{ij} = \frac{1}{2}\begin{pmatrix} 0 & 1 \\ -1 & 0 \end{pmatrix},$$

and

$$s_{ij} = \begin{pmatrix} 0 & 0 \\ 0 & 0 \end{pmatrix},$$

while from the figure it is clear that, in this case, the square region will simply rotate clockwise as a consequence of the flow, with no tendency toward being sheared or stretched in any way.

2.3.2 Relation between the Stress and Deformation-Rate Tensors

We can now discuss the relation between the *cause*, the stress tensor σ_{ij}, and its *effect*, the deformation-rate tensor d_{ij}. This is analogous to the cause being the force and the effect being the acceleration in Newton's Second Law.

In what follows, we will make the assumption that the relation between the stress tensor and the deformation-rate tensor is *linear*. Fluids that satisfy this condition are called *Newtonian fluids*. Non-Newtonian fluids are out of the scope of this book.

In order to determine σ_{ij}, we note a number of considerations that it must satisfy:

1. If the fluid is at rest, the only possible source of stress is the thermal pressure, because there are no relative bulk motions between different layers of the fluid that can exert friction.

2. The stress tensor must only depend on the deformation-rate tensor and on the external pressure.
3. Since there is no shearing action during solid-body rotation, the shear stresses for this type of motion must be zero.
4. Since there are no preferred directions in the fluid, the stress tensor must be isotropic.

Conditions 1 and 2 require σ_{ij} to be of the form

$$\sigma_{ij} = -P\delta_{ij} + \tau_{ij}, \tag{2.43}$$

where P is the thermal pressure, δ_{ij} is the Kroenecker delta, and τ_{ij} is the *shear-stress tensor*, which depends only on the motion of the fluid.

To apply condition 2, we note that both the shear-stress and the deformation-rate tensors are *rank-2* tensors, thus having nine elements each. A linear relation between them implies that each of the nine components of σ_{ij} will depend on a linear combination of each of the nine elements of d_{ij}, thus requiring $9^2 = 81$ coefficients, which we denote as a fourth-rank tensor β_{ijkl}. However, condition 3 requires that τ_{ij} be zero when the motion is a simple solid-body rotation, and thus it can only depend on the shearing rate. We then write the combined application of conditions 2 and 3 as

$$\tau_{ij} = \beta_{ijkl} s_{kl}. \tag{2.44}$$

This expression still leaves us with 81 elements of β_{ijkl} to determine. However, application of condition 4 provides an enormous simplification, since it can be shown (see, e.g., Currie 2012) that the most general isotropic tensor of rank 4 is of the form

$$\lambda\delta_{ij}\delta_{kl} + \mu_u(\delta_{ik}\delta_{jl} + \delta_{il}\delta_{jk}) + \gamma(\delta_{ik}\delta_{jl} - \delta_{il}\delta_{jk}),$$

where λ, μ_u, and γ are scalars. Thus, we are left with only three parameters to determine. To further reduce their number, we note that condition 3 requires that the coefficient β_{ijkl} be symmetric, as it must satisfy the same condition as the shearing rate tensor. Thus, the coefficient of the antisymmetric part of the above expression, γ, must be zero, and so we finally write

$$\tau_{ij} = \frac{1}{2}[\lambda\delta_{ij}\delta_{kl} + \mu_u(\delta_{ik}\delta_{jl} + \delta_{il}\delta_{jk})]\left(\frac{\partial u_k}{\partial x_l} + \frac{\partial u_l}{\partial x_k}\right). \tag{2.45}$$

Now, using the properties of the Kroenecker delta, we note that

$$\frac{1}{2}\lambda\delta_{ij}\delta_{kl}\left(\frac{\partial u_k}{\partial x_l} + \frac{\partial u_l}{\partial x_k}\right) = \lambda\delta_{ij}\frac{\partial u_k}{\partial x_k} \tag{2.46}$$

$$\frac{1}{2}\mu_u\delta_{ik}\delta_{jl}\left(\frac{\partial u_k}{\partial x_l} + \frac{\partial u_l}{\partial x_k}\right) = \frac{1}{2}\mu_u\left(\frac{\partial u_i}{\partial x_j} + \frac{\partial u_j}{\partial x_i}\right), \tag{2.47}$$

and

$$\frac{1}{2}\mu_u \delta_{il}\delta_{jk}\left(\frac{\partial u_k}{\partial x_l} + \frac{\partial u_l}{\partial x_k}\right) = \frac{1}{2}\mu_u\left(\frac{\partial u_j}{\partial x_i} + \frac{\partial u_i}{\partial x_j}\right). \tag{2.48}$$

Therefore, the expression for the shear-stress tensor finally becomes

$$\tau_{ij} = \lambda\delta_{ij}\frac{\partial u_k}{\partial x_k} + \mu_u\left(\frac{\partial u_i}{\partial x_j} + \frac{\partial u_j}{\partial x_i}\right), \tag{2.49}$$

and we are left with the *constitutive equation for a Newtonian fluid*

$$\sigma_{ij} = -P\delta_{ij} + \lambda\delta_{ij}\frac{\partial u_k}{\partial x_k} + \mu_u\left(\frac{\partial u_i}{\partial x_j} + \frac{\partial u_j}{\partial x_i}\right), \tag{2.50}$$

in which the remaining coefficients, λ and μ_u, must be determined empirically. As we shall see now, these are the viscosity coefficients.

Indeed, we see that the coefficient μ_u multiplies the shear rate in Equation (2.50). If we imagine a simple two-dimensional, incompressible ($\nabla \cdot \boldsymbol{u} = 0$) shear flow such that $u_x = u_x(y)$ and $u_y = 0$, then we see that $\sigma_{12} = \sigma_{21} = \mu_u du_x/dy$, while $\sigma_{11} = \sigma_{22} = -P$. Thus, the force in the x-direction is (cf. Equation (2.23)) $dF_x = \sigma_{xy} n_y dA$, where dA is the area element over which the force is being computed. We thus get $dF_x = \mu_u(du_x/dy)dA$, and the force is proportional to the shearing velocity gradient, with proportionality constant μ_u. Such a proportionality is the defining property of Newton's law of viscosity, in which case the proportionality constant is called the *dynamic viscosity*.

On the other hand, the parameter λ is usually referred to as the *second viscosity coefficient* or *bulk viscosity*, and, as we can see from Equation (2.50), it only acts when the flow is compressible ($\nabla \cdot \boldsymbol{u} \neq 0$), being related to the dissipation by viscosity in the propagation of sound and shock waves.

Using the explicit form of the stress tensor given by Equation (2.50), we can then calculate the form of the stress terms that enter the momentum equation, Equation (2.27). We find

$$\begin{aligned}\frac{\partial\sigma_{ij}}{\partial x_j} &= \frac{\partial}{\partial x_j}\left[-P\delta_{ij} + \lambda\delta_{ij}\frac{\partial u_k}{\partial x_k} + \mu_u\left(\frac{\partial u_i}{\partial x_j} + \frac{\partial u_j}{\partial x_i}\right)\right] \\ &= -\frac{\partial P}{\partial x_i} + \lambda\frac{\partial}{\partial x_i}\frac{\partial u_k}{\partial x_k} + \mu_u\left(\frac{\partial^2 u_i}{\partial x_j^2} + \frac{\partial}{\partial x_i}\frac{\partial u_j}{\partial x_j}\right),\end{aligned} \tag{2.51}$$

where in the last step we have used the fact that the partial derivatives with respect to x_i and x_j commute because the two are independent variables. In vector notation:

$$\nabla \cdot \overleftrightarrow{\sigma} = -\nabla P + \lambda\nabla\nabla\cdot\boldsymbol{u} + \mu_u(\nabla^2\boldsymbol{u} + \nabla\nabla\cdot\boldsymbol{u}).$$

Finally, substituting Equation (2.51) into the tensor momentum equation per unit mass, Equation (2.27), we obtain

$$\frac{\partial u_i}{\partial t} + u_j \frac{\partial u_i}{\partial x_j} = -\frac{1}{\rho}\frac{\partial P}{\partial x_i} + \frac{\lambda}{\rho}\frac{\partial}{\partial x_i}\frac{\partial u_k}{\partial x_k} + \frac{\mu_u}{\rho}\left(\frac{\partial^2 u_i}{\partial x_j^2} + \frac{\partial}{\partial x_i}\frac{\partial u_j}{\partial x_j}\right) + a_i, \tag{2.52}$$

which is known as the *Navier-Stokes (NS) equation*. In vector notation,

$$\frac{\partial \boldsymbol{u}}{\partial t} + \boldsymbol{u} \cdot \nabla \boldsymbol{u} = -\frac{1}{\rho} \nabla P + \frac{\lambda}{\rho} \nabla \nabla \cdot \boldsymbol{u} + \frac{\mu_u}{\rho}(\nabla^2 \boldsymbol{u} + \nabla \nabla \cdot \boldsymbol{u}) + \mathbf{a}. \tag{2.53}$$

Per unit volume, after canceling terms proportional to the full continuity equation, the NS equation takes the form,

$$\frac{\partial}{\partial t}(\rho u_i) + \frac{\partial}{\partial x_j}(\rho u_i u_j) = -\frac{\partial P}{\partial x_i} + \lambda \frac{\partial}{\partial x_i}\frac{\partial u_k}{\partial x_k} + \mu_u\left(\frac{\partial^2 u_i}{\partial x_j^2} + \frac{\partial}{\partial x_i}\frac{\partial u_j}{\partial x_j}\right) + f_i \tag{2.54}$$

in tensor notation, and

$$\frac{\partial}{\partial t}(\rho \boldsymbol{u}) + \nabla \cdot (\rho \boldsymbol{u}\boldsymbol{u}) = - \nabla P + \lambda \nabla \nabla \cdot \boldsymbol{u} + \mu_u(\nabla^2 \boldsymbol{u} + \nabla \nabla \cdot \boldsymbol{u}) + \mathbf{f}, \tag{2.55}$$

in vector notation.

Note that the terms in the velocity divergence, $\partial u_k/\partial x_k$, vanish for incompressible flows. Note also that the coefficient μ_u/ρ is often referred to as the *kinematic viscosity*, denoted ν. Finally, note that this equation, neglecting viscous terms, is known as the *Euler* equation.

Insight:

It is important to note that the viscous terms all involve second spatial derivatives of the velocity. This is a manifestation of the fact that viscosity is important in the presence of strong velocity gradients, while it is inactive in the case of smooth, uniform flows. Equivalently, we can think of the effect of viscosity as tending to produce uniform velocity fields. This is in direct competition with the role of the advection term, which describes the transport of momentum in a certain direction by the velocity component in another direction, producing net *mixing*. We will return to this in Chapter 3.

2.3.3 The Conductive and Radiative Heat Fluxes

A second important relation between external conditions and effects on the fluid is that between the heat flux vector and the temperature gradient. Fourier's law of heat conduction states that heat flows from high- to low-temperature regions in proportion to the negative temperature gradient, and so the *constitutive equation for the heat flux* is

$$q_j = -\kappa \frac{\partial T}{\partial x_j}, \tag{2.56}$$

where T is the temperature and κ is the *thermal conductivity*.

It is important to note that, in astrophysical applications, another important (and often dominant) form of heat transport is through radiation. It is customary to split the net effect of the radiative heat exchange into the *heating and cooling rates per unit volume* $\rho\Gamma$ and $n_{\rm H}^2\Lambda$, where $n_{\rm H} = \rho/m_{\rm H}$ is the hydrogen number density of the gas and $m_{\rm H}$ is the hydrogen atom mass. The *cooling efficiency* Λ is therefore the cooling rate per hydrogen atom per collision.

Thus, we write the internal energy conservation equation per unit mass, Equation (2.36) including the heat conduction and *radiative heating and cooling terms* as

$$\frac{\partial e}{\partial t} + \boldsymbol{u} \cdot \nabla e = \frac{1}{\rho}(\overleftrightarrow{\sigma} \cdot \nabla) \cdot \boldsymbol{u} - \frac{1}{\rho} \nabla \cdot (\kappa \nabla T) + \Gamma - \frac{n_{\rm H}\Lambda}{m_{\rm H}}. \tag{2.57}$$

Now, in view of the expression for the stress tensor, Equation (2.43), we rewrite the previous equation as

$$\frac{\partial e}{\partial t} + \boldsymbol{u} \cdot \nabla e = -\frac{P}{\rho} \nabla \cdot \boldsymbol{u} + \frac{1}{\rho}(\overleftrightarrow{\tau} \cdot \nabla) \cdot \boldsymbol{u} - \frac{1}{\rho} \nabla \cdot (\kappa \nabla T) + \Gamma - \frac{n_{\rm H}\Lambda}{m_{\rm H}}. \tag{2.58}$$

This form shows that the stress tensor makes two types of contribution to the internal energy equation: one from the thermal pressure and another from the shear-stress tensor τ_{ij}. These correspond to adiabatic heating from PdV work and to heating by friction, respectively. The former is very important in astrophysical processes, and so we now obtain an expression for it.

A first point to note is that, since it describes the PdV work done on the fluid parcel, this term represents the *adiabatic* contribution to the parcel's change in internal energy. Deviations from adiabaticity are described by the other (radiative, frictional, and conductive) terms in the equation. Therefore, we can assume an adiabatic process when dealing with this term, to write the pressure in terms of the density and internal energy. In this case, recalling that $1/\rho$ is the specific volume, or volume per unit mass in the gas, we can write the first law of thermodynamics per unit mass as

$$de = -Pd\left(\frac{1}{\rho}\right), \tag{2.59}$$

where we have taken the change in specific entropy $ds = 0$ due to adiabaticity. Also, for this type of process, we can write $P = A\rho^\gamma$, where A is a constant and γ is the specific heat ratio of the gas. We therefore find that

$$de = A\rho^{\gamma-2}d\rho, \tag{2.60}$$

and, integrating,

$$e = \frac{A\rho^{\gamma-1}}{\gamma - 1}, \tag{2.61}$$

so that we finally obtain, for an adiabatic process,

$$P = (\gamma - 1)\rho e. \tag{2.62}$$

Using Equation (2.43), inserting the above result into Equation (2.58), and incorporating the factor $1/m_{\rm H}$ into Λ, we finally obtain the equation of internal energy conservation, which is nothing other than the first law of thermodynamics:

$$\begin{aligned} \frac{\partial e}{\partial t} + \boldsymbol{u} \cdot \nabla e = &- (\gamma - 1)e \nabla \cdot \boldsymbol{u} + \frac{1}{\rho}(\boldsymbol{\tau} \cdot \nabla) \cdot \boldsymbol{u} \\ &- \frac{1}{\rho} \nabla \cdot (\kappa \nabla T) + \Gamma - n_{\rm H}\Lambda. \end{aligned} \tag{2.63}$$

In astrophysical applications, the transfer of heat by friction (second term on the right-hand side) is often neglected. On the other hand, in Chapter 4, we will discuss extensively the thermodynamic consequences of heat transport by radiation and thermal conduction in these flows.

2.4 The Fluid Equations from Boltzmann's Equation

The Boltzmann equation constitutes a more general conservation principle than the fluid equations. It can be applied to a wide variety of systems made up of microscopic particles, such as collisionless dark matter particles, atoms and molecules, and photons. When applied to ensembles of atoms or molecules, it allows the derivation of the fluid equations, while it leads to the radiative transfer equation in the case of photons. Here we state it from physical considerations, and describe the derivation of the fluid equations from it.

The Boltzmann equation considers *phase space*, the space of the position ($\boldsymbol{x}$) and momentum ($\boldsymbol{p}$) coordinates of an ensemble of particles. In particular, we will consider the case when the particles are all identical. We are interested in the conservation of the number of particles in a given six-dimensional volume element of phase space.

To this end, it is convenient to separate the types of forces that may act on the particles into two classes: *smooth* and *microscopic*. The former may be considered as being applied to the whole ensemble of particles, and to vary smoothly with position. These correspond to the *conservative forces* of classical mechanics, such as the gravitational or electrostatic forces. Instead, the microscopic forces may be considered as being applied to individual particles in a random way, as is the case of particle-particle collisions in the fluid. Moreover, these typically have durations much shorter than the characteristic timescales of the evolution of the particle ensemble.

We note then that the contribution of the smooth forces to the Lagrangian derivative of a physical quantity must be zero, since the motion of these particles *defines* the evolution of the Lagrangian volume element of the gas, since it is defined to move together *with* the gas. Therefore, we write

$$\left(\frac{df}{dt}\right)_{\rm sm} = 0. \tag{2.64}$$

On the other hand, collisions between particles may abruptly change the number of particles in the volume element, because these individual interactions do not affect its global evolution. These changes occur essentially instantaneously at any position in phase space, and can thus be described by $(\partial f/\partial t)_{\rm coll}$. Therefore, the total change of f is given by

$$\frac{df}{dt} = \left(\frac{\partial f}{\partial t}\right)_{\rm coll}. \tag{2.65}$$

We now expand this equation. Let $f(\boldsymbol{x}, \boldsymbol{p}, t)\, d^3x\, d^3p$ be the number of particles within a volume $d^3x\, d^3p$ around the position $(\boldsymbol{x}, \boldsymbol{p})$ in phase space. As we have done before, we consider that $\boldsymbol{x} = \boldsymbol{x}(t)$ and $\boldsymbol{p} = \boldsymbol{p}(t)$, noting that $\dot{\boldsymbol{x}} = \boldsymbol{v}$ and $\dot{\boldsymbol{p}} = \boldsymbol{F}$, where $\boldsymbol{v}$ is the velocity of the particles, and $\boldsymbol{F}$ is the applied force, which we assume depends only on position, as is the case for conservative forces. Then, the Boltzmann equation reads

$$\begin{aligned}\frac{df}{dt} &= \frac{\partial f}{\partial t} + \frac{\partial f}{\partial x_i}\frac{dx_i}{dt} + \frac{\partial f}{\partial p_i}\frac{dp_i}{dt}\\ &= \frac{\partial f}{\partial t} + v_i\frac{\partial f}{\partial x_i} + F_i\frac{\partial f}{\partial p_i} = \left(\frac{\partial f}{\partial t}\right)_{\rm coll}.\end{aligned}$$

One particular usefulness of the Boltzmann equation is that, since f is a distribution function (i.e., a density in phase space, or a number of particles per spatial and momentum interval), it allows us to compute average values of physical quantities of interest. For example, the particle number density is

$$n(x_i) = \int_{V_p} f(x_i) d^3p, \tag{2.66}$$

where V_p is the volume in the momentum coordinates. That is, the volume number density of particles, which depends on position x_i, is the density in phase space, integrated over all possible momenta. Essentially, Equation (2.66) gives the total number of particles in a volume element, regardless of how fast they are moving.

Also, the *velocity-averaged* value of any quantity α is given by the first *moment* of the distribution function:

$$\langle\alpha\rangle_v(x_i) = \frac{1}{n}\int_{V_p} \alpha f(x_i) d^3p. \tag{2.67}$$

For convenience, in what follows we will drop the subscript $\boldsymbol{v}$ from the average symbol, but it must be kept in mind that this is *not* a spatial average. Rather, averaging over velocity leaves us with a position-dependent quantity, such as, for example, the particle number density $n(x_i)$. For compactness, in what follows we will also drop the explicit dependence on x_i of the momentum-averaged quantities.

The expression of *spatial* conservation for any quantity α may then be obtained by multiplying the Boltzmann equation by α and then integrating over the momenta; i.e., by taking the moments of the Boltzmann equation:

$$\int_{V_p} \left(\alpha \frac{\partial f}{\partial t} + \frac{\alpha p_i}{m} \frac{\partial f}{\partial x_i} + \alpha F_i \frac{\partial f}{\partial p_i} \right) d^3p = \int_{V_p} \alpha \left(\frac{\partial f}{\partial t} \right)_{\rm coll} d^3p, \tag{2.68}$$

where we have used the fact that $v_i = p_i/m$, with m being the particle mass.

We will consider in particular the cases when α is the particle's mass m (a constant) and the linear momentum mv_i. The case when α equals the kinetic energy $mv_i^2/2$ will be left as an exercise for the reader. The first two quantities are conserved in any kind of collisions among the particles, while the third is only conserved if the collisions are elastic, which we will assume to be the case. Therefore, the right-hand-side of Equation (2.68) will be zero in all three cases. Moreover, these quantities are conserved (i.e., they are *integrals of motion*) during the collisionless parts of a particle's trajectory, and so they do not explicitly depend on time.

To proceed further, we note the following intermediate results:

1.

$$\int_{V_p} \alpha \frac{\partial f}{\partial t} d^3p = \int_{V_p} \frac{\partial(\alpha f)}{\partial t} d^3p = \frac{\partial}{\partial t} \int_{V_p} \alpha f d^3p = \frac{\partial}{\partial t}(n\langle\alpha\rangle),$$

where we have used the facts that α does not depend explicitly on time, since it is an integral of motion, and that the *partial* derivative with respect to time can be pulled out of the integral over p_i.

2.

$$\begin{aligned}\int_{V_p} \alpha \frac{p_i}{m} \frac{\partial f}{\partial x_i} d^3p &= \int_{V_p} \frac{\partial}{\partial x_i} \left(\frac{\alpha p_i f}{m} \right) d^3p - \int_{V_p} f \frac{p_i}{m} \frac{\partial \alpha}{\partial x_i} d^3p \\ &= \frac{\partial}{\partial x_i} \int_{V_p} \left(\frac{\alpha p_i f}{m} \right) d^3p - \int_{V_p} f \frac{p_i}{m} \frac{\partial \alpha}{\partial x_i} d^3p \\ &= \frac{\partial}{\partial x_i}(n\langle\alpha v_i\rangle) - n\left\langle v_i \frac{\partial \alpha}{\partial x_i} \right\rangle,\end{aligned}$$

where we have integrated by parts in the first step.

3.

$$\begin{aligned}\int_{V_p} \alpha F_i \frac{\partial f}{\partial p_i} d^3p &= \int_{V_p} \left[\frac{\partial}{\partial p_i}(\alpha F_i f) - \alpha f \frac{\partial F_i}{\partial p_i} - f F_i \frac{\partial \alpha}{\partial p_i} \right] d^3p \\ &= \int_{S_p} \alpha f F_i n_i \, dS_p - \int_{V_p} f F_i \frac{\partial \alpha}{\partial p_i} d^3p \\ &= -n \left\langle F_i \frac{\partial \alpha}{\partial p_i} \right\rangle,\end{aligned}$$

where S_p is the surface that bounds volume V_p, n_i is its unit normal vector, and furthermore we have used the following facts: In the first equality, the second term within the brackets vanishes because F_i does not depend on the momenta p_i. Also, we note that the first term is a volume integral of a divergence. Thus, in the second equality we have applied Gauss' theorem to replace it by a surface integral over the boundary of the volume in momentum space. If we assume that the distribution function goes to zero as $p_i \to \infty$,[6] then this surface integral vanishes as well.

Using these results, Equation (2.68) (recalling its right-hand side is zero for the quantities of interest) becomes

$$\frac{\partial}{\partial t}(n\langle\alpha\rangle) + \frac{\partial}{\partial x_i}(n\langle\alpha v_i\rangle) - n\left\langle v_i \frac{\partial \alpha}{\partial x_i}\right\rangle - \frac{n}{m}\left\langle F_i \frac{\partial \alpha}{\partial v_i}\right\rangle = 0. \quad (2.69)$$

In what follows, we derive the continuity and momentum equations using this formalism, leaving the derivation of the energy equation as an exercise for the reader.

2.4.1 The Continuity Equation

Let us first consider mass conservation. For this, we set $\alpha = m$, the particle mass, in Equation (2.69). Since $m=$ cst., we note that $\langle m\rangle = m$. Also, we define $u_i \equiv \langle v_i\rangle$. That is, the fluid velocity at each spatial position is the average over the particle velocities at that position. We thus find

$$\frac{\partial}{\partial t}(nm) + \frac{\partial}{\partial x_i}(nm\langle v_i\rangle) - n\left\langle v_i \frac{\partial m}{\partial x_i}\right\rangle - \frac{n}{m}\left\langle F_i \frac{\partial m}{\partial v_i}\right\rangle = 0.$$

In this equation, the third and fourth terms vanish because m is a constant, and note that $nm = \rho$, where ρ is the fluid mass density. We thus obtain the *equation of continuity*,

$$\frac{\partial \rho}{\partial t} + \frac{\partial}{\partial x_i}(\rho u_i) = 0. \quad (2.70)$$

2.4.2 The Momentum Equation

We now set $\alpha = mv_j$ in Equation (2.69), finding

$$\frac{\partial}{\partial t}(n\langle mv_j\rangle) + \frac{\partial}{\partial x_i}(n\langle mv_i v_j\rangle) - n\left\langle v_i \frac{\partial (mv_j)}{\partial x_i}\right\rangle = nF_i\left\langle \frac{\partial (mv_j)}{\partial (mv_i)}\right\rangle,$$

Noting that $\partial v_i/\partial v_j = \delta_{ij}$ and that $\partial v_j/\partial x_i = 0$ because v_j and x_i are independent coordinates in phase space, the previous equation becomes

$$\frac{\partial}{\partial t}(\rho u_j) + \frac{\partial}{\partial x_i}(\rho\langle v_i v_j\rangle) = nF_i\delta_{ij} = nF_j = nma_j, \quad (2.71)$$

where a_j is the j-th component of the acceleration.

[6] That is, that there are no particles moving at infinite speeds.

Now, defining the velocity *fluctuation* of particle i over the mean velocity as $w_i \equiv v_i - \langle v_i \rangle = v_i - u_i$, so that $\langle w_i \rangle = 0$, we can compute the average in the second term of the left-hand side as follows:

$$\begin{aligned}\langle v_i v_j \rangle &= \langle (u_i + w_i)(u_j + w_j) \rangle \\ &= \langle u_i u_j + u_i w_j + u_j w_i + w_i w_j \rangle \\ &= u_i u_j + u_i \langle w_j \rangle + u_j \langle w_i \rangle + \langle w_i w_j \rangle \\ &= u_i u_j + \langle w_i w_j \rangle .\end{aligned}$$

With this, Equation (2.71) becomes the equation of momentum conservation

$$\frac{\partial}{\partial t}(\rho u_j) + \frac{\partial}{\partial x_i}(\rho u_i u_j) = -\frac{\partial}{\partial x_i}(\rho \langle w_i w_j \rangle) + \rho a_j. \tag{2.72}$$

Comparing the first term on the right-hand side of Equation (2.72) with the corresponding term in Equation (2.26) we see that

$$-\rho \langle w_i w_j \rangle = \sigma_{ij} = -P\delta_{ij} + \lambda \delta_{ij} \frac{\partial u_k}{\partial x_k} + \mu_u \left(\frac{\partial u_i}{\partial x_j} + \frac{\partial u_j}{\partial x_i} \right), \tag{2.73}$$

where we have used Equation (2.50) in the second equality. This shows that both the effects of thermal pressure and of viscosity (cf. Equation (2.52)) arise from the velocity fluctuations of the particles around the mean flow velocity. These thermal motions "diffuse" the momentum from one fluid parcel to its neighbors.

2.5 The Gas Dynamics Equations

2.5.1 The Equations and Their Physical Interpretation

For convenience, we now collect the fluid equations, in their most suitable form for astrophysical applications, for future reference in the book. These are the evolution equations for the *fluid variables* (density, momentum, internal energy and magnetic field) over the volume occupied by the fluid. We include the Lorentz and gravitational forces in the momentum equation, and the equation of magnetic flux conservation, Equation (2.77), which we do not derive in this book. For its derivation, we refer the reader, for example, to Shu (1992). We also include Poisson's equation, Equation (2.78), which describes the fact that the gravitational potential arises from the mass distribution. The equations in cgs units are:

$$\frac{\partial \rho}{\partial t} + \boldsymbol{u} \cdot \nabla \rho = -\rho \nabla \cdot \boldsymbol{u}, \tag{2.74}$$

$$\begin{aligned}\frac{\partial \boldsymbol{u}}{\partial t} + \boldsymbol{u} \cdot \nabla \boldsymbol{u} = &- \frac{\nabla P}{\rho} - \nabla \phi + \frac{1}{4\pi\rho}(\nabla \times \boldsymbol{B}) \times \boldsymbol{B} \\ &+ \frac{\lambda}{\rho} \nabla \nabla \cdot \boldsymbol{u} + \frac{\mu_u}{\rho}(\nabla^2 \boldsymbol{u} + \nabla \nabla \cdot \boldsymbol{u})\end{aligned} \tag{2.75}$$

$$\frac{\partial e}{\partial t} + \boldsymbol{u} \cdot \nabla e = -(\gamma - 1)e \nabla \cdot \boldsymbol{u} + \Gamma - n\Lambda + \nabla \cdot (\kappa \nabla T), \tag{2.76}$$

$$\frac{\partial \boldsymbol{B}}{\partial t} = \nabla \times (\boldsymbol{u} \times \boldsymbol{B}), \tag{2.77}$$

$$\nabla^2 \phi = 4\pi G \rho, \tag{2.78}$$

where $\boldsymbol{u}$ is the velocity field of the gas, ρ is the density, e is the internal energy per unit mass (or *specific* internal energy), P is the thermal pressure, μ_u is the dynamic viscosity, such that $\mu_u/\rho = \nu$ is the kinematic viscosity; $\boldsymbol{B}$ is the magnetic field, $\gamma \equiv c_P/c_V$ is the heat capacity ratio, with c_P and c_V being the heat capacities at constant pressure and volume, respectively. Also, κ is the thermal conductivity, and ϕ is the gravitational potential, which is governed by Poisson's equation, Equation (2.78). Note that we have chosen to write Equations (2.75) and (2.76) *per unit mass*.

The gas dynamics equations need to be complemented by the *constitutive relations*, which relate physical properties of specific materials. In our case, besides the viscous and conductive equations discussed in Section 2.3, we have the *equation of state*, which relates the thermodynamic variables, and for which we will assume throughout this book the *ideal gas* equation of state, appropriate for the very low densities of the ISM, and the relation between temperature and internal energy, involving c_V, the specific heat per unit volume:

$$P = \frac{\rho k_B T}{\mu m_H} \equiv n k_B T, \tag{2.79}$$

$$e = c_V T. \tag{2.80}$$

where k_B is Boltzmann's constant, $n = \rho/(\mu m_H)$ is the number density, μ is the mean gas particle mass, and m_H the hydrogen atom mass.

Insight:

A bit of insight is convenient at this point. First, note that Equations (2.74), (2.75), and (2.76) are written *per unit mass*. In order to write them per unit volume, they must be multiplied by the mass density.

Second, note that the left-hand sides of Equations (2.74), (2.75), and (2.76) express the *total*[7] rate of change of the physical variables of the gas (density, velocity or internal energy) *following the flow*. This change is due both to the temporal change in the value of that variable at its instantaneous position (the first term on the LHS) and to the change in that property due to the *transport* of the gradient of the conserved physical variables into the present position due to the motion of the fluid, described by the second term on the LHS, also called the "advection" term. This term, of the form $\boldsymbol{u} \cdot \nabla \alpha$, where α is any of the physical variables, describes how the

[7] Recall that the *total* derivative of a function of time and space is given by $d/dt = \partial/\partial t + (\partial \mathbf{x}/\partial t) \cdot \nabla = \partial/\partial t + \boldsymbol{u} \cdot \nabla$.

projection of $\nabla\alpha$ onto the velocity field $\boldsymbol{u}$ is transported by the latter from one location to another.

The RHS of the equations, on the other hand, represents the sources and sinks of the otherwise conserved quantities. That is, advection alone only transports the conserved quantities, but does not cause any change in the total content of the conserved quantities in the flow, a property that is reflected in the fact that the total derivative is given in general by

$$\frac{d}{dt} = \frac{\partial}{\partial t} + \boldsymbol{u} \cdot \nabla, \tag{2.81}$$

so that if all the RHS terms in the equations vanish, the total rate of change of the quantity is identically zero.

Equation (2.77) is an exception to this general description, since in this case the conserved quantity is the magnetic flux, not the magnetic energy nor the field itself. A detailed derivation of this equation can be found in Chapter 21 of Shu (1992). It has the important implication that, in the absence of resistivity (the so-called *ideal MHD* limit), the magnetic field is "frozen" into the fluid: the gas can flow freely along magnetic field lines, but motions perpendicular to the magnetic field drag the magnetic field lines with them. However, this property is lost in the presence of diffusive or dissipative (resistive) effects.

In the mass conservation equation, Equation (2.74), also known as the *continuity equation*, we note that the source term is the negative divergence (i.e., the *convergence*) of the velocity field around the fluid parcel of density ρ. This means that the gas must flow (converge) toward the location of the fluid parcel in order to cause an increase of the parcel's density. Incidentally, in an incompressible flow, the density is constant everywhere and at all times, so that $\nabla \cdot \boldsymbol{u} = 0$. In this case, the velocity field is said to be purely *rotational*, also referred to as *solenoidal*, or *vortical*. Conversely, a flow that has $\nabla \times \boldsymbol{u} = 0$ everywhere is called *potential* or *irrotational*.

In Equation (2.75), which represents *Newton's Second Law of Motion*, it can be seen that the source terms determining the rate of change of linear momentum of the flow are (a) minus the pressure gradient (i.e., the fluid tends to move from high- to low-pressure regions); (b) minus the gradient of the gravitational potential (the fluid tends to move toward regions where the gravitational potential is deeper); (c) the Lorentz force,

$$\boldsymbol{f}_{\rm Lor} = \frac{1}{c}\boldsymbol{j} \times \boldsymbol{B}, \tag{2.82}$$

where

$$\boldsymbol{j} \equiv ne\boldsymbol{v}' = \frac{c}{4\pi\rho}(\nabla \times \boldsymbol{B})$$

is the electric current density, $\boldsymbol{v}'$ is the drift velocity between the electrons and the ions in a (partially) ionized gas, and c is the speed of light, and (d) the viscous forces, which transfer momentum from the bulk fluid motions in the flow to microscopic thermal motions.

In the internal energy conservation equation, Equation (2.76), which, as noted before, represents the *first law of thermodynamics*, we note that the source terms are, first, the PdV adiabatic work done on the fluid parcel. This can be seen by noting that, from the continuity equation, one can obtain the following equation for the volume per unit mass of fluid, $\upsilon \equiv 1/\rho$ (cf. Equation (2.21)),

$$\frac{d\upsilon}{dt} = \frac{\nabla \cdot \boldsymbol{u}}{\rho}, \tag{2.83}$$

and so $\nabla \cdot \boldsymbol{u}/\rho$ is the rate of change of the specific volume of the fluid parcel. In addition, using the First Law on an ideal gas, it follows that $P = (\gamma - 1)\rho e$ (cf. Section 2.3.3). Thus, the first term on the RHS of Equation (2.76) is indeed the PdV work per unit mass and per unit time exerted on the fluid parcel.

The terms Γ and $n\Lambda$ in Equation (2.76) collectively represent the heating and cooling processes acting on the fluid parcel. Note that both Γ and $n\Lambda$ have units of specific power; i.e., energy per unit mass per unit time. It is customary to refer to the quantity $\mathcal{L} = n\Lambda - \Gamma$ as the *net cooling*, which is positive when the gas cools down, and negative when it heats up. Note that the inclusion of these radiative heating and cooling rates in the internal energy equation implies that the flow in general does *not* behave adiabatically, but instead every fluid parcel gains or loses heat by emitting radiation. So, it can be said that the very fact that we can *observe* the ISM through various radiative processes means that it is undergoing radiative cooling.

2.5.2 Individual Process Rates and Rescalability

The fluid Equations (2.74)–(2.76), although ominous-looking at first sight, reflect simple conservation principles of the physical variables—mass, momentum and internal energy—, and express that the total rate of change of these quantities (left-hand side) for a fluid element is given by the sum of the rates due to individual processes, given by each of the terms on the right-hand side of the corresponding equation.

The relative importance of the individual rates is given by their relative magnitudes and so, when the rate associated with one of the processes is much larger than the others, the evolution of the physical variable will be dominated by this process. This means that what really matters for the evolution of the physical variables are the ratios of the various individual rates rather than their absolute magnitudes.

This leads to the property that the gas dynamics equations are in general *rescalable*, meaning that, if a physical variable A is written as

$$A = A_0 A', \tag{2.84}$$

where A_0 is a *unit* and A' is a *nondimensional* variable, then the equations retain their original form, except for the fact that some nondimensional coefficients appear next to the various rates, effectively giving them weights. When this is done, the equations are said to be in *nondimensional* form.

For example, let us consider the momentum equation, Equation (2.75), including only the thermal pressure-gradient term in the isothermal case, for which $P = c_s^2 \rho$, with the sound speed c_s being constant. In addition, for simplicity, we write it in Lagrangian form, as

$$\frac{d\boldsymbol{u}}{dt} = -\frac{\nabla P}{\rho}. \tag{2.85}$$

We then write the variables in terms of a nondimensional variable (denoted by a prime) and a unit as

$$\boldsymbol{x} = L\boldsymbol{x}'; \quad \boldsymbol{u} = u_0 \boldsymbol{u}'; \quad \rho = \rho_0 \rho'; \quad P = c_s^2 \rho_0 \rho'; \quad t = t_0 t' \equiv \frac{L}{u_0} t'.$$

This implies that the differential operators rescale as

$$\frac{\partial}{\partial x_i} = \frac{1}{L}\frac{\partial}{\partial x_i'}; \quad \frac{\partial}{\partial t} = \frac{u_0}{L}\frac{\partial}{\partial t'}.$$

Therefore, Equation (2.85) can be rewritten as

$$\frac{u_0^2}{L}\frac{d\boldsymbol{u}'}{dt'} = -\frac{c_s^2}{L}\frac{\nabla' P'}{\rho'}, \tag{2.86}$$

from which we obtain

$$\frac{d\boldsymbol{u}'}{dt'} = -\frac{1}{M_0^2}\frac{\nabla P'}{\rho'}, \tag{2.87}$$

where $M_0 = u_0/c_s$ is the *Mach number of the velocity unit*, and represents the relative weight of the two terms appearing in the equation depending on the characteristic values of the physical parameters (in this case, the characteristic flow velocity u_0 and the sound speed, c_s), which are collected in a single nondimensional parameter. The other rate terms can be similarly treated, and so each term has a weight that depends on the characteristic values of the variables appearing in it.

In particular, the property of rescalability is the reason why two flows at different size scales behave similarly, if the relative values of their characteristic parameters are the same. This allows, for example, the study of the aerodynamic flow around cars and airplanes using wind tunnels and small-scale models of the vehicles. It also allows using the same numerical simulation to describe two different-scale objects, so long as their nondimensional parameters are the same.

2.5.3 Characteristic Timescales

Another important point to notice is that, since the gas dynamics equations give the rate of change of the fluid variables, then dividing each equation by its corresponding fluid variable gives an expression with units of inverse time. In this case, the terms on the right-hand-side of each equation give the inverse of the *characteristic time* for each physical process. Therefore, the nondimensional parameters discussed in Section 2.5.2 can also be understood as the ratios of the characteristic timescales associated to the various source terms. For example, the parameter M_0^2 appearing in Equation (2.87) represents the ratio of the characteristic timescale of the pressure-gradient term to that of the advection term. Also, it can be seen that the processes with the *shortest* timescales will be the ones dominating the evolution of the flow, as they correspond to the largest values of the rates.

2.5.4 Self-Similarity

In addition, the solutions of the fluid Equations (2.74)–(2.76) often exhibit the important property known as *self-similarity*, which means that the flow variables at a certain spatial position and a certain moment in time take the same value as they did at another position and another time. This happens far from the *initial* or *boundary* conditions, where the flow has had a chance to become nearly *stationary*. In this case it is convenient to rewrite the equations in terms of a single independent variable, called the *similarity variable*, that combines the spatial and temporal dimensions, and both the spatial profile as well as the temporal evolution of the fluid along its path can be represented by a single plot. A simple example of this behavior is that of *traveling waves*. For example, for a wave traveling along the x direction, the fluid variables can in general be written as

$$A(x, t) = A_0 \sin(ikx - \omega t),$$

where A_0 is the amplitude, i is the imaginary unit, k is the *wavenumber* or spatial frequency, and ω is the temporal frequency. Then, at any time t, the value of A at position $x = \omega t/k$ is the same. Equivalently, it can be said that the value of A is *transported* downstream by the flow velocity. We will see an example of this type of treatment applied to gravitational collapse flow in Section 7.3.

2.5.5 Magnetic Concepts: Field Lines, Flux Tubes, and Field Freezing

A fundamental notion in magnetohydrodynamics (MHD) is that of *magnetic field lines*, which are graphical aids for visualizing the magnetic (or any other vector) field. They are defined as lines that are locally tangent to the magnetic field throughout their length. From these, the notion of a *flux tube* can also be constructed, which is another graphical aid defined as a tubular region such that its curved surface is everywhere parallel to the magnetic field lines.

Ideal (non-resistive) MHD is the regime described by Equation (2.77), the flux conservation equation, with no resistive nor diffusive additional terms. One of the fundamental features of this type of flow is that, if the velocity is parallel to the magnetic field, then the magnetic field remains unperturbed, as can be seen from setting $\boldsymbol{u} \| \boldsymbol{B}$ in Equation (2.77). Conversely, when $\boldsymbol{u} \perp \boldsymbol{B}$, the magnetic field lines are "dragged" and bent by the velocity, similarly to a bow's string being pulled by the archer's finger. More generally, when $\boldsymbol{u} \nparallel \boldsymbol{B}$, the Lorentz force tends to reorient the velocity so that it tends to become parallel to the magnetic field again. The net effect is that the gas can flow freely along field lines without distorting the field, while instead it drags the field lines and feels a restoring force due to the magnetic tension when its velocity is not parallel to the field. This property is known as *flux (or field) freezing*. Note, however, that the "freezing" is only with respect to motions perpendicular to the magnetic field. Motions parallel to the magnetic field remain "loose", in the sense that they are not opposed by the field nor the field is disturbed by them. When resistive or diffusive terms are included, then the magnetic field is not perfectly "frozen" in the gas.

Note also that Maxwell's equation describing the non-existence of magnetic monopoles, $\nabla \cdot \boldsymbol{B} = 0$, implies that magnetic field lines, and consequently flux tubes as well, do not end. They either extend to infinity or form closed loops.

2.5.6 The Cost of Simplifying the Equations of Motion

An important, somewhat historical perspective, is that, prior to the development of large computers that allowed solving the gas dynamics equations in their full form, solutions used to be found under simplifying assumptions, such as dropping terms in the equations, linearizing the equations by means of expanding the terms in Taylor series and keeping only up to first-order terms, or assuming similarity conditions under stationarity, etc. These strategies, although very powerful and helpful in understanding the onset of instabilities, or the nature of the flow under stationary situations, however missed the incredible richness and chaotic[8] nature of the full solutions of the equations, that is only revealed through numerical simulation.

The simplest example of this loss is that turbulence cannot be represented without including the full nonlinear advection term in the momentum equation, since it is this nonlinearity that gives rise to turbulent flow. This will be the subject of the following chapter. Here we can mention, however, that simplification is indeed useful and necessary, and should certainly be used to isolate the phenomena of interest, but one must be careful as to which phenomena may be eliminated by the simplifications assumed.

References

Currie, I. 2012, Fundamental Mechanics of Fluids (Civil and Mechanical Engineering Series) (4th ed; London: Taylor and Francis)

Shu, F. H. 1992, Physics of Astrophysics (Mill Valley, CA: Univ. Science Books), Vol II

[8] Here, we use the term "chaotic" in the sense of nonlinear dynamics and complex systems, meaning that infinitesimal perturbations to the initial conditions lead to large differences in the evolution of the physical variables after finite times—the so-called "butterfly effect".

AAS | IOP Astronomy

Interstellar Flow and Star Formation

Enrique Vázquez-Semadeni

Chapter 3

Turbulence

Reference: "Turbulence in Fluids", Lesieur, 2008

3.1 What Is Turbulence? The Reynolds Number

Before starting our discussion of turbulence, it is important to note that the vast majority of the theoretical developments on turbulent flows refer to the *incompressible* case, since this is the case for most terrestrial flows. As mentioned in Section 2.2.3.1, an incompressible flow is characterized by $d\rho/dt = 0$, which implies $\nabla \cdot \boldsymbol{u} = 0$. Thus, in this chapter, we will first study the results for this case, and then we will discuss some known extensions of those results to the compressible case, which is most relevant for astrophysical flows.

Turbulence is a highly disordered and chaotic[1] *flow regime* that occurs in fluids when the advection term in the momentum equation, Equation (2.75), is dominant. This is because this term is responsible for the transport of the momentum gradient in the direction of the velocity vector. This can be seen from the fact that the advection term, $\boldsymbol{u} \cdot \nabla \boldsymbol{u}$, in tensor notation is written as $u_j \partial_j u_i$, indicating that the gradient of the i-th component of the velocity is projected onto the velocity vector. That is, the spatial variation of the velocity (the specific momentum), described by the gradient, is transported by the velocity itself, causing a change of the momentum at a neighboring position. Therefore, the advection term tends to *stir* the fluid, causing *mixing*.

[1] In nonlinear dynamics, the term *chaotic* has the very precise mathematical meaning of *sensitivity to initial conditions*, as described below.

doi:10.1088/2514-3433/adcb28ch3

On the other hand, as mentioned at the end of Section 2.3.2, the viscous term has the opposite effect, of tending to homogenize the velocity field. Thus, a flow is expected to become turbulent when the ratio of the magnitude of the advection (mixing) term to that of the viscous (smoothing) term is sufficiently large. This ratio is known as the *Reynolds number*, *Re*, which is defined as

$$Re \equiv \frac{|\boldsymbol{u} \cdot \nabla \boldsymbol{u}|}{|\nu \nabla^2 \boldsymbol{u}|} \sim \frac{UL}{\nu}, \tag{3.1}$$

where U is a characteristic value of the velocity, L is the characteristic scale length of the flow, and ν is the kinematic viscosity, applicable in the incompressible case (cf. Equation 2.53). Experimentally, it is found that a flow becomes fully turbulent when $Re \gtrsim$ a few $\times\ 10^3$ (Figure 3.1). For example, a typical molecular clump may have a size scale $L \sim 1$ pc, a temperature $T \sim 10$ K, and a velocity dispersion $\sigma \sim$ a few km s^{-1}, which we could take as U. On the other hand, the kinematic viscosity is $\nu \approx c_s \lambda$, where c_s is the sound speed and $\lambda = (\pi n r^2)^{-1}$ is the mean free path, r being the radius of the gas particles, and n their number density. The radius of the hydrogen molecule is $r \sim 10^{-8}$ cm, and, for a molecular clump, $n \sim 10^3$cm^{-3}. Therefore, $\lambda \sim 3 \times 10^{12}$ cm $\sim 10^{-6}$ pc. Moreover, the hydrogen sound speed at $T = 10$ K is $c_s \approx 0.2$km s^{-1}. We thus find that a typical molecular cloud has a Reynolds number $Re \sim 10^7$, and is thus expected to be extremely turbulent.

Defining turbulence is an elusive task, but it is generally agreed that a turbulent flow is characterized by the following features:

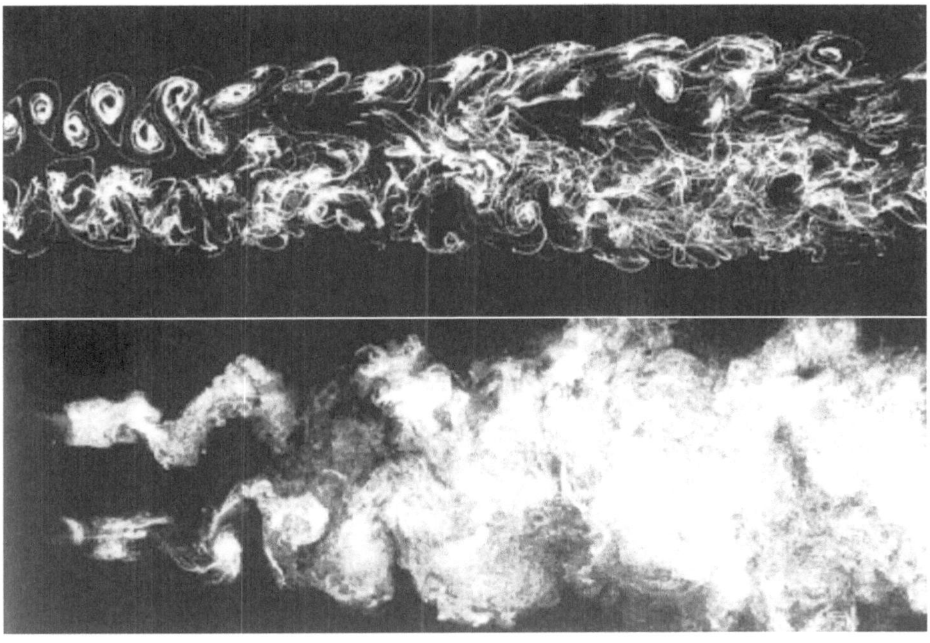

Figure 3.1. *Top:* A moderately turbulent flow at $Re \sim 240$. *Bottom*: A nearly fully turbulent flow at $Re \sim 1800$. Reproduced with permissions from Frisch (1995).

1. It contains *coherent* structures (generally, in the velocity field) on a wide range of scales. By "coherent" we mean that certain features of the flow maintain their identity over times long compared to typical timescales in the flow. Note, however, that by "coherent" we do *not* mean entire fluid patches moving ballistically through space as a single unit, but rather to regions of a given scale size across which there are well-defined velocity *differences*—for example, hurricanes, streams, vortices, shock fronts, etc.

   The existence of coherent structures over a wide range of scales is best illustrated by Richardson's poem

> "Big whirls have little whirls that feed on their velocity, and little whirls have lesser whirls, and so on to viscosity",[2]

   Mathematically, the wide range of scales is described by stating that turbulent flows contain an extremely large number of excited degrees of freedom (or *modes*). Of course, defining what is meant by "degrees of freedom" in this context is also an elusive task, but a reasonable estimate is the number of the smallest-scale fluid elements that fit in the system. In a numerical simulation, this is given by the number of computational cells used, or the number of Fourier modes used, when making a Fourier expansion of the flow.
2. The modes are able to nonlinearly exchange energy. This refers to the fact that the nonlinear advection $\boldsymbol{u} \cdot \nabla \boldsymbol{u}$ term in the momentum Equation (2.28) couples different modes and scales, as will be seen in Section 3.2.
3. The system is unpredictable in the sense that it exhibits *sensitivity to initial conditions*. This means that even initial conditions that lie arbitrarily close in phase space diverge exponentially, so that any uncertainty in the initial conditions leads to an unpredictability of the evolution, because the uncertainties grow exponentially.

   This property is often referred to as the "butterfly effect", a term coined by the pioneer of chaos theory, Edward N. Lorenz, which states that the flapping of the wings of a butterfly introduces fluctuations that are amplified by the flow, possibly making the difference between a tornado forming at some distant place several weeks later or not. The sensitivity to initial conditions is a fundamental property of *chaotic* systems, and implies that chaotic systems are predictable only in a statistical sense, while predictability in a detailed sense (i.e., predicting the value of the fluid variables everywhere and at all times) would require an arbitrarily large precision in both the initial conditions and in the computation, which are unattainable in practice.

[2] This is an adaptation of the poem by Jonathan Swift

> "Great fleas have little fleas upon their backs to bite 'em, and little fleas have lesser fleas, and so ad infinitum."

This is the reason why, for example, the weather forecast is stated in terms of probabilities, rather than precise predictions, and why it often turns out to be wrong: a fully certain prediction is impossible because it is impossible to densely sample the instantaneous physical conditions in the atmosphere and the ocean, and even if it were, it is impossible to solve the equations with infinite precision. As weather measurements and computer facilities improve, the predictions become more accurate, but prediction with 100% accuracy is impossible.

3.2 The Incompressible Momentum Equation in Fourier Space

Item 1 above states that turbulence contains coherent motions on a wide variety of scales. In the incompressible case, these coherent motions are usually thought of as vortices, or *eddies*, although in the compressible case, the coherent motions can also be compressive, and these are sometimes referred to as *shocklets*.

A natural way of describing a field (density, velocity, etc.) is to write it as a superposition of Fourier modes, since each of these is characterized by a length scale, given by its wavelength λ. This allows us to describe features of the field of size comparable to the wavelength of the corresponding Fourier mode.[3] We thus take a Fourier decomposition of the fields of the form

$$A(\boldsymbol{x}, t) = \int A_{\boldsymbol{k}}(t) e^{i\boldsymbol{k}\cdot\boldsymbol{x}} d^3k, \tag{3.2}$$

where A is the physical field (velocity components or pressure; recall that the density is constant in the incompressible case), i here is the imaginary unit, such that $i^2 = -1$, $\boldsymbol{k}$ is the wavevector of the Fourier mode, with associated wavenumber $k \equiv |\boldsymbol{k}| = 2\pi/\lambda$, and $A_{\boldsymbol{k}}$ is the corresponding Fourier *amplitude*, a complex number, which we assume depends only on time. The wavenumber k represents the angular spatial frequency of the wave; that is, the number of radians that fit in the spatial domain. We absorb all necessary normalization coefficients into the amplitude $A_{\boldsymbol{k}}$.

Under the assumption of periodic boundary conditions in a domain of size L, only modes that fit an integer number of times in the domain are allowed. Thus, the wavevectors $\boldsymbol{k}$ have discrete components, multiples of $2\pi/L$, so that the integration becomes a summation over all wavevectors. Choosing a box length $L = 2\pi$ thus gives wavevectors with integer components. Considering only the pressure and viscous terms on the right-hand side (RHS), the momentum equation, Equation (2.28) then becomes (in the incompressible case)

[3] Note that, strictly speaking, a single Fourier mode does not represent any specific spatial scale, as it is simply an infinite wave train of a certain wavelength. The well-known result from Fourier analysis, that the Fourier transform of a Gaussian function is another Gaussian function whose standard deviation is inversely proportional to that of the Gaussian pulse, implies that, in order to represent a pulse (a wave packet) of a given width, it is necessary to add Fourier modes with wavelengths in the *vicinity* of the pulse's width, with the extent of this vicinity in Fourier space being inversely proportional to the pulse's width. Informally, then, it is common to discuss as if the Fourier mode's wavelength defines a spatial scale represented by that mode.

$$\frac{\partial}{\partial t}\sum_k \boldsymbol{u}_k(t)e^{i\boldsymbol{k}\cdot\boldsymbol{x}} + \left[\sum_p \boldsymbol{u}_p(t)e^{i\boldsymbol{p}\cdot\boldsymbol{x}}\right]\cdot\nabla\left[\sum_q \boldsymbol{u}_q(t)e^{i\boldsymbol{q}\cdot\boldsymbol{x}}\right] = -\frac{1}{\rho}\sum_k \nabla\,[P_k\, e^{i\boldsymbol{k}\cdot\boldsymbol{x}}] + \nu\sum_k \nabla^2\,[\boldsymbol{u}_k(t)e^{i\boldsymbol{k}\cdot\boldsymbol{x}}],$$

where the last term corresponds to the viscous term in the incompressible case.

Note that, in the second term, we have used a different summation index for each factor of the velocity, since the entire sum from one of the factors needs to be multiplied by that from the other factor. Also, because of the linearity of the derivative operators, we can interchange them with the summation operator to get

$$\sum_k \frac{d\boldsymbol{u}_k}{dt}e^{i\boldsymbol{k}\cdot\boldsymbol{x}} + i\sum_p\sum_q(\boldsymbol{u}_p\cdot\boldsymbol{q})\boldsymbol{u}_q e^{i(\boldsymbol{p}+\boldsymbol{q})\cdot\boldsymbol{x}} = -\frac{i}{\rho}\sum_k \boldsymbol{k}P_k e^{i\boldsymbol{k}\cdot\boldsymbol{x}} - \nu\sum_k k^2\boldsymbol{u}_k e^{i\boldsymbol{k}\cdot\boldsymbol{x}},$$

where we have used the property that

$$\nabla[A_k(t)e^{i\boldsymbol{k}\cdot\boldsymbol{x}}] = i\boldsymbol{k}A_k(t)e^{i\boldsymbol{k}\cdot\boldsymbol{x}}.$$

That is, we have the highly convenient property that *the differentiation operation in real physical space, becomes a multiplication by the wavevector in Fourier space*. Also, in the above equation, we have written the temporal derivative as a total derivative, since the Fourier amplitudes only depend on time.

In order to obtain the equation for the $\boldsymbol{k}'$-th Fourier component, we compute the inverse Fourier transform of the above equation, multiplying by $e^{-i\boldsymbol{k}'\cdot\boldsymbol{x}}$ and integrating over all space. We obtain

$$\int_x \sum_k \frac{d\boldsymbol{u}_k}{dt}e^{i(\boldsymbol{k}-\boldsymbol{k}')\cdot\boldsymbol{x}}d^3x + i\int_x \sum_p\sum_q(\boldsymbol{u}_p\cdot\boldsymbol{q})\boldsymbol{u}_q e^{i(\boldsymbol{p}+\boldsymbol{q}-\boldsymbol{k}')\cdot\boldsymbol{x}}d^3x = -\int_x\sum_k\left[\frac{i}{\rho}\boldsymbol{k}P_k + \nu k^2\boldsymbol{u}_k\right]e^{i(\boldsymbol{k}-\boldsymbol{k}')\cdot\boldsymbol{x}}d^3x. \tag{3.3}$$

Now we note that, according to the definition of the delta function,

$$\int_x e^{i(\boldsymbol{k}-\boldsymbol{k}')\cdot\boldsymbol{x}}d^3x = 2\pi\ \delta(\boldsymbol{k}-\boldsymbol{k}'),$$

and so, reversing the order of the integration and summation, and dropping the prime from $\boldsymbol{k}'$, Equation (3.3) becomes (again incorporating normalization coefficients into the amplitudes)

$$\frac{d\boldsymbol{u}_k}{dt} + i\sum_{\boldsymbol{p}+\boldsymbol{q}=\boldsymbol{k}}(\boldsymbol{q}\cdot\boldsymbol{u}_p)\boldsymbol{u}_q = -\frac{i}{\rho}\boldsymbol{k}P_k - \nu k^2\boldsymbol{u}_k. \tag{3.4}$$

Note that the second term in the left-hand side is a *convolution*, which is the Fourier transform of a product, and involves pairs of wavevectors $\boldsymbol{p}$ and $\boldsymbol{q}$ such that $\boldsymbol{p}+\boldsymbol{q}=\boldsymbol{k}$. Moreover, this equation can be further manipulated by noting that, because $\boldsymbol{p}$ and $\boldsymbol{q}$ are dummy summation indices, we have the property

$$\sum_{p+q=k} (\boldsymbol{q} \cdot \boldsymbol{u}_p)\boldsymbol{u}_q = \sum_{p+q=k} (\boldsymbol{p} \cdot \boldsymbol{u}_q)\boldsymbol{u}_p,$$

and therefore

$$\sum_{p+q=k} (\boldsymbol{q} \cdot \boldsymbol{u}_p)\boldsymbol{u}_q = \frac{1}{2} \sum_{p+q=k} [(\boldsymbol{q} \cdot \boldsymbol{u}_p)\boldsymbol{u}_q + (\boldsymbol{p} \cdot \boldsymbol{u}_q)\boldsymbol{u}_p],$$

so that Equation (3.4) becomes

$$\frac{d\boldsymbol{u}_k}{dt} + \frac{i}{2} \sum_{p+q=k} [(\boldsymbol{q} \cdot \boldsymbol{u}_p)\boldsymbol{u}_q + (\boldsymbol{p} \cdot \boldsymbol{u}_q)\boldsymbol{u}_p] = -\frac{i}{\rho}\boldsymbol{k}P_k - \nu k^2 \boldsymbol{u}_k. \quad (3.5)$$

In the incompressible case, an important further simplification can be made, because in this case, the pressure amplitude can be solved for in terms of the velocity amplitudes. Indeed, taking the dot product of Equation (3.5) with wavevector $\boldsymbol{k}$ (equivalent to taking the divergence of the momentum equation), we obtain

$$P_k = - \sum_{p+q=k} \frac{\rho}{k^2} (\boldsymbol{k} \cdot \boldsymbol{u}_q)\ (\boldsymbol{k} \cdot \boldsymbol{u}_p), \quad (3.6)$$

where the time derivative term has been eliminated thanks to the incompressibility condition, Equation (2.22), which in Fourier space reads

$$\boldsymbol{k} \cdot \boldsymbol{u}_k = 0. \quad (3.7)$$

Note also that what would be a double integration in physical ($\boldsymbol{x}$) space, in Fourier space it amounts to a division by the factor k^2.

Thus, collecting all terms within the summation, the incompressible momentum equation finally reads

$$\frac{d\boldsymbol{u}_k}{dt} + \frac{i}{2} \sum_{p+q=k} \left[(\boldsymbol{q} \cdot \boldsymbol{u}_p)\boldsymbol{u}_q + (\boldsymbol{p} \cdot \boldsymbol{u}_q)\boldsymbol{u}_p - \frac{2\boldsymbol{k}}{k^2} (\boldsymbol{k} \cdot \boldsymbol{u}_q)\ (\boldsymbol{k} \cdot \boldsymbol{u}_p) \right] = -\nu k^2 \boldsymbol{u}_k. \quad (3.8)$$

The following points are worth noting:

- The incompressibility condition in Fourier space, Equation (3.7), expresses the fact that the velocity amplitude must have a zero component along the direction of its wavevector $\boldsymbol{k}$. Recalling that multiplication by the wavevector represents taking the gradient of the quantity, this means that the incompressibility condition requires that there is no *longitudinal* gradient of the velocity field, in agreement with the fact that, according to the continuity equation, Equation (2.20), the variation of the Lagrangian density depends directly on the velocity divergence (cf. Section 2.2.3).
- The momentum equation, which is a partial differential equation in real space, becomes a set of coupled *ordinary* differential equations in Fourier space, one for each mode $\boldsymbol{k}$.
- The nonlinear advection term in the momentum equation in real space is seen to become a convolution term of the form $\sum_{p+q=k}(\boldsymbol{q} \cdot \boldsymbol{u}_p)\boldsymbol{u}_q$, involving a summation over pairs of wavevectors $\boldsymbol{p}$ and $\boldsymbol{q}$ such that $\boldsymbol{p} + \boldsymbol{q} = \boldsymbol{k}$, in the

equation for mode $\boldsymbol{k}$. The sets of vectors $\boldsymbol{k}$, $\boldsymbol{p}$ and $\boldsymbol{q}$ are called *triads*. This term explicitly shows the *coupling* between different modes; i.e., the fact that the evolution of mode $\boldsymbol{k}$ is affected by other modes of different scales. This is a direct consequence of the *nonlinear* character of the advective term.

- Noting that, in the incompressible case, $\boldsymbol{k} \cdot \boldsymbol{u}_p = (\boldsymbol{p} + \boldsymbol{q}) \cdot \boldsymbol{u}_p = \boldsymbol{q} \cdot \boldsymbol{u}_p$, we see that the third term in the brackets in Equation (3.8), which is the pressure gradient term in terms of the velocity amplitudes, effectively removes the projection of each of the previous two terms onto the direction of wavevector $\boldsymbol{k}$. That is, the pressure gradient term removes the longitudinal (compressible) component of the contributions from modes $\boldsymbol{p}$ and $\boldsymbol{q}$ to the change of mode $\boldsymbol{k}$, maintaining incompressibility.
- This instantaneous effect of the pressure is possible because in the incompressible case the sound speed is assumed to approach infinity. This in turn implies that the internal energy equation is "bypassed". Indeed, in the full compressible case, the effect of compressive velocity fluctuations is to produce density fluctuations via the continuity equation. In turn, these produce fluctuations of the internal energy, and thus of the pressure, via the internal energy equation. The pressure then feeds back on the momentum equation via the $\nabla P/\rho$ term, and these pressure fluctuations propagate in the medium at the sound speed. Only when the sound speed approaches infinity is the pressure gradient capable of instantaneously removing all compressible motions from the velocity field (cf. Section 3.9).

3.3 Energy Conservation and Dissipation

3.3.1 Energy Conservation in Physical Space

A fundamental property of turbulence is that it is a dissipative process. By this it is meant that the kinetic energy in the turbulent motions is quickly dissipated by viscosity, and it is necessary to continuously inject fresh energy into the turbulence in order to keep it stationary. In the Earth, the main source of energy for the turbulent motions in the ocean and atmosphere is the Sun. In the ISM, plenty of available sources exist.

Therefore, an important quantity to compute is the rate of kinetic energy dissipation. In real space this is accomplished by taking the dot product of the incompressible NS equation in per-unit-mass form (Equation (2.52)) with the velocity vector to obtain an evolution equation for the kinetic energy:[4]

$$u_i \frac{\partial u_i}{\partial t} = -u_i u_j \frac{\partial u_i}{\partial x_j} - \frac{1}{\rho} u_i \frac{\partial P}{\partial x_i} + u_i \frac{\mu}{\rho} \frac{\partial^2 u_i}{\partial x_j^2} + u_i a_i.$$

Using the incompressibility condition, Equation (2.22), or $\partial u_i/\partial x_i = 0$ in tensor notation, to bring the velocities inside the spatial derivatives, this equation can be rewritten as

[4] Note that here i denotes a vector or tensor index, not to be confused with the imaginary unit i, which generally appears as a multiplicative factor, not as an index.

$$\frac{1}{2}\frac{\partial u_i^2}{\partial t} = -\frac{1}{2}\frac{\partial (u_i^2 u_j)}{\partial x_j} - \frac{1}{\rho}\frac{\partial (u_i P)}{\partial x_i} + u_i \frac{\mu}{\rho}\frac{\partial^2 u_i}{\partial x_j^2} + u_i a_i.$$

To compute the total rate of change of the specific kinetic energy in a volume V of the flow, we integrate over all space and apply Gauss' theorem, to get

$$\frac{dE_{\rm tot}}{dt} \equiv \frac{1}{2}\int_V \frac{\partial u_i^2}{\partial t} dV = -\frac{1}{2}\oint_S u_i^2 u_j dS_j - \frac{1}{\rho}\oint_S u_i P dS_i + \int_V u_i \frac{\mu}{\rho}\frac{\partial^2 u_i}{\partial x_j^2} dV + \int_V u_i a_i dV,$$

where S is the surface enclosing volume V. Letting $V \to \infty$ and assuming that the velocity and the pressure vanish at infinity, both surface terms vanish, and we get

$$\frac{dE_{\rm tot}}{dt} = \int_V u_i \frac{\mu}{\rho}\frac{\partial^2 u_i}{\partial x_j^2} dV + \int_V u_i a_i dV. \tag{3.9}$$

In the absence of viscosity (i.e., in *inviscid* flows) and absence of external forces, this equation reduces to

$$\frac{dE_{\rm tot}}{dt} = 0, \tag{3.10}$$

which is the expression of total specific kinetic energy conservation in the absence of viscosity and external forces.

Insight: The vanishing of the integrated advection and pressure gradient terms implies that they conserve the specific kinetic energy in the incompressible case. This result is the basis for the notion of an *energy cascade: energy that is injected by some force at large scale is only transferred through the flow, but conserved in total by the advection and pressure terms*. The only terms that can induce a change in the energy content of the flow are the viscous term and the power injected by the external forces.

3.3.2 Conservation in Fourier Space: Detailed Conservation by Triads

Dotting the momentum equation in the absence of viscosity and external forces with the (complex conjugate of the) velocity vector in Fourier space, $\boldsymbol{u}_k^*$, we can obtain the evolution equation for the kinetic energy per unit mass,

$$\frac{1}{2}\frac{du_k^2}{dt} + \Re\left\{\frac{i\boldsymbol{u}_k^*}{2}\cdot \sum_{p+q=k}\left[(\boldsymbol{q}\cdot\boldsymbol{u}_p)\boldsymbol{u}_q + (\boldsymbol{p}\cdot\boldsymbol{u}_q)\boldsymbol{u}_p - \frac{2\boldsymbol{k}}{k^2}(\boldsymbol{k}\cdot\boldsymbol{u}_q)\,(\boldsymbol{k}\cdot\boldsymbol{u}_p)\right]\right\} = 0, \tag{3.11}$$

where $u_k \equiv (\boldsymbol{u}_k \cdot \boldsymbol{u}_k^*)^{1/2} \equiv |\boldsymbol{u}_k|$ is the magnitude of the Fourier velocity amplitude $\boldsymbol{u}_k$, and $\Re$ denotes the real part, which must be taken since $|u_k|^2$ is real. Using the incompressibility condition $\boldsymbol{u}_k \cdot \boldsymbol{k} = 0$, which eliminates the third term within the brackets and allows writing, for example, $\boldsymbol{q}\cdot\boldsymbol{u}_p = (\boldsymbol{u}+\boldsymbol{p})\cdot\boldsymbol{u}_p = \boldsymbol{k}\cdot\boldsymbol{u}_p$, and noting that $\Re(iz) = -\Im(z)$ for an arbitrary complex number z, where $\Im$ denotes the imaginary part, the above expression reduces to

$$\frac{1}{2}\frac{du_k^2}{dt} + \frac{1}{2}\Im\left\{\sum_{p+q=k}\left[(\boldsymbol{k}\cdot\boldsymbol{u}_p)\left(\boldsymbol{u}_k^*\cdot\boldsymbol{u}_q\right) + (\boldsymbol{k}\cdot\boldsymbol{u}_q)\left(\boldsymbol{u}_k^*\cdot\boldsymbol{u}_p\right)\right]\right\} = 0, \tag{3.12}$$

In order to obtain the total conservation of energy, we sum over all wavevectors, to get

$$\frac{1}{2}\frac{d}{dt}\sum_k u_k^2 + \frac{1}{2}\sum_k\sum_{p+q=k} T(\boldsymbol{k}|\boldsymbol{p},\boldsymbol{q}) = 0, \tag{3.13}$$

where we have defined the *detailed energy transfer term* as

$$T(\boldsymbol{k}|\boldsymbol{p},\boldsymbol{q}) \equiv \Im\left[(\boldsymbol{k}\cdot\boldsymbol{u}_p)\left(\boldsymbol{u}_k^*\cdot\boldsymbol{u}_q\right) + (\boldsymbol{k}\cdot\boldsymbol{u}_q)\left(\boldsymbol{u}_k^*\cdot\boldsymbol{u}_p\right)\right]. \tag{3.14}$$

A fundamental property of the interaction between triads of wavevectors is that they conserve the kinetic energy. This can be seen by noting that $\boldsymbol{p} = \boldsymbol{k} - \boldsymbol{p}$ and $\boldsymbol{q} = \boldsymbol{k} - \boldsymbol{q}$, and that, upon performing the summation over $\boldsymbol{k}$ indicated in Equation (3.13), we will encounter in particular the sum $T(\boldsymbol{k}|\boldsymbol{p},\boldsymbol{q}) + T(\boldsymbol{q}|\boldsymbol{k},-\boldsymbol{p}) + T(\boldsymbol{p}|\boldsymbol{k},-\boldsymbol{q})$. Using the incompressibility condition, the result that, for any complex variable z, $\Im(z) = (z - z^*)/2i$, Equation (3.7), and the *reality condition* for complex variables, which states that, if $u_i(\boldsymbol{x})$ is a real field, then its Fourier transform satisfies

$$u_i(-\boldsymbol{k}) = u_i^*(\boldsymbol{k}), \tag{3.15}$$

we can write

$$\begin{aligned} T(\boldsymbol{k}|\boldsymbol{p},\boldsymbol{q}) = \frac{1}{2}\Im\Big\{&\left[(\boldsymbol{k}\cdot\boldsymbol{u}_p)\left(\boldsymbol{u}_k^*\cdot\boldsymbol{u}_q\right) + (\boldsymbol{k}\cdot\boldsymbol{u}_q)\left(\boldsymbol{u}_k^*\cdot\boldsymbol{u}_p\right)\right] \\ &- \left[\left(\boldsymbol{k}\cdot\boldsymbol{u}_p^*\right)\left(\boldsymbol{u}_k\cdot\boldsymbol{u}_q^*\right) + \left(\boldsymbol{k}\cdot\boldsymbol{u}_q^*\right)\left(\boldsymbol{u}_k\cdot\boldsymbol{u}_p^*\right)\right]\Big\}. \end{aligned} \tag{3.16}$$

Finally, from this expression, it can readily be shown that

$$T(\boldsymbol{k}|\boldsymbol{p},\boldsymbol{q}) + T(\boldsymbol{q}|\boldsymbol{k},-\boldsymbol{p}) + T(\boldsymbol{p}|\boldsymbol{k},-\boldsymbol{q}) = 0. \tag{3.17}$$

This result shows that, similarly to the result in real space, the advection term in the momentum equation *conserves kinetic energy*. However, in Fourier space, Equation (3.17) implies that the conservation is accomplished on a *triad-by-triad* basis. That is, each triad of wavevectors conserves energy individually. This is the fundamental physical mechanism underlying the energy cascade in the incompressible case. This proof, however, makes intensive use of the incompressibility condition, and thus cannot be used in the compressible case. We return to this in Section 3.9.

In view of the above, the conservation of specific kinetic energy in Fourier space reads

$$\frac{1}{2}\frac{d}{dt}\sum_k u_k^2 = 0. \tag{3.18}$$

This equation is the Fourier-space version of the real-space expression of specific kinetic energy conservation, Equation (3.10). Although we have arrived to it by direct calculation in Fourier space, it is noteworthy that it can also be derived by transformation of the real-space equation, using the standard result from Fourier analysis known as *Parseval's theorem*, which states that, for a real field $u(\boldsymbol{x})$, its Fourier transform satisfies

$$\int_{V_x} u^2(\boldsymbol{x})\, d^3x = \int_{V_k} u_{\boldsymbol{k}}^2\, d^3k, \tag{3.19}$$

so that total energy conservation in real space requires conservation in Fourier space as well. What is not so clearly seen in real space, however, is the property of detailed conservation by triads in Fourier space, which is the mechanism that allows the transfer of energy from one mode (say, $\boldsymbol{k}$) to other modes (say $\boldsymbol{p}$ and $\boldsymbol{q}$) of different wavelengths and directions, thus providing the underlying physical process for the energy cascade.

3.3.3 The Dissipation of Energy and the Singular Nature of the Inviscid Case

When viscosity cannot be neglected, but still in the absence of external forces (or *driving*, or *forcing*), Equation (3.9) becomes

$$\frac{dE_{\rm tot}}{dt} = \int_V \nu u_i \frac{\partial^2 u_i}{\partial x_j^2} dV = \int_V \nu \boldsymbol{u} \cdot \nabla^2\, \boldsymbol{u} dV, \tag{3.20}$$

where $\nu \equiv \mu/\rho$ is the kinematic viscosity, and in the second equality we have written the vector form of the RHS. The Fourier-space version of this equation reads

$$\frac{1}{2}\frac{d}{dt}\sum_{\boldsymbol{k}} u_{\boldsymbol{k}}^2 = -\nu \sum_{\boldsymbol{k}} k^2 u_{\boldsymbol{k}}^2. \tag{3.21}$$

Since ν and $u_{\boldsymbol{k}}^2$ are positive-definite quantities, Equation (3.21) clearly implies that, in the viscous but undriven case, the kinetic energy decays monotonically.

If we include driving as well, then the full energy conservation equation reads

$$\begin{aligned}\frac{dE_{\rm tot}}{dt} &= \int_V \nu u_i \frac{\partial^2 u_i}{\partial x_j^2} dV + \int_V u_i a_i dV \\ &= \int_V \nu \boldsymbol{u} \cdot \nabla^2\, \boldsymbol{u} dV + \int_V \boldsymbol{u} \cdot \mathbf{a} dV\end{aligned} \tag{3.22}$$

in real space, and

$$\frac{1}{2}\frac{d}{dt}\sum_{\boldsymbol{k}} u_{\boldsymbol{k}}^2 = -\nu \sum_{\boldsymbol{k}} k^2 u_{\boldsymbol{k}}^2 + \sum_{\boldsymbol{k}} \boldsymbol{u}_k \cdot \mathbf{a}_k \tag{3.23}$$

in Fourier space.

If we now consider a stationary regime, in which the kinetic energy remains constant, the left-hand side (LHS) of the above equation vanishes, and we can write

$$\sum_k \boldsymbol{u}_k \cdot \mathbf{a}_k = \nu \sum_k k^2 u_k^2. \tag{3.24}$$

This expression has a profound implication: under stationary conditions, the viscous dissipation, given by the RHS, *must adjust* to whatever the energy injection rate (the LHS) may be, *even at constant viscosity*. Conversely, at a given injection rate, the dissipation rate must be the same regardless of the value of the kinematic viscosity ν. Since the kinematic viscosity ν is constant, and the RHS contains only the wavenumber and the Fourier velocity mode at that wavenumber, the adjustment can only occur if the scales (or wavenumbers) at which dissipation mainly occurs are different for different energy injection rates.

This can be understood as follows. Assume, for example, that u_k does not decay with k faster than k^{-2}. In this case, the terms in the summation of the RHS of Equation (3.24) increase with increasing k, meaning that dissipation occurs preferentially at larger wavenumbers, or smaller scales. The wavenumber $k_{\rm d}$ at which

$$\sum_k \boldsymbol{u}_k \cdot \mathbf{a}_k \approx \nu \sum_{k \ni k_{\rm d} - \frac{1}{2} < |\boldsymbol{k}| < k_{\rm d} + \frac{1}{2}} k^2 u_k^2 \tag{3.25}$$

is called the *dissipation wavenumber*. Its associated size scale, $\sim 1/k_{\rm d}$, is referred to as the *dissipation scale*. If the forcing is applied preferentially at large scales, as is the case in most turbulent flows, then the fact that most of the dissipation occurs at small scales implies that there exists a range of scales, called the *inertial range*, over which energy is not dissipated, but just transferred from the *injection scale* to the dissipation scale, as described in Section 3.3.2. This is at the basis of the theory behind the distribution of energy in wavenumber intervals, i.e., the turbulent energy spectrum, described in Section 3.5.

Moreover, the fact that dissipation occurs preferentially at the dissipation scale also has the important implication that, if the viscosity ν is varied, the dissipation scale varies accordingly so that the dissipation rate continues to balance the injection rate. This means that, in a stationary regime, *the dissipation rate is independent of the viscosity, and, for smaller viscosities, the dissipation scale becomes smaller*. This also implies that, as the viscosity becomes smaller, the inertial range becomes larger, while the dissipation rate remains constant, equal to the injection rate. Instead, in the strictly inviscid case, the dissipation rate is necessarily zero, and so a continuously driven inviscid flow would accumulate energy without limit. Thus, it is said that the inviscid case, in which $\nu = 0$, is *singular*, since it is different from the limit $\nu \to 0$.

3.4 The Energy Spectrum

The description of the flow in Fourier space allows a straightforward treatment of the amount of kinetic energy contained in coherent motions at each size scale. Let us start by writing the total, instantaneous kinetic energy *per unit mass* (or *specific* kinetic energy) in the flow as

$$
\begin{aligned}
E(t) &= \frac{1}{2}\int_{V_x} u^2(\boldsymbol{x}, t)\, d^3x \\
&= \frac{1}{2}\int_{V_k} |\boldsymbol{u_k}(t)|^2\, d^3k \\
&= \frac{1}{2}\int_{k=0}^{k=\infty}\int_{\theta_k}\int_{\phi_k} |\boldsymbol{u_k}(t)|^2\, k^2 \sin\theta_k\, dk\, d\theta_k d\phi_k,
\end{aligned}
\tag{3.26}
$$

where $u^2 = \boldsymbol{u}(\boldsymbol{x}) \cdot \boldsymbol{u}(\boldsymbol{x})$ in the first equality, and the integral is performed over the whole flow. The second equality follows from *Parseval's theorem*. Note that, physically, this expression describes the fact that the total energy is the integral of the square of the velocity either in real or in Fourier space. Finally, the third equality expands the k-integral in spherical coordinates.

The *turbulent energy spectrum* is then defined as the total kinetic energy per unit mass contained in all modes of magnitude between k and $k + dk$ (Figure 3.2):

$$
E(k) \equiv \frac{1}{2}\int_{\theta_k}\int_{\phi_k} |\boldsymbol{u_k}|^2 k^2 \sin\theta_k d\theta_k d\phi_k, \tag{3.27}
$$

so that

$$
E = \int_{k=0}^{k=\infty} E(k)\, dk, \tag{3.28}
$$

where we have omitted the explicit dependence on time. The quantity $|\boldsymbol{u_k}|^2$ is sometimes referred as the *velocity power spectrum*.

Note that all of this mathematical construction has been developed for the incompressible regime, in which $\rho =$ cst., and so the kinetic energy is simply proportional to the specific kinetic energy. This, however, is not true for the compressible case, in which there exist density fluctuations as well, and so the total

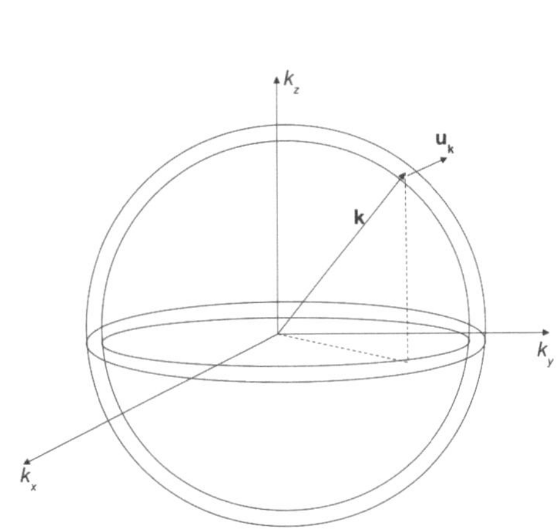

Figure 3.2. A spherical shell in Fourier (or $k-$) space showing a wavevector lying in a shell of radius k. The energy spectrum, $E(k)$, is the total kinetic energy of the amplitudes of all wavevectors in this shell.

kinetic energy needs to be calculated by integrating the kinetic energy *density* (i.e., *per unit volume*) over the volume of the flow. We will discuss this in Section 3.9.

3.5 Kolmogorov–Obukhov Theory for the Energy Spectrum and the Turbulent Energy Cascade

In 1941, Kolmogorov and Obukhov independently derived the expected functional form of the energy spectrum (and the structure functions, which we will discuss in Section 3.8) for *homogeneous* turbulence (i.e., whose statistical properties are independent of position). This result is a milestone in turbulence theory, and is often referred to as the K41 theory, which we now discuss.[5] It is based on the following assumptions:

- A very large range of scales of "coherent" motion is active in the flow. Again, by "coherent" we mean parcels of the flow that are recognizable as unique structures, such as vortices (or "eddies"), streams, etc. Because these structures have a certain characteristic scale l, they can be associated with Fourier modes of wavenumber $k \approx 2\pi/l$.
- Energy is injected primarily at large scales (which are usually referred to as the *energy injection* or *energy-containing* scales). Conversely, it is dissipated (i.e., converted to thermal energy of the fluid molecules) at small scales (as can be seen from the k^2 dependence of the viscous term in Equation (3.8), which causes this term to be important at large k, or small scales).
- Energy is *transferred* among the modes (or scales), and the transfer is *local* in Fourier space; i.e., it occurs mainly among similar wavenumbers (in triads with $p \sim q \sim k$). Together with the two previous assumptions, this implies that there is an intermediate range of wavenumbers in which energy can only *cascade* from large to small scales (small to large k). This transfer-only range is called the *inertial range* of wavenumbers.
- The system is in a stationary state, so that the rates of energy injection at large scales, of transfer at intermediate scales and dissipation at small scales are equal. In consequence, the transfer rate in the inertial range, where no energy is injected nor dissipated, must be independent of wavenumber k.

With these assumptions, the form of the spectrum in the inertial range can be estimated from dimensional analysis as follows. The (constant) energy transfer rate for eddies of scale size l can be written as

$$\varepsilon \sim \frac{u_l^2}{\tau_l}, \tag{3.29}$$

where u_l is the characteristic velocity difference across the eddy, and

$$\tau_l = \frac{l}{u_l} \tag{3.30}$$

[5] Part of the discussion in this section is based on Vázquez-Semadeni (1999).

is the characteristic, or *eddy turnover*,[6] time at scale l. Thus, the velocity differences scale with separation l as

$$u_l = (\varepsilon l)^{1/3}. \tag{3.31}$$

Now, the characteristic velocity difference at scale l can be identified with the rms specific kinetic energy contained in scales $l' < l$, which is given by the integrated energy spectrum:

$$u_l^2 = 2\int_{2\pi/l}^{\infty} E(k)dk. \tag{3.32}$$

Assuming that the spectrum has a power-law dependence on wavenumber, $E(k) \propto k^{-n}$, and substituting the scaling relation (3.31), we obtain

$$l^{2/3} \propto \int_{2\pi/l}^{\infty} k^{-n}dk \propto l^{n-1}, \quad (n > 1)$$

implying $n = 5/3$. This is the famous Kolmogorov "$-5/3$ law", and has been verified experimentally in terrestrial flows (e.g., Grant et al. 1962, Figure 3.3) and in astrophysical flows that are expected to behave nearly incompressibly (Armstrong et al. 1995).

Insight:

The Kolmogorov spectral slope is unique because it is the only one that causes the energy transfer among scales to be *independent of scale (or wavenumber)*. It is easy to check that, if the slope were steeper, the energy transfer rate would *decrease* with wavenumber, and so the small wavenumbers would transfer less energy than they receive from the large ones, increasing their energy, and flattening the spectrum. The opposite would happen for a slope shallower than the Kolmogorov value, and so the Kolmogorov slope is an *attractor* for the spectrum.

It should be noted that the K41 theory involves two fundamental concepts, which sometimes are even considered as the signature of turbulent flows: a $k^{-5/3}$ energy spectrum and the concept of an energy cascade from large to small scales. We will see in Section 3.9, however, that compressible turbulence is not necessarily characterized by these properties, and also, *inverse* cascades (from small to large scales) can occur in MHD and two-dimensional flows. For example, hurricanes in the Earth's atmosphere and Jupiter's Great Red Spot are consequences of the inverse cascade in nearly-two-dimensional flows, which tends to form large, coherent, long-lived structures out of smaller-scale perturbations. Thus, K41 theory is only expected to apply for incompressible, three-dimensional, non-magnetic turbulent flows.

---

[6] The concepts of "eddy" and "turnover" arise from the fact that the most readily identifiable velocity structures in incompressible turbulence are vortices, or "eddies", whose associated characteristic time is the time for a fluid parcel to spin from one side to the other of the vortex—a "turnover time". Note that in the compressible case, compressive motions can generate a completely different type of structure, a density fluctuation. These velocity structures are sometimes called "shocklets", although, more simply, the concept of eddy can just be generalized to include all types of coherent velocity structures. For shocklets, the characteristic time is the so-called *turbulent crossing time*.

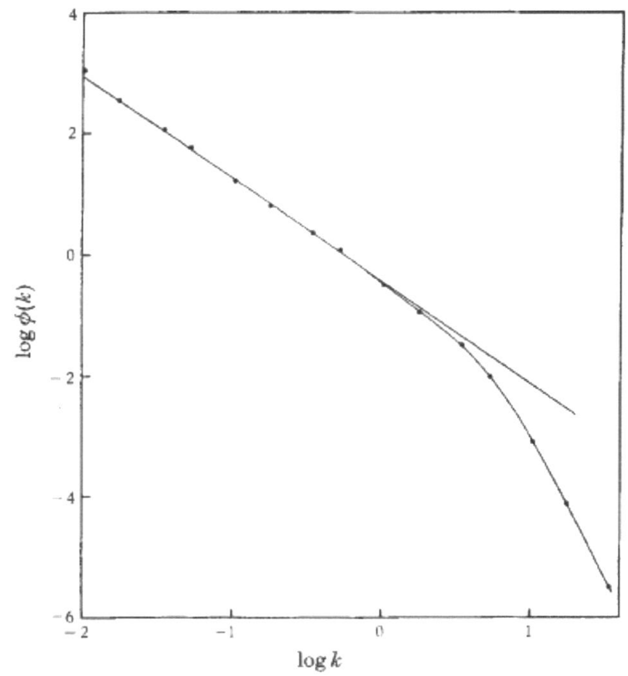

Figure 3.3. The first measurement of the Kolmogorov–Obukhov spectrum in a tidal maritime channel. Reproduced with permissions from Grant et al. (1962) © 1962 Cambridge University Press.

Finally, it should be remarked that the energy spectrum loses a considerable amount of information due to the averaging over angles in k-space. Armi & Flament (1985) have shown that the information concerning the spatial distribution of the physical fields resides not in the energy spectrum, but in the complex phases of the Fourier modes, which are lost upon squaring and angle-averaging. Thus, the energy spectrum is by no means completely determinant of the type of flow present in the fluid.

3.5.1 The Dissipation Timescale

The dimensional analysis of the turbulent cascade from the last section allows also an estimation of the decay rate and characteristic timescale of the turbulence, in the absence of driving. According to the K41 theory, the energy transfer rate is independent of scale, and equals the energy injection rate and the energy dissipation rate. Moreover, one of the fundamental assumptions of the K41 theory is that the characteristic timescale for the transfer rate at scale l is the eddy turnover (or more generally, the turbulent crossing) time, $\tau_l = l/u_l$. At the largest, energy-containing scale L, where most of the kinetic energy resides, we can write

$$\frac{dE_{\rm tot}}{dt} \sim \frac{u_L^2}{\tau_L} = \frac{u_L^3}{L} \sim \frac{u_{\rm rms}^3}{L} \sim \frac{E_{\rm tot}^{3/2}}{L}. \tag{3.33}$$

Integrating this equation, we obtain

$$E_{\rm tot}^{1/2} \propto u_{\rm rms} \propto Lt^{-1}. \tag{3.34}$$

Insight:

Equations (3.33) and (3.34) have important implications. First, Equation (3.33) implies that the dissipation rate is a strongly increasing function of the velocity dispersion, meaning that it is increasingly difficult to maintain larger velocity dispersions in the medium, while a small velocity dispersion persists easily.

Second, noting that Equation (3.34) is equivalent to that for the turbulent crossing time (Equation (3.30)) for the energy-containing scale, and that the injection, transfer, and dissipation rates are equal as required by the K41 theory, we have that the dissipation timescale for the turbulence (in the case of decaying turbulence, after the energy injection is turned off) is

$$\tau_{\rm diss} \sim \frac{L}{u_{\rm rms}}, \tag{3.35}$$

i.e., *the dissipation time for the system is of the order of the crossing time.*

3.6 The Range of Excited Scales

As mentioned in Section 3.1, an extremely important feature of turbulent flows is that they contain a wide range of scales of the excited modes. An estimation of the dynamic range of wavenumbers can be obtained as follows. From Equation (3.8), it is seen that the viscous term contains a factor of k^2. This means that, all other factors being equal, the relative importance of the viscous term increases quadratically with the wavenumber. So, even for flows with very large global Reynolds numbers, there is a certain (small) scale $l_{\rm d}$ for which its associated Reynolds number, $Re_{l_{\rm d}} \equiv u_{\rm d} l_{\rm d}/\nu \sim 1$. If the flow follows a K41 scaling, then, according to Equation (3.31), we can write

$$u_{\rm d} = \left(\frac{l_{\rm d}}{L}\right)^{1/3} U,$$

where $U \equiv u_L$ is the characteristic large-scale velocity of the flow. Taking the ratio of the Reynolds numbers at scales L and $l_{\rm d}$, we then get

$$\frac{L}{l_{\rm d}} = Re^{3/4}. \tag{3.36}$$

That is, for fully developed, incompressible, three-dimensional turbulence, for which the Kolmogorov scaling applies, the dynamic range between the largest and the smallest scales in a turbulent flow scales as $Re^{3/4}$. Moreover, for a three-dimensional flow, the total number of degrees of freedom (the number of smallest-scale eddies) scales as

$$\left(\frac{L}{l_{\rm d}}\right)^3 = Re^{9/4}, \tag{3.37}$$

exhibiting the need for very high resolution for numerical simulations of fully developed turbulence (for which, as mentioned in Section 3.1, $Re \gtrsim$ a few $\times 10^3$). The scale l_d is known as the *dissipative* or *Kolmogorov* scale, and represents the end of the inertial range. The range L/l_d represents the extent of the inertial range. Note that l_d is in general still much larger than the molecular mean free path in the fluid, since the continuum approximation still applies at l_d.

3.7 Moments of the Velocity and Correlation Function

The velocity field in a turbulent flow behaves, in many aspects, as a random variable, since its exact value at a certain spatial location and moment in time is in principle unpredictable. It is therefore relevant to consider its various *moments*, which are averages of products of the velocity components at different times and positions. In general, the n-th *order moment* of the velocity is defined as (using tensor notation)

$$\left\langle u_i(\boldsymbol{x}_1, t_1)\, u_j(\boldsymbol{x}_2, t_2)\ \ldots\ u_k(\boldsymbol{x}_n, t_n)\right\rangle, \tag{3.38}$$

where the symbol $\langle\rangle$ denotes an *ensemble average*, in a statistical mechanical sense.[7]

A turbulent flow is said to be *homogeneous* if this average is independent of position; that is, if

$$\begin{aligned}&\left\langle u_i(\boldsymbol{x}_1, t_1)\, u_j(\boldsymbol{x}_2, t_2)\ \ldots\ u_k(\boldsymbol{x}_n, t_n)\right\rangle\\ &\quad = \left\langle u_i(\boldsymbol{x}_1 + \boldsymbol{r}, t_1)\, u_j(\boldsymbol{x}_2 + \boldsymbol{r}, t_2)\ \ldots\ u_k(\boldsymbol{x}_n + \boldsymbol{r}, t_n)\right\rangle,\end{aligned} \tag{3.39}$$

where $\boldsymbol{r}$ is any constant vector. For homogeneous flows, the so-called *ergodic hypothesis* states that the ensemble average equals the volume average over the entire flow; that is,

$$\langle u_i\rangle(t) = \frac{1}{V}\int u_i(\boldsymbol{x}, t)\, d^3x. \tag{3.40}$$

The first-order moment $\langle u_i\rangle$ is simply the average of the velocity field. The second-order moment $\left\langle u_i(\boldsymbol{x}_1, t_1) u_j(\boldsymbol{x}_2, t_2)\right\rangle$ is known as the *velocity correlation tensor*, a simplification of which is the *autocorrelation function* of the velocity, given by

$$U(\boldsymbol{r}, t) = \langle u_i(\boldsymbol{x}, t)\, u_i(\boldsymbol{x} + \boldsymbol{r}, t)\rangle = \langle \boldsymbol{u}(\boldsymbol{x}, t) \cdot \boldsymbol{u}(\boldsymbol{x} + \boldsymbol{r}, t)\rangle. \tag{3.41}$$

Physically, the autocorrelation function of the velocity can be interpreted as the probability that the velocity at two points separated by a vector $\boldsymbol{r}$ have the same value.

An important property of the autocorrelation function is that its Fourier transform is the velocity spectrum. Indeed, taking the Fourier transform of Equation (3.41) and applying the spatial averaging operator:

---

[7] In statistical mechanics, an *ensemble* is the set of N-particle systems with different macroscopic configurations that share some macroscopic property, like the volume, number of particles, total energy, etc. In the context of a turbulent flow, it can be understood as the set of possible velocity fields that have the same value of a global quantity, like the velocity dispersion, for example.

$$
\begin{aligned}
U(\boldsymbol{k}, t) &\equiv \int_{V_r} U(\boldsymbol{r}, t)\, e^{-i\boldsymbol{k}\cdot\boldsymbol{r}} d^3r \\
&= \int_{V_r} \langle u_i(\boldsymbol{x}, t)\, u_i(\boldsymbol{x}+\boldsymbol{r}, t)\rangle e^{-i\boldsymbol{k}\cdot\boldsymbol{r}} d^3r \\
&= \frac{1}{V}\int_{V_r}\int_{V_x} u_i(\boldsymbol{x}, t)\, u_i(\boldsymbol{x}+\boldsymbol{r}, t)\, e^{-i\boldsymbol{k}\cdot\boldsymbol{r}} d^3x\, d^3r.
\end{aligned}
\tag{3.42}
$$

Now, defining the variable $\boldsymbol{x}'$ so that $\boldsymbol{x} = \boldsymbol{x}' - \boldsymbol{r}$, we get

$$
\begin{aligned}
VU(\boldsymbol{k}, t) &= \int_{V_x'}\int_{V_x} u_i(\boldsymbol{x}, t)\, u_i(\boldsymbol{x}', t)\, e^{i\boldsymbol{k}\cdot(\boldsymbol{x}'-\boldsymbol{x})} d^3x\, d^3x' \\
&= \int_{V_x'} u_i(\boldsymbol{x}', t)\, e^{-i\boldsymbol{k}\cdot\boldsymbol{x}'} d^3x' \int_{V_x} u_i(\boldsymbol{x}, t)\, e^{-i(-\boldsymbol{k})\cdot\boldsymbol{x}} d^3x \\
&= u_i(\boldsymbol{k}, t)\, u_i^*(\boldsymbol{k}, t) \equiv |\boldsymbol{u}(\boldsymbol{k}, t)|^2,
\end{aligned}
\tag{3.43}
$$

where in the last step we have used the reality condition, Equation (3.15). Equation (3.43) then shows that the Fourier transform of the autocorrelation function equals the power spectrum of the velocity, modulo a constant factor.

3.8 Structure Functions

Another class of statistical measures of the velocity are the so-called *structure functions of order p*, defined as

$$
S_p(\boldsymbol{r}, t) = \langle [\boldsymbol{u}(\boldsymbol{x}, t) - \boldsymbol{u}(\boldsymbol{x}+\boldsymbol{r}, t)]^p \rangle. \tag{3.44}
$$

In particular, the second-order structure function is related to the velocity autocorrelation as follows (dropping the explicit dependence on time):

$$
\begin{aligned}
S_2(\boldsymbol{r}) &= \langle [\boldsymbol{u}(\boldsymbol{x}) - \boldsymbol{u}(\boldsymbol{x}+\boldsymbol{r})]^2 \rangle \\
&= \langle \boldsymbol{u}^2(\boldsymbol{x}) - 2\boldsymbol{u}(\boldsymbol{x})\cdot\boldsymbol{u}(\boldsymbol{x}+\boldsymbol{r}) + \boldsymbol{u}^2(\boldsymbol{x}+\boldsymbol{r}) \rangle \\
&= \langle \boldsymbol{u}^2(\boldsymbol{x}) \rangle - 2\langle \boldsymbol{u}(\boldsymbol{x})\cdot\boldsymbol{u}(\boldsymbol{x}+\boldsymbol{r}) \rangle + \langle \boldsymbol{u}^2(\boldsymbol{x}+\boldsymbol{r}) \rangle \\
&= 2[\langle \boldsymbol{u}^2(\boldsymbol{x}) \rangle - \langle \boldsymbol{u}(\boldsymbol{x})\cdot\boldsymbol{u}(\boldsymbol{x}+\boldsymbol{r}) \rangle] \\
&= 2[\langle \boldsymbol{u}^2 \rangle - U(\boldsymbol{r})],
\end{aligned}
\tag{3.45}
$$

where, in going from the third to the fourth equality, we have used the assumption of homogeneity, so that the average of $\boldsymbol{u}^2$ is independent of position.

Also, the Fourier transform of the second-order structure function is related to the velocity spectrum as follows. From Equation (3.45), using the definition of the averaging operator and noting that $U(\boldsymbol{r})$ is the inverse Fourier transform of $U(\boldsymbol{k})$ (cf. Equation (3.42)), we obtain

$$
\begin{aligned}
S_2(\boldsymbol{r}) &= 2[\langle \boldsymbol{u}^2 \rangle - U(\boldsymbol{r})] \\
&= 2\left[\frac{1}{V}\int_x u^2(\boldsymbol{x})\, d^3x - \int_k U(\boldsymbol{k})\, e^{i\boldsymbol{k}\cdot\boldsymbol{r}} d^3k\right].
\end{aligned}
$$

Now, applying Parseval's theorem to the first term on the right, and using Equation (3.43) in the second term, we get

$$S_2(\boldsymbol{r}) = \frac{2}{V}\int_{\boldsymbol{k}} |\boldsymbol{u}(\boldsymbol{k})|^2(1 - e^{i\boldsymbol{k}\cdot\boldsymbol{r}})d^3k. \tag{3.46}$$

In general, the results above show that the structure function is directly related to the correlation function and to the velocity spectrum. Its usefulness lies in that it is often more convenient to experimentally measure the structure function than the correlation function (Schulz-Dubois & Rehberg 1981). It is also important to remark that the K41 theory predicts that

$$S_p(l) \propto l^{p/3}, \tag{3.47}$$

as can be seen if we identify the velocity difference across scale l with u_l in Equation (3.31) with the velocity difference appearing in the structure function of order p, Equation (3.44).

3.9 The Compressible Case

3.9.1 Fully Thermodynamic and Barotropic Flow

The discussion in the previous sections has in general been carried under the condition of incompressibility, which corresponds to the vast majority of terrestrial flows. However, in many astrophysical applications, the flow is highly compressible, and so in this section we now discuss some basic extensions and modifications of the previous results when referring to the compressible case.

Obviously, an immediate consequence of non-incompressibility is that we need to consider the continuity equation in its full extent, rather than in the incompressible form, $\nabla \cdot \boldsymbol{u} = 0$. Slightly less obvious is that, in principle, the internal energy equation, Equation (2.76), must also be considered, since it determines the internal energy (essentially, the temperature), and, through the equation of state, the pressure, which feeds back on the momentum equation through the pressure gradient term. However, an intermediate step is possible, which is the assumption of a *barotropic* flow, in which the thermal pressure depends only on the density, $P = P(\rho)$, so that the energy equation may still be bypassed. A frequently used approximation is that of *polytropic* flows, where the dependence is assumed to be a power law of the form $P \propto \rho^{\gamma_e}$ (cf. Section 4.2), where γ_e is the *effective polytropic exponent*, and is a free parameter, that represents any net heating and cooling acting on the gas. This case includes the extremely common assumption of *isothermal* flows, for which $\gamma_e = 1$.

3.9.2 Magnitude of the Density Fluctuations

In the compressible case, the density of a fluid parcel is no longer constant, as was required by the incompressibility condition $\nabla \cdot \boldsymbol{u} = 0$. A first question in this case is what is the relative amplitude of the density fluctuations. To respond to this question, we again consider the ratio of the two relevant terms in the momentum equation. As mentioned in Section 3.2, the term opposing compressions is the pressure gradient term, while the term causing mixing (which, if we remove the incompressibility condition, may very well cause compressions), is the nonlinear advection term. Note, of course, that in the compressible case it is not possible to solve for the pressure gradient term like we did in Section 3.2 because now the incompressibility condition $\boldsymbol{k} \cdot \boldsymbol{u}_k = 0$ does not apply anymore, and we cannot eliminate the time derivative term in the momentum equation by dotting the equation with the wavevector $\boldsymbol{k}$.

So, let us now consider the ratio of the advection and the pressure gradient terms, and assume it is of order unity, so that both terms contribute equally to the momentum equation. That is,

$$\frac{|\boldsymbol{u} \cdot \nabla \boldsymbol{u}|}{\left|\frac{\nabla P}{\rho}\right|} \sim 1. \tag{3.48}$$

Let us now express ∇P in terms of the density field in two common cases: an adiabatic and an isothermal flow. First, for an adiabatic case, we know that

$$P = A\rho^{\gamma} \quad \text{(Adiabatic)},$$

where $\gamma = c_P/c_V$ is the ratio of specific heats and A is a constant. Thus,

$$\frac{dP}{d\rho} = \gamma A \rho^{\gamma - 1} = \frac{\gamma P}{\rho} = \frac{\gamma k_B T}{m} \equiv c_{s,a}^2, \tag{3.49}$$

where k_B is the Boltzmann constant and m is the mean particle mass in the gas. In the third equality, we have used the ideal gas equation of state, and the last equality defines the adiabatic sound speed. Thus, in the adiabatic case, we have

$$\nabla P = c_{s,a}^2 \nabla \rho. \quad \text{(Adiabatic)} \tag{3.50}$$

Now, in the isothermal case, $T =$ cst., we simply have, from the equation of state,

$$P = \rho k_B T/m \equiv c_{s,i}^2 \rho,$$

where the last equality defines the *isothermal* sound speed. Thus, in the isothermal case, we have

$$\nabla P = c_{s,i}^2 \nabla \rho. \quad \text{(Isothermal)} \tag{3.51}$$

Thus, we can write, in general, $\nabla P \approx c_s^2 \nabla \rho$, where c_s is the sound speed appropriate to the kind of thermodynamic process, and so we can rewrite Equation (3.48) as

$$|\boldsymbol{u} \cdot \nabla \boldsymbol{u}| \sim \left| \frac{c_s^2 \nabla \rho}{\rho} \right|,$$

To order of magnitude, this gives

$$\frac{U^2}{L} \sim \frac{c_s^2}{\rho} \frac{\delta \rho}{L},$$

or, equivalently,

$$\frac{\delta\rho}{\rho} \sim \frac{U^2}{c_s^2} \equiv \mathcal{M}_s^2, \qquad (3.52)$$

where $\mathcal{M}_s$ is the *sonic Mach number*. We see then that the typical relative density fluctuation is of the order of the Mach number squared. Note, however, that in the adiabatic case, the sound speed also increases upon compression, and, as is well known from shock jump theory, the maximum density enhancement in this case is by a factor of 4. In the isothermal case, the sound speed is constant, and thus the relative density fluctuation is not bounded. Finally, it has been shown by Vázquez-Semadeni et al. (1996) that, for polytropic flows of the form $P \propto \rho^{\gamma_e}$, with γ_e a free parameter, $\delta\rho/\rho \to e^{\mathcal{M}_s^2}$ when $\gamma_e \to 0$, implying that, in this case, the relative density fluctuation is much larger. The limit $\gamma_e \to 0$ corresponds to the isobaric limit, because P becomes independent of the density.

3.9.3 Rotational and Compressible Modes

In the compressible case, the velocity field is no longer subject to the incompressibility condition, $\boldsymbol{k} \cdot \boldsymbol{u_k} = 0$. However, it is convenient to still decompose the field in Fourier space into a rotational (also known as *incompressible* or *solenoidal*) part, denoted $\boldsymbol{u}_{k,s}$, and a compressible (or *potential*) one $\boldsymbol{u}_{k,c}$, such that

$$\boldsymbol{k} \cdot \boldsymbol{u}_{k,s} = 0 \quad \boldsymbol{k} \times \boldsymbol{u}_{k,c} = 0. \qquad (3.53)$$

These components of the velocity field can be constructed as follows. The compressible component is obtained simply by taking the projection of the Fourier velocity amplitude onto its corresponding wavevector:

$$\boldsymbol{u}_{k,c} \equiv \frac{(\boldsymbol{k} \cdot \boldsymbol{u_k})\boldsymbol{k}}{k^2}, \qquad (3.54)$$

while the solenoidal component is given by

$$\boldsymbol{u}_{k,s} \equiv \boldsymbol{u_k} - \boldsymbol{u}_{k,c}. \qquad (3.55)$$

Note that these two components are orthogonal to each other in Fourier space. An immediate consequence of this is that they can be considered as independent degrees of freedom, and in fact, separate equations can be written for the evolution of the *vorticity*, $\omega \equiv \nabla \times \boldsymbol{u}$ and of the *dilatation*, $\nabla \cdot \boldsymbol{u}$ by respectively taking the curl and the divergence of the momentum equation. The *vorticity equation* is

$$\frac{\partial\omega}{\partial t} + (\boldsymbol{u} \cdot \nabla)\,\omega = (\omega \cdot \nabla)\,\boldsymbol{u} - \omega(\nabla \cdot \boldsymbol{u}) - \frac{\nabla P \times \nabla\rho}{\rho^2} + \nabla \times \left(\frac{\nabla \cdot \overleftrightarrow{\tau}}{\rho}\right), \qquad (3.56)$$

where $\overleftrightarrow{\tau}$ is the shear-stress tensor (cf. Section 2.3.2), and use has been made of the vector calculus identities

$$\begin{aligned}\nabla(\boldsymbol{A} \cdot \boldsymbol{B}) &= (\boldsymbol{A} \cdot \nabla)\boldsymbol{B} + (\boldsymbol{B} \cdot \nabla)\boldsymbol{A} + \boldsymbol{A} \times (\nabla \times \boldsymbol{B}) + \boldsymbol{B} \times (\nabla \times \boldsymbol{A}),\\ \nabla \times (\boldsymbol{A} \times \boldsymbol{B}) &= \boldsymbol{A}(\nabla \cdot \boldsymbol{B}) - \boldsymbol{B}(\nabla \cdot \boldsymbol{A}) + (\boldsymbol{B} \cdot \nabla)\boldsymbol{A} - (\boldsymbol{A} \cdot \nabla)\boldsymbol{B},\end{aligned}$$

where $\boldsymbol{A}$ and $\boldsymbol{B}$ are arbitrary vectors, and noting that $\nabla \cdot (\nabla \times \boldsymbol{u}) = 0$.

In turn, the *divergence*, or *dilatation equation* is

$$\frac{\partial \ \nabla \cdot \boldsymbol{u}}{\partial t} + \boldsymbol{u} \cdot (\nabla \times \omega) = - \nabla^2 \frac{u^2}{2} + \omega^2 - \frac{\nabla^2 P}{\rho} + \frac{\nabla P \cdot \nabla \rho}{\rho^2} + \nabla \cdot \left(\frac{\nabla \cdot \overleftrightarrow{\tau}}{\rho} \right), \quad (3.57)$$

where we have made use of the vector identity

$$\nabla \cdot (\boldsymbol{A} \times \boldsymbol{B}) = \boldsymbol{B} \cdot (\nabla \times \boldsymbol{A}) - \boldsymbol{A} \cdot (\nabla \times \boldsymbol{B}).$$

3.9.3.1 Sources and Transfer among the rotational and Compressive Modes

Several remarks are in order about these equations:

- The left-hand side (LHS) of the vorticity equation, Equation (3.56), can be written as the total, or Lagrangian, derivative $d\omega/dt$. This implies that, if the terms on the RHS are zero under some circumstances, then the vorticity of a Lagrangian fluid parcel is conserved.
- The first term on the RHS of the vorticity equation is the so-called *vortex--stretching* term. It represents the projection of the vorticity along the velocity gradient, and it enters the equation with a positive sign. This means that, if there is a component of the vorticity aligned with the velocity gradient, the vorticity is amplified. In two dimensions, where the vorticity is necessarily perpendicular to $\nabla \boldsymbol{u}$, this term is identically zero.
- The second term in the RHS of the vorticity equation amplifies the vorticity when there is a compression. This is the well-known phenomenon of objects spinning up when their radius contracts, as a consequence of angular momentum conservation.
- The third term is the so-called *baroclinic* term, and represents the generation of vorticity when the density and pressure gradients are *not* aligned. This term is identically zero in barotropic flows, where the two gradients are necessarily parallel (cf. Section 3.9.1), since the pressure depends only on density in this case. On the other hand, in fully thermodynamic flows, where the internal energy equation is solved together with the mass and momentum conservation equations, the pressure depends on both the density and the temperature, through the equation of state, and thus $\nabla P \nparallel \nabla \rho$ in general. This means that fully thermodynamic flows generate vorticity from the misalignment between the thermal pressure and density gradients.
- The vorticity and dilatation equations are clearly non-symmetric. In particular, the LHS of the dilatation equation, Equation (3.57), cannot be written as a Lagrangian derivative, and thus the dilatation is not a conserved Lagrangian quantity under any circumstance. Furthermore, the terms on the RHSs of both equations are very different from each other, and thus the generation of dilatation proceeds differently than that of vorticity.
- The dilatation equation contains a source term proportional to ω^2. This means that the mere presence of vorticity generates dilatation. However, the opposite is not true: no term proportional to the dilatation alone is present in

the vorticity equation, so dilatation by itself does not generate vorticity. The term $\omega(\nabla \cdot \boldsymbol{u})$ *amplifies* the vorticity if it is already present, but does not *generate* it if it is not previously present.

- In view of the above, it appears that the transfer is predominantly from rotational to compressible modes, so that rotational energy tends to decay if no external sources of vorticity (from the baroclinic, viscous or magnetic [not shown] terms) are present. This conclusion is supported by numerical simulations of both moderately (Kida & Orszag 1990b, 1990a) and highly (Vázquez-Semadeni et al. 1996) compressible turbulence.

3.9.4 The Burgers and Fleck/Kritsuk Energy Spectra

The energy spectrum is one of the fundamental and most studied properties of turbulence. However, the derivation of the spectral slope in the compressible case is far from being as clear-cut as for the incompressible case. In the latter, the derivation of the Kolmogorov (1941) spectrum (cf. Section 3.5) was based on an energy conservation argument: since in this case there is only one type of velocity modes (the solenoidal modes), it is reasonable to assume that, if energy is injected at large scales and dissipated at small scales, with a wide range of "inertial" scales in-between, then the energy can only "cascade" from the large to the small scales, being transferred in a loss-less manner through the inertial range. However, in the compressible case, new modes appear, from the compressible modes of the velocity itself, to the internal energy modes, when the internal energy equation is being solved together with the continuity and momentum equations. Thus, the energy can be transferred "horizontally" among the different modes at fixed wavenumber, and it is not clear that an energy cascade from large to small scales can occur along the inertial range. This is still an open subject of research, and here we just mention a few important efforts in this direction.

3.9.4.1 The Burgers Spectrum

Burgers (1974) presented a simple equation in one dimension with the pressure gradient term removed, with the intention to isolate the effects of the nonlinear advection term. The *Burgers equation* reads

$$\frac{\partial u}{\partial t} + u\frac{\partial u}{\partial x} = \nu\frac{\partial^2 u}{\partial x^2}. \tag{3.58}$$

The solutions to this equation develop shocks (velocity discontinuities) in a finite time, and, if a density field is attached to it via a continuity equation, the density develops spikes approaching delta functions. It is expected that the velocity field of highly compressible turbulence, in which the pressure gradient term is negligible, behaves in a similar way, and the velocity field consists essentially of a superposition of planar shock fronts. In this case, the spectrum of the compressible component of the velocity can be derived from the spectral signature of a shock. Indeed, a shock can be represented by a Heaviside step function of position, which in one dimension is defined as

$$u(x) \propto H(x) \equiv \begin{cases} 1 & \text{if} x \leqslant 0; \\ 0 & \text{if} x > 0. \end{cases}$$

Therefore, its derivative is given by a delta function,

$$u'(x) \propto \delta(x),$$

and so the Fourier transform of the velocity spatial derivative is then

$$\mathcal{F}_k(u'(x)) \propto \int \delta(x) e^{ikx} dx = \text{cst}.$$

The Fourier transform of the velocity is then obtained by dividing by the wavenumber k (equivalent to integrating over space in physical space):

$$u(k) \propto k^{-1},$$

and the velocity power spectrum is finally given by

$$u^2(k) \propto k^{-2}. \tag{3.59}$$

This gives the velocity power spectrum for a shock. The derivation is one-dimensional, but, assuming that shocks are essentially one-dimensional structures, then integration over angles of the squared Fourier amplitude to obtain the energy spectrum (cf. Equation (3.27)) does not bring in any additional dependence on wavenumber because there are no modes in other directions, and the energy spectrum can be assumed to have the same form:

$$E(k) \propto k^{-2}. \quad \text{(Burgers)} \tag{3.60}$$

Numerical simulations of isothermal supersonic turbulence indeed suggest that, as the Mach number is increased, the slope of the (specific) energy spectrum shifts from the Kolmogorov value of $-5/3$ to the Burgers one, of -2 (Kritsuk et al. 2007).

It is important to note that, numerically, the slope (-2) of the Burgers spectrum is not very different from that of Kolmogorov ($-5/3$). However, physically, it has a *very* different meaning: while the Kolmogorov spectrum is the spectral slope that maintains a wavenumber-independent transfer rate across the inertial range (i.e., it is essentially an energy-conservation argument), the slope of the Burgers spectrum follows simply from a geometrical argument, based on the spectral signature of a shock jump, and involves no energy considerations whatsoever.

A final remark is that, applying Equation (3.32), we see that the typical velocity difference across scale ℓ for Burgers' turbulence is

$$u_\ell \propto \ell^{1/2}, \tag{3.61}$$

in contrast to the $u_\ell \propto \ell^{1/3}$ scaling that arises in the case of Kolmogorov turbulence (Equation (3.31)).

3.9.4.2 An Energy Density Cascade: The Fleck/Kritsuk Spectrum

There have been attempts in the literature to extend the energy cascade concept from the incompressible to the compressible case. For example, based on a model by

Fleck (1996), Kritsuk et al. (2007) define the spectrum of kinetic energy *density*, $\mathcal{E}(k)$, such that

$$\int \mathcal{E}(k)dk = \frac{1}{2}\int \rho u^2 d^3x,$$

and hypothesize that it is the kinetic energy density that cascades down the wavenumber range. They thus write

$$\frac{\rho_l u_l^2}{\left(\frac{l}{u_l}\right)} = \text{cst.} \Rightarrow w_l \equiv \rho_l^{1/3} u_l \propto l^{1/3}, \tag{3.62}$$

where the identity defines the density-weighted velocity w. Thus, they propose that Equation (3.62) is the compressible counterpart of Equation (3.31), so that w_l should exhibit a Kolmogorov-like scaling, and $\mathcal{E}(k) \propto k^{-5/3}$. Their numerical simulations partially support this, as they exhibit a scaling $\mathcal{E}(k) \propto k^{-3/2}$ roughly independently of Mach number. This scaling is close, but not exactly equal to the Kolmogorov scaling. The tendency toward maintaining a Mach number-independent slope is expected since, while the velocity spectrum steepens with increasing Mach number, the density power spectrum flattens as the density jumps become more pronounced (Kritsuk et al. 2007; Gazol & Kim 2010), and so the effects of the density and of the velocity in the spectrum of the combined quantity w tend to cancel each other out. However, it remains to be shown (via an energy transfer study) whether there is really a conservative cascade of the kinetic energy density. This is not obvious, because in shocks all Fourier modes are in phase, in order to generate a near-discontinuity. In this case, energy is transferred directly from the largest scales to all smaller scales (Kadomtsev & Petviashvili, 1973). Nevertheless, it has been suggested that the energy spectrum should still exhibit a $k^{-5/3}$ slope at small scales, where the velocity dispersion is subsonic, even in the globally supersonic case (Galtier & Banerjee, 2011).

3.10 Supersonic Turbulent Fragmentation: The Probability Density Function (PDF) of the Density Fluctuations

As discussed in Section 3.9.2, for supersonically turbulent flow, an ensemble of density fluctuations is expected to arise in the flow, as a consequence of a corresponding population of traveling shock waves, each of which produces a density jump. A theory for the development of the probability density function (PDF; essentially, the *histogram* of the density values in a certain region) of the density fluctuations in polytropic flows where the thermal pressure depends only on the density, and with a power-law form, $P \propto \rho^{\gamma_e}$, where γ_e is a parameter, was developed by Passot & Vázquez-Semadeni (1998 hereafter, PV98), and later refined by Federrath et al. (2008). Here we review the derivation for isothermal and, more generally, polytropic flows. The fundamental hypothesis behind the derivation of the density PDF is that the density fluctuations arise from a succession of shock wave passages through any given location in the flow (Vázquez-Semadeni, 1994), and that each shock passage is an independent random event.

3.10.1 The Isothermal Case

In the isothermal case, where $\gamma_e = 1$, it is well known that the density jump across a shock is

$$\rho_2 = \mathcal{M}_s^2 \rho_1, \tag{3.63}$$

where the subindices "1" and "2" represent the pre- and post-shock conditions, and $\mathcal{M}_s \equiv u_1/c_s$ is the sonic Mach number, u being the velocity, and $c_{s,iso}$ the isothermal sound speed. That is, the density shock is *multiplicative*. This in turn implies that the jump in the natural logarithm of the normalized density, $s \equiv \ln(\rho/\rho 0)$, where $\rho 0$ is the mean density of the flow, is *additive*:

$$s_2 = s_1 + \ln \mathcal{M}_s^2 = s_1 + 2 \ln \mathcal{M}_s. \tag{3.64}$$

If we now assume that the density at a given location in the flow is the result of the successive passages of a large number n of shock waves, then the final log of the density, s_n, will be related to the original, s_1, by

$$s_n = s_1 + \sum_i \ln M_{s,i}, \tag{3.65}$$

where $M_{s,i}$ can be considered a random variable with a certain, possibly unknown, distribution. However, this distribution must be the same for all jumps. This can be seen writing the one-dimensional (1D) continuity and momentum equations in terms of s, and including only the pressure term in the momentum equation. We obtain

$$\frac{Ds}{Dt} = -\frac{\partial u}{\partial x} \tag{3.66}$$

$$\frac{Du}{Dt} = -c_s^2 \frac{\partial s}{\partial x}, \tag{3.67}$$

where $D/Dt \equiv \partial/\partial t + u\partial/\partial x$ and u is the 1D velocity in the direction of x. It is seen that these equations are invariant to shifts $s \to s + s_0$, where s_0 is a constant, implying that the sequence of density jumps caused by the successive passage of shocks is independent of the previous density, and thus all jumps are governed by the same equations. Therefore, all jumps should belong to the same probability distribution.

At this point, the Central Limit Theorem of probability theory can be invoked. This theorem states that the distribution of the sum of independent random variables drawn from the same distribution converges to a normal (Gaussian) distribution. Therefore, the variable s is expected to have a normal distribution, which has the form

$$P(s)ds = \frac{1}{\sqrt{2\pi\sigma_s^2}} \exp\left[-\frac{(s - s_0)^2}{2\sigma_s^2}\right] ds, \tag{3.68}$$

where s_0 is the mean value of s and σ_s is its standard deviation. Equation (3.68) implies that the density itself has a *lognormal* distribution. The parameters s_0 and σ_s need to be evaluated by physical considerations.

Concerning the mean, s_0, it is convenient to recall the standard condition for distributions of related variables, $P(\rho/\rho 0)d(\rho/\rho 0) = P(s)ds$, in order to calculate the mean normalized density as

$$\langle \rho/\rho 0 \rangle \equiv 1 = \int_{-\infty}^{\infty} e^s P(s) ds. \tag{3.69}$$

From here, inserting Equation (3.68), it can be shown that

$$s_0 = -\frac{\sigma_s^2}{2}. \tag{3.70}$$

Concerning the standard deviation of the distribution, PV98 showed, in high-resolution numerical simulations of 1D turbulence, that

$$\sigma_s \approx \mathcal{M}_s, \tag{3.71}$$

(Figure 3.4). Also, owing to the standard relationship between the variance of the linear density and that of its logarithm for the lognormal distribution, which reads

$$\sigma_\rho^2 = e^{\sigma_s^2} - 1, \tag{3.72}$$

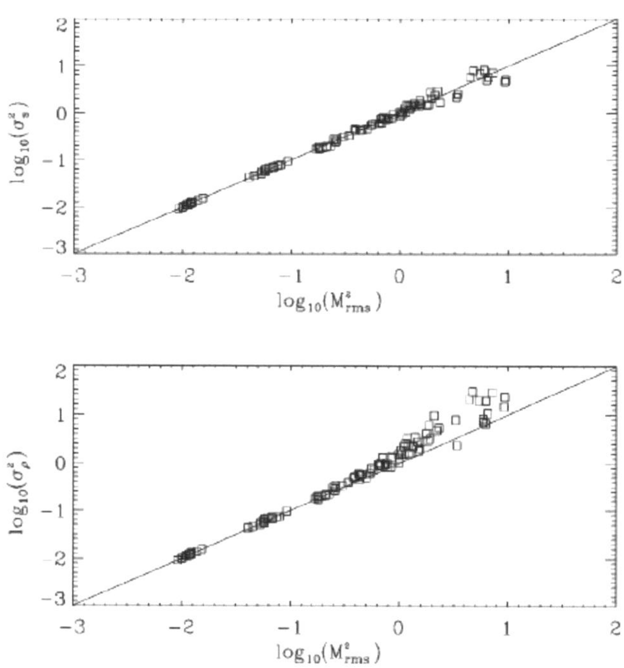

Figure 3.4. *Top:* Scaling between σ_s and the rms Mach number in one-dimensional simulations of purely compressible turbulence by PV98. The scaling is seen to be nearly linear for *rms* Mach numbers up to $\mathrm{Ms} \sim \sqrt{10}$. *Bottom:* Scaling between σ_s and the rms Mach number. A deviation from linearity begins to be apparent for $\mathrm{Ms} \gtrsim 1$. Reprinted figure with permission from Passot and Vázquez-Semadeni (1998). Copyright 1998 by the American Physical Society.

it is seen in Figure 3.4 that σ_ρ appears to also scale linearly with $\mathcal{M}_s$ up to Mach numbers of several, while for larger values, the relationship curves up slightly.

The above results imply that the width of the PDF increases while its peak shifts to lower densities as the Mach number is increased. The latter is a consequence of mass conservation: when higher densities are produced by the turbulence, more mass is deposited in the density peaks, removing mass from the low-density regions, which however occupy a larger volume. Thus, the most probable value of the volume-weighted density becomes smaller. This is illustrated in Figure 3.5.

However, since the simulations of PV98 were one-dimensional, the velocity field contained exclusively compressible modes. In three dimensions, nonlinear transfer between the compressible and solenoidal modes implies that compressible modes are generated from solenoidal ones and vice versa. From 3D numerical simulations forced with various ratios of compressible and solenoidal energy, Federrath et al. (2008) have suggested that it is the variance of the *linear* density that scales linearly with the *rms* Mach number:

$$\frac{\sigma_\rho}{\rho_0} \approx b\mathcal{M}_s, \tag{3.73}$$

so that

$$\sigma_s^2 = \ln\left(1 + b^2\mathcal{M}_s^2\right), \tag{3.74}$$

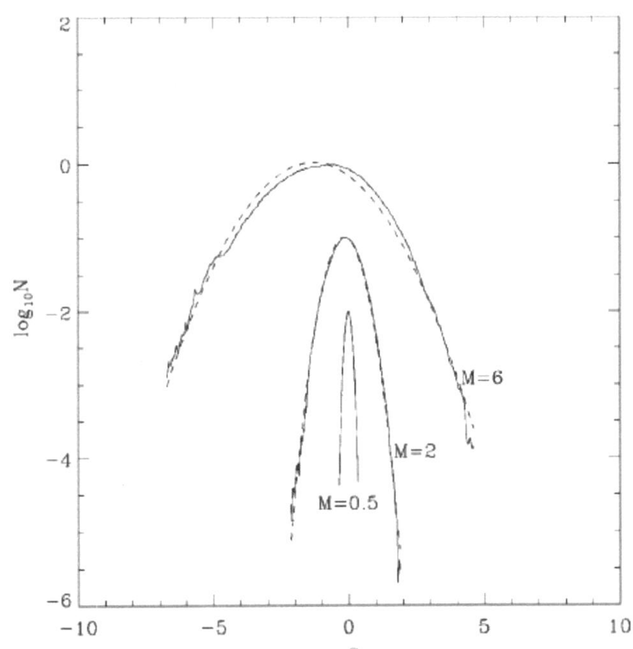

Figure 3.5. The density PDF in 1D numerical simulations at different rms Mach numbers. The width increases and the peak shifts to lower densities as the Mach number is increased. Reprinted figure with permission from Passot and Vázquez-Semadeni (1998). Copyright 1998 by the American Physical Society.

where b is the so-called *forcing parameter*, which can take values from $1/3$ in the case of purely solenoidal driving, to 1 in the case of purely compressible driving. The difference between Equations (3.71) and (3.73) may arise from the fact that in 1D all the energy must reside exclusively in compressible modes throughout the turbulent cascade, while in 3D, some excitation of solenoidal modes necessarily occurs even in the case of purely compressible driving.

3.10.2 The General Polytropic Case

PV98 extended the argument for the isothermal case to the general polytropic case by writing the 1D equations in the polytropic case, making the substitution $z \equiv (1 - \gamma_e)s = \ln(\rho/\rho_0)^{1-\gamma_e}$, and writing the polytropic equation as $P = P_0(\rho/\rho_0)^{\gamma_e}$, so that

$$\frac{Dz}{Dt} = -(1 - \gamma_e)\frac{\partial u}{\partial x} \tag{3.75}$$

$$\frac{Du}{Dt} = \frac{c_{s,0}^2}{1 - \gamma_e}\frac{\partial e^{-z}}{\partial x}, \tag{3.76}$$

where $c_{s,0} \equiv \sqrt{\gamma_e P_0/\rho_0}$ is the effective sound speed (cf. Equation (3.50)). PV98 noted that these equations return to the form of Equations (3.66) and (3.67) upon the density-dependent substitution $c_{s,0} \to c_{s,0}e^{(1-\gamma_e)s/2}$, and thus proposed the ansatz that this replacement can be made also in the density PDF, to obtain its form in the general polytropic case. Making the above substitution, and moving the term in s_0 from inside the exponential to a multiplicative constant C, they wrote

$$P(s) = C \exp\left[-\frac{s^2 c_{s,0}{}^2 e^{(\gamma_e - 1)s}}{2} - \alpha(\gamma_e)s\right], \tag{3.77}$$

where $\alpha(\gamma_e)$ is a parameter that also must be constrained by imposing the condition that $\int_{-\infty}^{\infty} e^s P(s) ds = \langle\rho\rangle$. Equation (3.77) shows that, for $(\gamma_e - 1)s < 0$, the first term in the exponential becomes small at large values of $|s|$, and the exponential is dominated by the second term, so that the PDF approaches an exponential in s, or a power law in ρ. On the other hand, for $(\gamma_e - 1)s > 0$, the first term in the exponential dominates, and so the PDF drops faster than a Gaussian in s, or a lognormal in ρ. This is illustrated in Figure 3.6, which shows the density PDF for values of γ_e smaller and larger than unity.

Insight:
It is important to note that the lognormal density PDF is *not* the general form of the PDF for any supersonically turbulent flow, contrary to some statements in the literature. It applies *only* to the isothermal case. The polytropic case with $\gamma_e \neq 1$, which represents non-isothermal flows, already develops a power-law side of the PDF. Also, it is worth noting that a power-law tail at high densities is known to

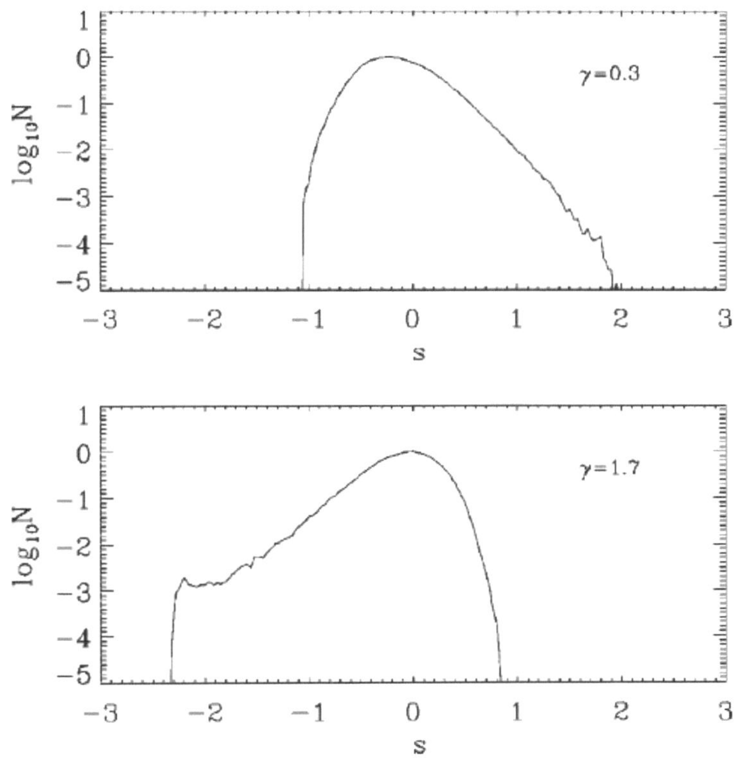

Figure 3.6. PDFs of two 1D numerical simulations by PV98, one with $\gamma_e = 0.3$ and the other with $\gamma_e = 1.7$, both with rms Mach number $\mathcal{M}_s \sim 3$. An exponential form in s (a power law in ρ) is seen to appear at low densities for $\gamma_e > 1$ and at high densities for $\gamma_e < 1$. Reprinted figure with permission from Passot and Vázquez-Semadeni (1998). Copyright 1998 by the American Physical Society.

occur in self-gravitating flows (Ballesteros-Paredes et al. 2011; Kritsuk et al. 2011). Essentially, both a self-gravitating flow and one with $0 < \gamma_e < 1$ are *more compressible at high densities*, thus generating a larger fraction of high-density gas parcels.

3.11 Turbulence Amplification During Gravitational Contraction and the Thermodynamic Behavior of Turbulence

As mentioned in Section 3.9.3.1, the second term on the right-hand side of Equation (3.56) implies that vorticity is amplified upon compressions. This is of particular interest in the case of a gravitationally-induced compression, which could then drive turbulent amplification. This problem can be investigated phenomenologically, by treating the turbulent velocity dispersion as an "internal energy" of the turbulence, and thus investigating the thermodynamics of the turbulent motions. In this section we follow the treatment by Robertson & Goldreich (2012) and Murray & Chang (2015). Note, however, that treating the turbulent velocity dispersion as an additional internal energy of the flow loses all information of its multi-scale nature, and therefore this analogy must be used with great caution (see Section 3.12 below).

The increase in the turbulent velocity caused by a reduction in size can be estimated phenomenologically by considering conservation of the angular momentum $\boldsymbol{j} = \boldsymbol{r} \times m\boldsymbol{v}$ of a fluid element within a spherical clump, moving with turbulent velocity $\boldsymbol{v}$ in the direction perpendicular to the radius $\boldsymbol{r}$ from the clump's center. The conservation of angular momentum then implies that $rv=$ cst., where $r = |\boldsymbol{r}|$ and $v = |\boldsymbol{v}|$. Therefore, we can write

$$v(r) = \frac{v_0 r_0}{r}, \tag{3.78}$$

where $v_0 = v(r_0)$, and r_0 is the initial radial distance from the fluid element to the clump's center. Therefore, if the clump contracts at velocity $u \equiv dr/dt$, the change in the turbulent velocity is

$$\left(\frac{dv}{dt}\right)_{\text{heat}} = -\frac{v_0 r_0}{r^2}\frac{dr}{dt} = -v\frac{u}{r}. \tag{3.79}$$

Note that $u < 0$ since the motion is contracting, and thus it is in the opposite direction of increasing radius.

The change in the turbulent velocity given by Equation (3.79) can be considered the analogue of the adiabatic heating suffered by a gas parcel upon being compressed by PdV work, and is thus often referred to as the "adiabatic heating" of the turbulence by the contraction.

On the other hand, the dissipation of turbulence, which, in our thermodynamic analogy, corresponds to the "cooling" of the turbulence, is given by Equation (3.33), from which we can write

$$\frac{1}{2}\left(\frac{dv^2}{dt}\right)_{\text{cool}} = v\frac{dv}{dt} = -\eta\frac{v^2}{r}, \tag{3.80}$$

where η is a constant of order unity, which we have introduced to take care of the "$\sim$" sign in Equation (3.33). Also, we have included a minus sign because the energy is *lost* from the turbulence, and we have homologated the distance scale L in Equation (3.33) to the radius r in our contracting clump.

We can thus write a model "internal energy" equation for the turbulence as

$$\frac{dv}{dt} = -\left(1 + \eta\frac{v}{u}\right)\frac{vu}{r}, \tag{3.81}$$

where the first term within the parentheses corresponds to the driving (the "heating") and the second corresponds to the dissipation (the "cooling").

As pointed out by Robertson & Goldreich (2012), it can be seen from Equation (3.81) that, if the turbulent velocity v is much larger than the infall speed $|u|$, then the dissipation term dominates, and the turbulent velocity tends to decrease (recall that $u < 0$, so that, in this case, $dv/dt < 0$). Conversely, if $|u| \gg v$, then the driving dominates, and the turbulent velocity tends to increase. Therefore, those authors concluded that the turbulent velocity should become comparable to the infall speed. Indeed, in numerical simulations of turbulent amplification by gravitational contraction, Guerrero-Gamboa & Vázquez-Semadeni (2020) found that the system approached a state where $v \sim 40\%$ of $|u|$.

An interesting consequence of this value is that it implies that the specific (i.e., per unit mass) kinetic energy in the turbulent velocity dispersion, $E_{\rm k,t}$, satisfies $E_{\rm k,t} \approx 1/2E_{\rm g}$, where $E_{\rm g} \approx \beta GM(r)/r$ is the specific gravitational energy, with G the gravitational constant, $M(r)$ the mass contained within radius r, and β a constant of order unity that depends on the mass distribution within the sphere. This can be seen as follows. Since we have considered the turbulent velocity to be in a direction perpendicular to the radius, it can be associated with the one-dimensional turbulent velocity dispersion, $\sigma_{\rm 1D}$. The total kinetic energy in the turbulent velocity dispersion, including all three directions, is then

$$E_{\rm k,t} = \frac{3}{2}\sigma_{\rm 1D}^2. \tag{3.82}$$

On the other hand, the specific kinetic energy in the infall motion, $E_{\rm k,i}$, assuming that the infall speed is the free-fall speed, must satisfy

$$E_{\rm k,i} = \frac{1}{2}u^2 = E_{\rm g} = \frac{\beta GM(r)}{r}. \tag{3.83}$$

Therefore, if $\sigma_{\rm 1D} \sim 0.4|u|$, then $E_{\rm k,t} \approx 3/2(0.4u)^2 \approx 1/4u^2 = 1/2E_{\rm g}$. This is a similar condition to that expected in virial equilibrium (see Section 5.2.2), although the physical condition is completely different, corresponding to free-fall collapse rather than to equilibrium, and therefore it can be considered a *pseudo-virial* state. This is discussed further in Section 5.3.3.

3.12 How Turbulent Pressure Differs from Thermal Pressure

Turbulence is often invoked in the star formation community as an additional nonthermal velocity dispersion similar to the thermal one, with an associated *turbulent pressure*, which is typically larger than the thermal pressure, because the turbulence is often supersonic, especially in molecular clouds. However, this analogy suffers from the serious problem that the turbulent motions are essentially multi-scale in nature, while the thermal motions of atoms and molecules are not.

As mentioned in Section 2.1, a prerequisite for applying the fluid approach is that the system scales must be much larger than the mean free path of the particles making up the gas (atoms or molecules), so that meaningful averages can be constructed over large numbers of particles to define the fluid properties. The particle mean free path is precisely the size scale of the gas particle motions, so the particle motions are several orders of magnitude smaller than the fluid system's scale. This also implies that the velocity dispersion of the thermal motions is independent of the size of the region in which it is measured, and that the particle motions are isotropic within any fluid element, so that the pressure satisfies Pascal's law that the pressure in a fluid is the same on all planes passing through any given point; i.e., it is *isotropic*.

In order for turbulence to generate an isotropic pressure, it would have to consist of motions on much smaller scales than the characteristic size of the flow (i.e., the turbulence should be *microscopic*, often referred to as *microturbulence*). This is explicitly stated in early treatments of support against self-gravity (e.g., Chandrasekhar 1951), in which it was assumed that the autocorrelation function of the turbulent motions vanishes at large separations. However, contrary to this assumption, turbulent motions are characterized by larger velocity differences as larger separations are considered, as implied by the negative slope of the turbulent energy spectrum in both incompressible and compressible flows, up to the scale at which energy is injected (*the energy injection scale*). Since the kinetic energy in molecular clouds is typically injected at the largest scales (in case of gravitational contraction) or even from the outside (if the clouds are formed by turbulent compressions driven from outside of them), this means that the turbulent velocity dispersion depends on the size scale of the region considered, and that the correlation length of a turbulent flow is of the order of the size scale of the flow itself, contradicting the hypothesis.

Although a scale-dependent support has been considered by a number of authors (e.g., Bonazzola et al. 1987; Vázquez-Semadeni & Gazol 1995), these attempts still assumed that turbulence generates an isotropic pressure at each scale. However, this is clearly impossible, because it would imply that the turbulent pressure measured at one point in space would depend on the size of the region around that point where it is measured. i.e., the turbulent pressure would be multi-valued at that point.

The solution to this conundrum lies in the fact that the scale-dependence of the velocity dispersion within structures of different size scales arises from the typical velocity *differences* Δv between points separated by a certain distance (or "lag") ℓ, not from the typical velocities of the fluid parcels with respect to a single reference frame, as thermal motions do. Specifically, for Kolmogorov incompressible turbulence, $\Delta v \propto \ell^{1/3}$, while for strongly compressible, Burgers turbulence, $\Delta v \propto \ell^{1/2}$. The only possible way to accomplish this is that larger fluid parcels move coherently with larger velocities than smaller ones, as illustrated in Figure 3.7, where the velocity difference between regions separated by a certain distance ℓ is depicted as a rotation—i.e., as a rotating *eddy*. This figure then illustrates that the velocity difference across a larger size scale involves larger fluid parcels moving coherently (i.e., as a single unit).

However, this property implies that only a handful of large-scale fluid parcels exist moving with the velocity corresponding to that scale, and therefore, it is not possible to make a meaningful ensemble average over them to obtain an isotropic pressure. Instead, each large-scale fluid parcel behaves as a *stream*, that causes a *ram pressure* with well-defined direction, analogous to the splash of an ocean wave against a cliff, or the jet from a firehose. That is, *the large-scale turbulent motions cannot produce an isotropic pressure on that same size scale*. This is a crucial consideration, for example, for the problem of whether clouds can be stabilized against gravitational contraction by turbulent motions. For example, Brunt et al. (2009) have shown that the principal component of the velocity dispersion in molecular clouds is a dipole-like component, meaning that the dominant motions

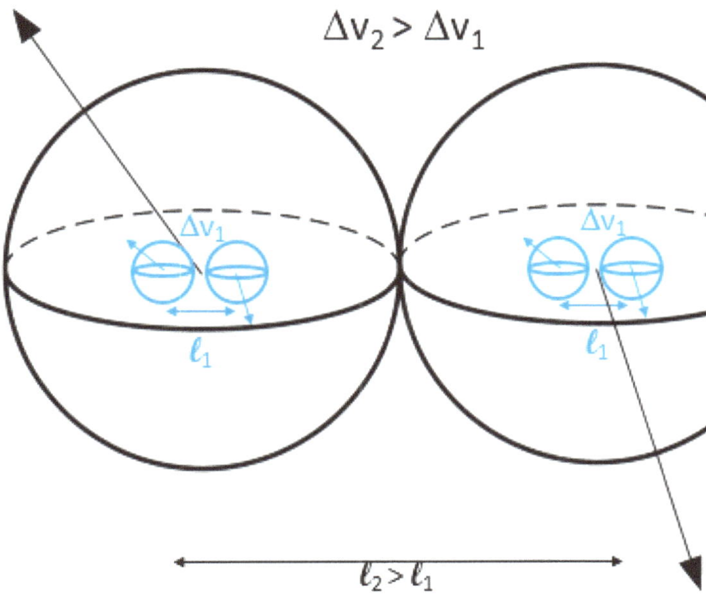

Figure 3.7. Schematic illustration of the multi-scale nature of the velocity differences in a turbulent flow. The small-scale eddies and quantities referring to them are denoted by blue symbols. The velocity difference between the small-scale eddies is denoted Δv_1, and the separation between them is denoted ℓ_1. The large-scale eddies and their corresponding properties are denoted by black lines. The difference between their velocities is $\Delta v_2 > \Delta v_1$, and the separation between them is $\ell_2 > \ell_1$.

are cloud-scale velocity gradients, rather than microscopic turbulent motions. This implies that the main effect of the turbulent motions on a cloud of a given scale is to distort it (compress, stretch, spin, or shear), rather than provide an isotropic pressure that can "support" the cloud against gravity.

Therefore, turbulence should not be modeled simply as an "extra" velocity dispersion that provides a pressure additional to the thermal one in a certain region of a flow, unless it can be safely shown that the largest turbulent motions are much smaller than the size scale of the system considered. However, this is not the case for individual molecular clouds. Alternatively, the statement can be made in a statistical sense, if temporal averages of the turbulent motions can be computed over timescales much larger than the turbulent crossing time across the system's scale, but again, this does not describe the evolution of an individual cloud over its lifetime, but rather the stationary state of a cloud ensemble. Finally, note, however, that the analysis in Section 3.11 is not invalidated by this consideration, since only the amplification of turbulent motions by contraction was discussed, without making any assumption that those motions could provide an isotropic pressure.

References

Armi, L., & Flament, P. 1985, JGR, 90, 11
Armstrong, J. W., Rickett, B. J., & Spangler, S. R. 1995, ApJ, 443, 209

Ballesteros-Paredes, J., Vázquez-Semadeni, E., Gazol, A., Hartmann, L. W., Heitsch, F., & Colín, P. 2011, MNRAS, 416, 1436
Bonazzola, S., Heyvaerts, J., Falgarone, E., Perault, M., & Puget, J. L. 1987, A&A, 172, 293
Brunt, C. M., Heyer, M. H., & Mac Low, M. M. 2009, A&A, 504, 883
Burgers, J. M. 1974, The Nonlinear Diffusion Equation: Asymptotic Solutions and Statistical Problems (Dordrecht: Springer)
Chandrasekhar, S. 1951, RSPSA, 210, 26
Federrath, C., Klessen, R. S., & Schmidt, W. 2008, ApL, 688, L79
Fleck, R. C. 1996, ApJ, 458, 739
Frisch, U. 1995, Turbulence: The Legacy of A.N. Kolmogorov (Cambridge: Cambridge Univ. Press)
Galtier, S., & Banerjee, S. 2011, PhRvL, 107, 134501
Gazol, A., & Kim, J. 2010, ApJ, 723, 482
Grant, H. L., Stewart, R. W., & Moilliet, A. 1962, JFM, 12, 241
Guerrero-Gamboa, R., & Vázquez-Semadeni, E. 2020, ApJ, 903, 136
Kadomtsev, B. B., & Petviashvili, V. I. 1973, SPhD, 18, 115
Kida, S., & Orszag, S. A. 1990a, JSCom, 5, 85
Kida, S., & Orszag, S. A. 1990b, JSCom, 5, 1
Kolmogorov, A. 1941, DoSSR, 30, 301
Kritsuk, A. G., Norman, M. L., Padoan, P., & Wagner, R. 2007, ApJ, 665, 416
Kritsuk, A. G., Norman, M. L., & Wagner, R. 2011, ApL, 727, L20
Murray, N., & Chang, P. 2015, ApJ, 804, 44
Passot, T., & Vázquez-Semadeni, E. 1998, PhRvE, 58, 4501
Robertson, B., & Goldreich, P. 2012, ApL, 750, L31
Schulz-Dubois, E. O., & Rehberg, I. 1981, ApPhy, 24, 323
Vázquez-Semadeni, E. 1994, ApJ, 423, 681
Vázquez-Semadeni, E. 1999, Millimeter-Wave Astronomy: Molecular Chemistry & Physics in Space: Proc. of the 1996 INAOE Summer School of Millimeter-Wave Astronomy, Vol. 241, ed. W. F. Wall, A. Carramiñana, & L. Carrasco (Dordrecht: Springer) 161
Vázquez-Semadeni, E., & Gazol, A. 1995, A&A, 303, 204
Vázquez-Semadeni, E., Passot, T., & Pouquet, A. 1996, ApJ, 473, 881

AAS | IOP Astronomy

Interstellar Flow and Star Formation

Enrique Vázquez-Semadeni

Chapter 4

ISM Thermodynamics and Compressibility

As mentioned in Chapter 1, the ISM is subject to a large number of cooling and heating processes. The net effect of these is to cause a unique thermodynamic behavior of the ISM, quite different from the nearly adiabatic or isothermal regimes frequently considered in thermodynamics. Specifically, the ISM at large exhibits a roughly uniform thermal pressure, as can be seen in Figure 4.1, which shows the temperature as a function of density for various types of regions, implying that the pressure is roughly uniform, as indicated by the solid lines, which represent pressures of 10^2, 10^3, and 10^4 K cm^{-3}. The only types of regions that appear to have pressures in excess of 10^4 K cm^{-3} are the HII regions and the molecular structures (dark clouds, globules, and molecular clouds). The HII regions are overpressured because they are heated by the photoionizing radiation from massive stars within them, while molecular clouds are likely overpressured because they have been compressed by gravity or some external mechanical agent, like the collision of expanding shells in the ISM. In this chapter we will discuss the origin of this behavior, and how it gives rise to the so-called *phases* of the ISM.

4.1 The Heating and Cooling Functions in the ISM

The heating and cooling rates Γ and $n\Lambda$ include a large number of processes at the atomic, molecular, and dust level, which we will not discuss here. It suffices, for our purposes, to know that these processes can have a strong dependence on the density and temperature of the gas. Detailed discussions on the processes acting on the diffuse neutral hydrogen can be found in the classical papers by Dalgarno & McCray (1972) and Wolfire et al. (1995); Sutherland & Dopita (1993) discuss the processes acting on the diffuse ionized gas, and Koyama & Inutsuka (2000) (and references therein) discuss the processes acting on the dense, cold atomic and molecular gas. The end result of all of these heating and cooling processes is that the net rate of temperature change of the gas depends strongly on its temperature, density and metallicity. As classical examples, Figure 4.2 illustrates the density

doi:10.1088/2514-3433/adcb28ch4

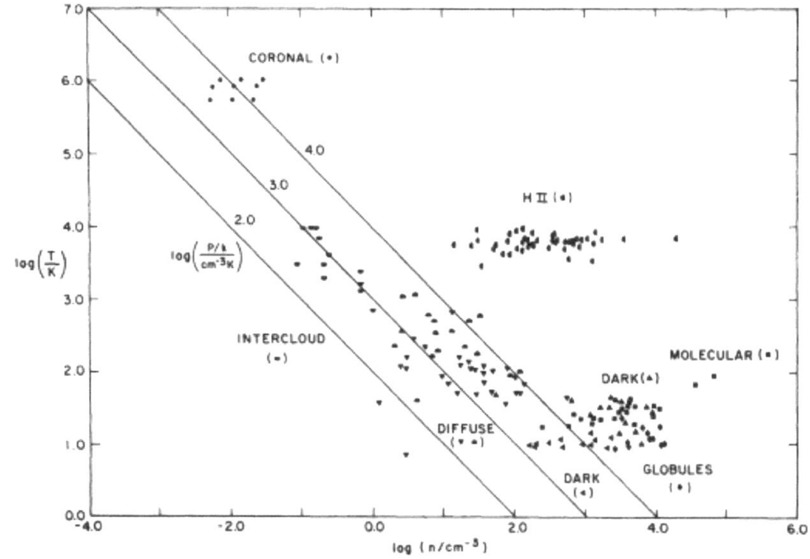

Figure 4.1. Temperature *versus* density of various types of regions in the ISM. The diagonal lines indicate *loci* of constant pressure, for $P/k_{\rm B} = 10^2$, 10^3 and 10^4 K cm^{-3}. Reproduced from Myers (1978).

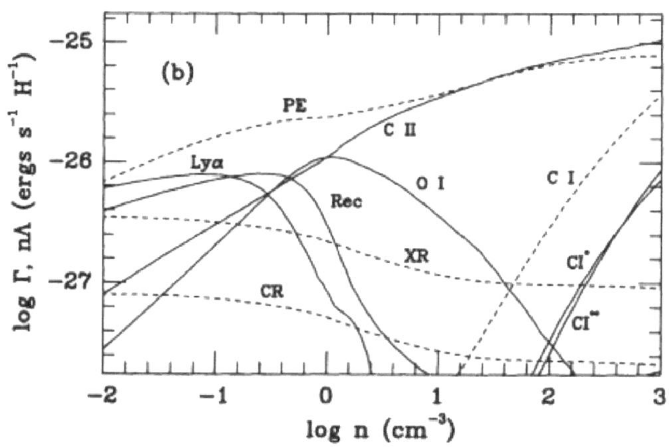

Figure 4.2. Density dependence of the cooling and heating processes acting on the neutral atomic gas in the ISM for the "standard" model of Wolfire et al. (1995). The dashed lines represent heating processes, while the solid lines represent cooling processes. The labels have the following meanings: "PE", photoelectric heating on dust grains and poly-aromatic hydrocarbons ("PAHs"); "XR", heating from X-rays; "CR": heating by cosmic rays' "CI", coiling by photoionization of neutral Carbon; "Lyα", cooling by Lyα emission; "Rec", cooling by recombination on dust grains and PAHs; "CI*", cooling by emission of the fine-structure 609 μm line of CI; "CI**", cooling by emission of the fine-structure 370 μm of CI' "OI", cooling by fine-structure emission of OI. Reproduced from Wolfire et al. (1995).

dependence of the cooling and heating rates produced by the main processes acting on the atomic gas (taken from Wolfire et al., 1995), and Figure 4.3 illustrates the *cooling function* Λ (with units of erg cm^3 s^{-1}) as a function of temperature in the range $10 < T/{\rm K} < 10^8$ (taken from Dalgarno & McCray, 1972). Multiplying this

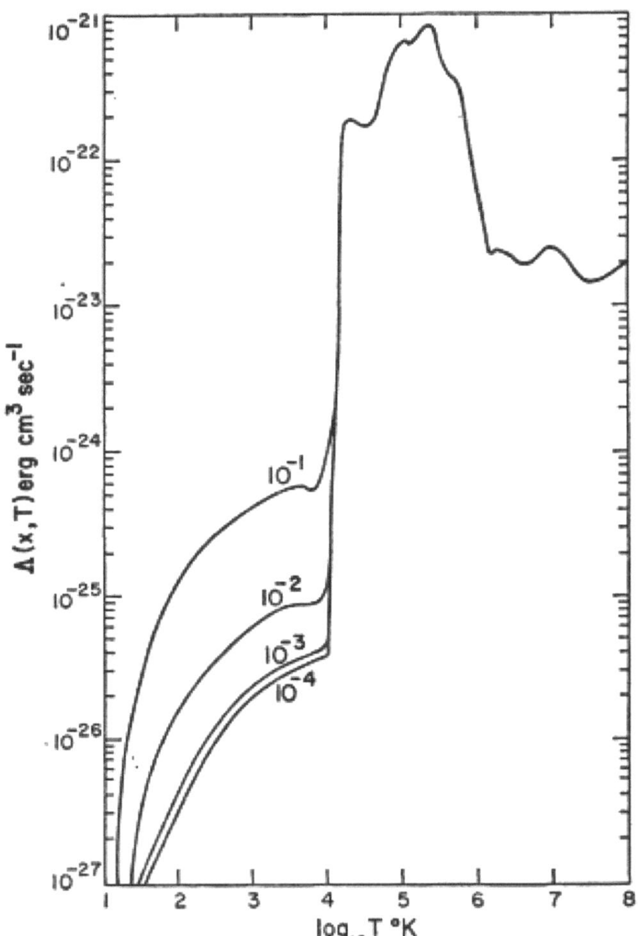

Figure 4.3. Cooling function Λ as a function of temperature and of the ionization fraction x. The numbers next to the curves indicate the value of x. Reproduced with permission from Dalgarno & McCray, (1972). © Annual Reviews.

function by the number density n (with units of cm^{-3}) gives the cooling rate per hydrogen atom, and multiplying Λ times n^2 gives the cooling rate per cubic cm. As can be seen from this figure, the cooling function is an extremely nonlinear function of the temperature. As a consequence, the cooling rate per unit volume, $n^2\Lambda$ is also a strongly nonlinear function of the density.

Instead, the heating rate does not vary as strongly with density, as illustrated by the line labeled "PE" in Figure 4.2, which shows the heating rate per unit volume due to photoelectric heating of dust grains and polycyclic aromatic hydrocarbon molecules ("PAHs") caused by the UV background radiation originating from massive stars. We see that this heating process, which is the dominant one for the atomic medium, varies only by roughly one order of magnitude over a five-order of magnitude range in density. Thus, a useful first-order approximation for this heating rate is to take it as a constant.

4.2 Fast Net Cooling and Barotropic Flow

It is often the case that the radiative heating (Γ) and cooling ($-n\Lambda$) rates are much larger than the adiabatic heating rate ($-[\gamma - 1]e\,\nabla \cdot \boldsymbol{u}$) in the internal energy equation, Equation (2.76). This condition is equivalent to the condition that the *net cooling time*,[1]

$$\tau_{\rm c} \approx \frac{e}{|n\Lambda - \Gamma|}, \tag{4.1}$$

is much shorter than the *sonic crossing time*,

$$\tau_{\rm s} = \frac{L}{c_{\rm s}}, \tag{4.2}$$

where L is the size of the fluid parcel and $c_{\rm s}$ is the sound speed. In this case, the Lagrangian derivative of the internal energy e is dominated by the competition between the two radiative rates, so that

$$\frac{de}{dt} \approx \Gamma - n\Lambda \equiv -\mathcal{L}, \tag{4.3}$$

where $\mathcal{L} \equiv n\Lambda - \Gamma$ is referred to as the *net cooling function*. If $\mathcal{L} = 0$, the gas is said to be in *thermal equilibrium*. Since Γ and Λ are in general functions of the density and temperature, then the condition of thermal equilibrium gives a relation between these two variables, of the form

$$\Gamma(n, T) = n\Lambda(n, T), \tag{4.4}$$

which in principle allows writing the temperature as a function of the density, $T_{\rm eq} = T(n)$, or, equivalently, $T_{\rm eq} = T(\rho)$. Inserting this dependence in the ideal-gas equation of state, Equation (2.79), one can eliminate the temperature, and write the pressure as a function of the density only, $P = P(\rho)$, a condition called *barotropic flow*, governed by an *effective barotropic equation of state*.[2] We also refer to $P(\rho)$ as the *thermal equilibrium pressure*, and denote it by $P_{\rm eq}(\rho)$.

Furthermore, if over some ranges of density and temperature, the cooling and heating functions may be written as power laws of these variables, then it is easy to show that the pressure can be written as a power law of the density, which we refer to as an *effective polytropic equation of state*, of the form

$$P_{\rm eq} = P_0\left(\frac{\rho}{\rho_0}\right)^{\gamma_{\rm e}}, \tag{4.5}$$

---

[1] This time can be understood as the time required by the temperature of a fluid parcel to decrease by a factor of e at the net cooling rate $n\Lambda - \Gamma$, under static conditions.

[2] Etymologically, the word "barotropic", from the Greek *baros* (pressure, weight) and *tropos* (way, form, manner), means "in the manner of pressure".

where γ_e is known as the *effective polytropic exponent*. Note that γ_e is different from the actual heat capacity ratio of the gas, γ, which refers to the internal degrees of freedom of the particles making up the gas. Instead, γ_e is a parameter representing the dependence of the radiative heating and cooling rates on the density and temperature.

Insight: It is worth noting that, informally, a flow is sometimes said to be *soft* if it has a small γ_e, or to have a "soft" effective equation of state. In particular, flows with $\gamma_e < 1$ are said to be "softer than isothermal".

Also, it is important to remark that, strictly speaking, barotropic flow, and in particular the polytropic case, apply when $\tau_c \to 0$ and, in this case, the internal energy equation, Equation (2.76), can be omitted altogether, and the thermal equilibrium pressure, given by Equation (4.5) can be inserted directly into the momentum equation, Equation (2.75). Therefore, the polytropic equation, which includes isothermal flow as a special case, implicitly implies infinitely fast cooling.

4.3 Thermal Instability and the Condensation Mode

The complex functional dependence of the cooling function with temperature and density is at the origin of a peculiar thermodynamical behavior of the ISM known as *thermal instability* (TI). Intuitively, TI can be understood as the tendency of the gas to drift away from the equilibrium between heating and cooling upon a perturbation. The linear analysis for the problem was first studied by Field (1965, hereafter F65). Here we discuss it first in qualitative form and then we present the basics of the linear instability analysis. Other discussions may be found in the monographs by Shu (1992) and Krumholz (2017), and in the review by Vázquez-Semadeni et al. (2003).

4.3.1 The Isochoric Criterion

For convenience, let us assume that, to first order, $\Gamma \approx$ cst. Then, for instance, let us consider the temperature range $10^{5.5} \lesssim T \lesssim 10^6$ K in Figure 4.3, in which we observe that the condition

$$\left(\frac{\partial \Lambda}{\partial T}\right)_\rho = \left(\frac{\partial \mathcal{L}}{\partial T}\right)_\rho < 0 \tag{4.6}$$

is satisfied. We write the derivative as a partial derivative because Λ is a function of T and ρ, although the plot shows only the dependence of Λ on temperature. Also for this reason, we write the derivative at constant density.

Condition (4.6) implies that, if the medium is initially in thermal equilibrium and a gas parcel suffers a small increment in its temperature ($\Delta T > 0$) at constant density, Λ decreases, and therefore now $\Gamma > n\Lambda$; that is, the heating is now larger than the cooling. In turn, from Equation (4.3), we see that this implies that $de/dt > 0$, and thus the temperature will continue to increase, and Λ will continue to decrease, causing the medium to drift away from equilibrium even more. The same happens, but in the opposite direction, for a perturbation such that the parcel's temperature decreases by a small amount. In this case, the cooling increases, becoming larger than the heating, and the parcel will continue to cool.

This evolution away from equilibrium ends when the slope of the curve $\Lambda(T)$ becomes positive again, a condition that happens at sufficiently lower ($\log T \lesssim 5.3$) or higher ($\log T \gtrsim 6$) temperature. The net effect is that the gas tends to segregate in a hot ($T \gtrsim 10^6$ K) and a warm phase ($T \lesssim 10^5$ K).

Because this mode of the instability tends to produce *condensations* (i.e., colder, denser regions embedded within warmer, more diffuse substrate), it is called the *condensation mode*. In turn, this mode can operate under different conditions. The condition (4.6) is referred to as the *isochoric criterion* for the condensation mode, since it implies that the perturbation may evolve isochorically (i.e., at roughly constant density), if its size scale λ is sufficiently large, as we now discuss.

To determine the meaning of "sufficiently large", consider the following: when the fluid parcel begins to cool, its pressure drops, and it begins to be compressed by its surroundings. However, this compression occurs via a sound wave from the external medium, which travels at the sound speed in that medium. Therefore, the compression takes one *sonic crossing time*, $\tau_s = \lambda / c_s$, to occur. The condition for the perturbation to remain at roughly its initial density, then, is that λ is large enough that

$$\tau_c \ll \tau_s, \tag{4.7}$$

where τ_c is the *cooling time*, given by Equation (4.1).

Given the range of temperatures in which the isochoric criterion (Equation (4.6)) is satisfied ($10^5 \lesssim T/K \lesssim 10^6$; cf. Figure 4.3) and the fact that the nearly isochoric character requires the scale of the perturbations to be large enough for relation (4.7) to be fulfilled, this mode of the instability is responsible for the formation of large patches of hot gas ($T \gtrsim 10^6$ K) that can last long times, since, additionally, the cooling time given by Equation (4.1) is very long at low densities. This is the case, for example, of hot bubbles produced by supernova explosions, which apply extremely large heating rates over brief periods, which in turn can overwhelm the large cooling "barrier" at $T \sim 10^4$–10^5 K (the steep rise in the cooling function in this temperature range), and leave the gas above these temperatures.

4.3.2 The Isobaric Criterion

Another, less stringent, criterion for the development of the condensation mode of thermal instability occurs when the condition

$$\left(\frac{\partial \mathcal{L}}{\partial T}\right)_P < 0 \tag{4.8}$$

is satisfied. This is referred to as the *isobaric criterion*.

To understand its origin and physical meaning, let us write explicitly the dependence of Γ and Λ on the density and temperature in the thermal equilibrium ($\mathcal{L} = 0$) condition:

$$\Gamma(\rho, T) = n\Lambda(\rho, T). \tag{4.9}$$

This is a relation between ρ and T that allows, in principle, to solve for T as a function of ρ, $T = T(\rho)$, so that the temperature can be eliminated in the ideal-gas equation of state, Equation (2.79). One then obtains a so-called *barotropic equation*,

$$P = P_{\rm eq}(\rho), \tag{4.10}$$

which indicates that the pressure depends only on the density, and the notation $P_{\rm eq}$ denotes that this is the value of the pressure that corresponds to thermal equilibrium.

The curve $P = P_{\rm eq}(\rho)$ can be understood as the locus of $\mathcal{L} = 0$ in the (P, ρ) plane of Figure 4.4. The region above the curve ($P > P_{\rm eq}$) corresponds to net cooling ($\mathcal{L} > 0$) because, on one hand, the cooling function is a monotonically increasing function of the temperature for $T \lesssim 10^5$ K, and on the other, according to the ideal-gas equation of state, Equation (2.79), an increase of pressure at constant density requires an increase of the temperature. Thus, $P > P_{\rm eq}$ implies $\mathcal{L} > 0$. Conversely, the region below the curve ($P < P_{\rm eq}$) corresponds to net heating ($\mathcal{L} < 0$).

Figure 4.4 shows $P_{\rm eq}(\rho)$ for the atomic gas in the ISM, as calculated by Wolfire et al. (1995) using the heating and cooling processes shown in Figure 4.2. It is noteworthy that $P_{\rm eq}$ *decreases* with increasing density over the density interval $1 \lesssim n/{\rm cm}^{-3} \lesssim 10$. That is, $dP_{\rm eq}/d\rho < 0$ in this density range. We can thus see that the condition $dP_{\rm eq}/d\rho < 0$ is equivalent to the isobaric condition (4.8) because, for example, a density enhancement at constant pressure corresponds to a horizontal displacement toward higher density in the (P, ρ) plane. In the region where $dP_{\rm eq}/d\rho < 0$, such a perturbation puts the perturbed fluid parcel above the $P_{\rm eq}(\rho)$ curve, and thus in a net cooling regime, with $\mathcal{L} > 0$. That is, $(\partial\mathcal{L}/\partial\rho)_P > 0$ for this process. However, at constant pressure, (say $P = P_0$), we note that the ideal-gas equation of state, Equation (2.79), implies $T = \mu m_{\rm H} P_0/k_{\rm B}\rho$. Thus

$$0 < \left(\frac{\partial\mathcal{L}}{\partial\rho}\right)_P = -\frac{\mu m_{\rm H} P_0}{k_{\rm B}\rho^2}\left(\frac{\partial\mathcal{L}}{\partial T}\right)_P,$$

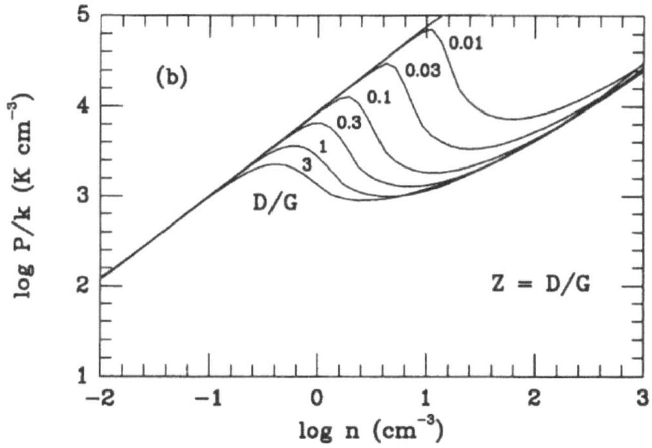

Figure 4.4. Thermal equilibrium pressure $P_{\rm eq}$ as a function of density for the atomic ISM. The different curves correspond to different values of the metallicity. Reproduced from Wolfire et al. (1995).

so that $(\partial\mathcal{L}/\partial\rho)_P > 0$ implies $(\partial\mathcal{L}/\partial T)_P < 0$, and therefore the isobaric criterion is satisfied when $dP_{\rm eq}/d\rho < 0$.

We can see, then, that the isobaric perturbation discussed above will cause the gas to cool, and thus reduce its pressure. But then it will be at a lower pressure than its surroundings, and continue to be squeezed by them, thus undergoing a runaway compression, which continues until the parcel leaves the density range where $dP_{\rm eq}/d\rho < 0$. The same occurs, albeit in the opposite sense, if the perturbation implies a decrease in the parcel's density.

However, the subsequent evolution of the perturbation depends on its wavelength, or size scale, λ. This is because the development of a pressure gradient between the perturbed parcel and its surroundings generates sound waves that tend to restore uniform pressure. The timescale for pressure to be restored is the *sound crossing time* $\tau_{\rm s} = \lambda/c_{\rm s}$. On the other hand, the cooling tends to increase the gradient, on the *cooling timescale*, $\tau_{\rm c}$ given by Equation (4.1). Then, if λ is small enough that $\tau_{\rm s} \ll \tau_{\rm c}$, pressure balance will be nearly fully restored before the parcel has time to cool further, and the parcel will thus cool and contract at *nearly* constant pressure.[3] Conversely, if λ is large enough that $\tau_{\rm s} \gg \tau_{\rm c}$, then the fluid parcel will cool long before sound waves have had time to induce much compression, and the parcel will cool nearly isochorically. This also implies that the growth rate $(1/\rho)d\rho/dt$ of the instability tends to zero as $\lambda \to \infty$. This is because the propagation speed of the compression induced by the environment on the parcel is essentially the speed of sound. Therefore, the larger the size of the parcel, the longer the time needed for the compression to traverse it, and the lower the relative increment of the density per unit time $((1/\rho)d\rho/dt)$.

Of course, over timescales much longer than the cooling time, the parcel will enter a nonlinear evolutionary regime, and deviations from the nearly isobaric or isochoric regimes will occur. This is the regime where numerical simulations become necessary to follow the evolution. Also, as will be discussed in Section 4.5, TI can be nonlinearly triggered by strong compressions (shocks) in the *linearly stable* density ranges.

On the other hand, if the isochoric criterion (which is more stringent than the isobaric one) is also satisfied, then the growth rate of the instability can remain finite even for arbitrarily large parcel scales, since in this case the mechanism involved is a direct cooling of the parcel, with no need for it to be compressed by its surroundings.

4.3.3 Thermally Bistable Media and Multi-phase Models of the ISM

A medium is called *thermally bistable* when its $P_{\rm eq}(\rho)$ curve implies the existence of a thermally unstable density range, in which $dP_{\rm eq}/d\rho < 0$, bounded by two thermally stable ranges where $dP_{\rm eq}/d\rho > 0$, as in Figure 4.4. A remarkable feature of the Galactic atomic ISM is that it is thermally bistable over a large range of galactocentric radii, and the mean density at each radius lies in the unstable range

[3] Note, that the parcel cannot remain at *exactly* the same pressure, since it would then not continue to be compressed.

(e.g., Pikel'Ner 1968; Field et al. 1969; Wolfire et al. 2003; Dickey et al. 2009). This implies that the medium will have an inherent tendency to spontaneously segregate into a warm-diffuse phase, called the *warm neutral medium*, WNM, and a cold-dense one, called the *cold neutral medium* (CNM), giving rise to the notion of a two-phase atomic ISM (Pikel'Ner 1968; Field et al. 1969). In particular, at the Solar Galactocentric radius, the mean density is $n \sim 1\ \mathrm{cm}^{-3}$, the mean temperature $T \sim 3000$ K, and the thermal pressure $P/k_\mathrm{B} \sim 3000\ \mathrm{K\ cm}^{-3}$. In quasistatic conditions, these mean values, however, would be attained through a mixture of CNM cloudlets[4] with $n \sim 30\ \mathrm{cm}^{-3}$ and $T \sim 100$ K, embedded in a WNM substrate with $n \sim 0.3\ \mathrm{cm}^{-3}$ and $T \sim 10^4$ K, both at the same pressure $P/k_\mathrm{B} \sim 3000\ \mathrm{K\ cm}^{-3}$.

Some years after the development of the two-phase model for the atomic ISM, McKee & Ostriker (1977) proposed that an additional, hot phase should also exist, due to the continued energy injection by supernova explosions, which creates bubbles of gas at $n \sim 10^{-2}\ \mathrm{cm}^{-3}$ and $T \sim 10^6$ K. The transition to and from this phase should be mediated by the isochoric criterion of TI, as it can be seen from Figure 4.3 that the cooling function Λ satisfies $(\partial\Lambda/\partial T)_\rho < 0$ in the temperature range $10^5 \lesssim T/\mathrm{K} \lesssim 10^6$. This became known as the *three-phase model* of the ISM. More recently, however, numerical simulations have shown that, in the presence of turbulence, mixing between the phases generates a substantial amount of gas transiting the unstable regime, as it flows between the stable phases, generating a continuous distribution of densities and temperatures in the ISM. This is discussed in detail in Section 4.5.1.

4.4 Linear Instability Analysis

As should be clear from the previous discussion, the ISM, subject to radiative heating and cooling, has an effective thermal behavior very different from that of an adiabatic ideal gas. In particular, under some circumstances, it can remain isothermal or even cool down upon compression. We now briefly review the linear instability analysis performed by F65.

Subject to these heating and cooling processes, and neglecting magnetic fields, self-gravity and other agents, Equations (2.74)–(2.76) reduce to (Field 1965; Meerson 1996)

$$\frac{d\rho}{dt} + \rho \nabla \cdot \boldsymbol{u} = 0, \tag{4.11}$$

$$\rho\frac{d\boldsymbol{u}}{dt} + \nabla P = 0, \tag{4.12}$$

$$\frac{1}{\gamma - 1}\frac{dP}{dt} + \frac{\gamma}{\gamma - 1}P \nabla \cdot \boldsymbol{u} + \rho\mathcal{L}(\rho, T) - \nabla \cdot (\kappa \nabla T) = 0, \tag{4.13}$$

[4] The reason that the CNM was thought to consist of small cloudlets was that, as discussed in Section 4.4, the fastest-growing modes of the instability have small wavelengths. Thus, the most abundant structures produced by the linear development of the instability would have the size scale of these fastest-growing modes.

where $d/dt = \partial/\partial t + \boldsymbol{u} \cdot \nabla$ is the total or Lagrangian derivative (cf. Equation (2.12)), $\boldsymbol{u}$ is the fluid velocity, P is the thermal pressure, T is the temperature, $\gamma = c_P/c_V$ is the heat capacity ratio, c_P and c_V are respectively the specific heats at constant pressure and at constant volume, $\kappa = \kappa(T)$ is the thermal conductivity, and the ideal-gas equation of state, $P = \rho RT = nk_B T$, is assumed, with R being the universal gas constant, $n = \rho/m$ the particle number density, and m the mean particle mass. It is worth recalling that $R = N_A k_B = c_P - c_V$, where k_B is the Boltzmann constant, $N_A \approx (m_H/g)^{-1}$ is Avogadro's number, and m_H/g is the mass of the hydrogen atom expressed in grams.[5] Note that Equation (4.13) is equivalent to Equation (2.76), because, for an adiabatic process (applicable to the first two terms in Equation (2.76)), we have $P = (\gamma - 1)\rho e$.

The linear analysis leading to TI was first performed in full in the classic paper by F65, and later generalized to the case of a flow in motion (Hunter & James, 1970). Other useful, more recent presentations may be found in Elmegreen (1991), Shu (1992), and Meerson (1996). To obtain the linearized equations, we assume small-amplitude perturbations for all variables, so that the latter are written generically as $A = A_0 + \delta A$, where $A_0 =$ cst. is the equilibrium value of the variable and $\delta A(\boldsymbol{x}, t) \ll A_0$ is the perturbation. Note that, for the velocity and the net cooling function, we have $\boldsymbol{u}_0 = \mathcal{L}_0 = 0$. This is because, for the velocity, a Galilean transformation is always possible to a frame in which the mean velocity is zero, and, for the net cooling, in thermal equilibrium, it equals zero. We thus obtain the linearized equations

$$\frac{d\delta\rho}{dt} + \rho_0 \, \nabla \cdot \delta\boldsymbol{u} = 0, \tag{4.14}$$

$$\rho_0 \frac{d\delta\boldsymbol{u}}{dt} + \nabla \, \delta P = 0, \tag{4.15}$$

$$\frac{1}{\gamma - 1}\frac{d\delta P}{dt} - \frac{\gamma}{\gamma - 1}\frac{P_0}{\rho_0}\frac{d\delta\rho}{dt} + \rho_0(\mathcal{L}_\rho \delta\rho + \mathcal{L}_{T,\rho}\delta T) - \kappa_0 \, \nabla^2 \, \delta T = 0, \tag{4.16}$$

where

$$\mathcal{L}_\rho \equiv \left(\frac{\partial\mathcal{L}}{\partial\rho}\right)_{T,0}, \quad \mathcal{L}_{T,\rho} \equiv \left(\frac{\partial\mathcal{L}}{\partial T}\right)_{\rho,0}, \tag{4.17}$$

and the subindices "0" in Equation (4.17) and in $\kappa 0$ indicate that these quantities are to be evaluated at the equilibrium state.

Next, we choose the perturbations to be ondulatory (complex exponential) functions of the form

$$\delta A = A_1 \exp(\omega t + i\boldsymbol{k} \cdot \boldsymbol{x}), \tag{4.18}$$

[5] Strictly speaking, $\mathcal{A} = (\text{Da}/\text{g})^{-1}$, where Da (the Dalton) is the atomic mass unit (AMU) expressed in grams. However, since $m_H = 1.007825$ Da, a very convenient, and much more intuitive relation, is that Avogadro's number is, to within 0.8%, the number of hydrogen atoms in one gram.

where ω is the frequency and $\boldsymbol{k}$ is the wavenumber. This turns the differential Equations (4.14)–(4.16) into the algebraic equations

$$\omega\rho_1 + \rho_0 i\boldsymbol{k}\cdot\boldsymbol{u}_1 = 0, \tag{4.19}$$

$$\omega\rho_0\boldsymbol{u}_1 + i\boldsymbol{k}P_1 = 0, \tag{4.20}$$

$$\frac{\omega}{\gamma-1}\delta P - \frac{\omega\gamma}{\gamma-1}\frac{P_0}{\rho_0}\delta\rho + \rho_0(\mathcal{L}_\rho\delta\rho + \mathcal{L}_{T,\rho}\delta T) + \kappa_0 k^2\delta T = 0, \tag{4.21}$$

This system of equations can be manipulated to obtain the cubic *dispersion relation*

$$\omega^3 + \omega^2\left[\frac{\mathcal{L}_{T,\rho}}{c_V} + \frac{ck^2}{k_K}\right] + \omega c^2k^2 + c^2k^2\left[\frac{\mathcal{L}_{T,P}}{c_P} + \frac{ck^2}{\gamma k_K}\right] = 0, \tag{4.22}$$

where ω is the growth rate, k is the wavenumber, $c \equiv \sqrt{\gamma kT/\mu}$ is the adiabatic sound speed, and

$$\mathcal{L}_{T,P} \equiv \left(\frac{\partial\mathcal{L}}{\partial T}\right)_{P,0} = \left(\mathcal{L}_{T,\rho} - \frac{\rho_0}{T_0}\mathcal{L}_\rho\right), \tag{4.23}$$

and

$$k_K \equiv \frac{Ak_\mathrm{B}}{\mu}\frac{c\rho_0}{(\gamma-1)\kappa_0}. \tag{4.24}$$

As pointed out by F65, k_K is the wavenumber corresponding to mean free path of the gas particles.

Being cubic, the dispersion relation (4.22) has three roots, one of them being always real, and the other two being either a complex conjugate pair, or a pair of real numbers. There is growth of the perturbation's amplitude whenever $\Re(\omega) > 0$, where $\Re$ denotes the real part. If the real root is positive, there is exponential growth without oscillation, and thus F65 referred to it as the *condensation mode*. On the other hand, the pair of complex roots corresponds to oscillatory behavior, whose amplitude grows when the real part of those roots is positive, and F65 thus referred to it as the *wave mode*. Finally, he described the case when the three roots are real and negative as the sound waves being *overdamped* (i.e., they decay without oscillating).

Through a detailed analysis of the regions of instability and overstability in the $\mathcal{L}_\rho$-$\mathcal{L}_{T,\rho}$ parameter space, which we shall not repeat here, F65 concluded that the condensation mode is unstable when the isobaric criterion (cf. Equation (4.8)), namely

$$\mathcal{L}_{T,P} < 0, \tag{4.25}$$

is satisfied. However, when only this criterion is satisfied, the growth rate vanishes as $k \to 0$. For the growth rate to remain non-zero at long wavelengths, it is necessary to also satisfy the isochoric criterion (cf. Equation (4.6))

$$\mathcal{L}_{T,\rho} < 0. \tag{4.26}$$

The corresponding instability criterion for the wave mode is the so-called *isentropic* criterion, reading[6]

$$\left(\frac{\partial \mathcal{L}}{\partial T}\right)_s = \frac{\mathcal{L}_{T,\rho}}{c_V} - \frac{\mathcal{L}_{T,P}}{c_P} < 0, \tag{4.28}$$

where s is the specific entropy. Note that

$$c_V = \frac{R}{(\gamma - 1)\mu}. \tag{4.29}$$

By way of illustration, let us consider a power-law cooling function of the form

$$\Lambda = CT^\beta, \tag{4.30}$$

where C and β are constants, and a constant heating rate

$$\Gamma = \Gamma_0. \tag{4.31}$$

Therefore, noting that partial derivatives at constant P ($=P_0$) must be taken using $\rho(T) = P_0/RT$, the isobaric, isochoric, and isentropic criteria, respectively, require $\beta < 1$, $\beta < 0$ and $\beta < 1/(1 - \gamma) = -3/2$, the latter equality being applicable for a monoatomic gas with $\gamma = 5/3$. Looking at Figure 4.3, we see that, in general $\beta \gtrsim 0$ for $T \lesssim 10^4$ K, so that the neutral atomic medium is linearly stable against both the wave mode and the condensation mode under the isochoric criterion.

On the other hand, for the condensation mode under the isobaric criterion, it is useful to rewrite the criterion in terms of the slope of the $\log P_{\rm eq}$ *vs.* $\log \rho$ curve. This can be done noting that the thermal equilibrium condition $\Gamma = n\Lambda$ implies $T_{\rm eq} = (\Gamma_0/C)^{1/\beta} n^{-1/\beta}$. Substituting this in the ideal-gas equation of state implies $P_{\rm eq} \propto \rho^{\gamma_{\rm e}}$, where

$$\gamma_{\rm e} \equiv 1 - \frac{1}{\beta} \tag{4.32}$$

is the *effective polytropic exponent*, which represents the effective dependence of the pressure on density under the action of the radiative heating and cooling processes alone in the limit of fast cooling. Therefore, in the range $0 < \beta < 1$, where the condensation mode is stable under the isochoric criterion, but unstable under the isobaric condition, we see that $\gamma_{\rm e} < 0$. That is, a sufficient condition for instability under the isobaric criterion is $\gamma_{\rm e} < 0$, or a negative slope of the curve $\log P_{\rm eq}$ *vs.* $\log \rho$.

---

[6] The derivation of this equation involves considering the ideal-gas equation of state $P = \rho RT$, with R the universal gas constant, and the facts that the specific entropy is given by $s = c_V \ln(P\rho^{-\gamma})$, and that $\mathcal{L}$ can be written as a function of any pair of thermodynamic variables, so that, in particular, it can be written as either $\mathcal{L}(s, T)$ or $\mathcal{L}(\rho, T)$. Thus, its total rate of change, $d\mathcal{L}$, satisfies

$$d\mathcal{L} = \left(\frac{\partial \mathcal{L}}{\partial T}\right)_s dT + \left(\frac{\partial \mathcal{L}}{\partial s}\right)_T ds = \left(\frac{\partial \mathcal{L}}{\partial T}\right)_\rho dT + \left(\frac{\partial \mathcal{L}}{\partial \rho}\right)_T d\rho, \tag{4.27}$$

and finally expressing $d\rho$ in terms of P and T through the equation of state.

It also is worth noting that $\beta < 0$ implies a "reversed" dependence of Λ on the temperature, since in this case, Λ *decreases* upon an increment of the temperature, and this is sufficient to trigger instability under the isochoric mode, even though $\gamma_e > 1$ for all $\beta < 0$. The same happens for the adiabatic criterion, for which the condition on the effective polytropic exponent is $1 < \gamma_e < 5/3$, but with the same reversed dependence of Λ on T, since the instability requirement on β in this case is $\beta < -3/2$ for a monatomic gas with $\gamma = 5/3$.[7]

We conclude this section by noting that isobaric and isochoric perturbations (which can be implemented in simulations, for example, by setting up density fluctuations at constant pressure or temperature fluctuations at constant density, respectively), are generically referred to as *entropy* perturbations, since they imply a variation of the ratio P/ρ^γ, which remains constant for reversible isentropic processes. On the other hand, *isentropic*, or *adiabatic*, perturbations can be set up by *dynamic* compressions with short enough characteristic timescales (*sound crossing times*) that the medium does not have time to cool, and so the fluctuation remains adiabatic. In the following sections, we discuss the evolution of the various kinds of perturbations in slightly greater detail.

4.4.1 Entropy Perturbations

Figure 4.5 shows a typical plot of the growth rate ω as a function of wavenumber $k = 2\pi/\lambda$. In the case of the pure development of the instability, without any external forcing processes, there are three clearly distinct scale ranges for the wavelength λ that arise from the presence of three characteristic timescales (Meerson, 1996): the dynamical time, in this case given by the sound crossing time, $\tau_s = c_s/\lambda$; the cooling time, which for isobaric processes is given by[8]

$$\tau_c = \frac{\gamma R}{(\gamma - 1)\mu|\mathcal{L}_{T,P}|}, \tag{4.33}$$

but which, to order of magnitude, is in general $\tau_c \approx c_V T/(\rho\Lambda)$; and the conductive time, $\tau_\kappa = \lambda^2/\kappa_0'$, where $\kappa_0' = \kappa_0/\rho_0$ is the thermal diffusivity. In the remainder of this chapter we will consider the ratio of the cooling time to the crossing time,

$$\eta \equiv \frac{\tau_c}{\tau_s}, \tag{4.34}$$

as an important governing parameter for the evolution of the perturbations.

[7] Recall, however, that the isochoric mode requires large-scale perturbations, so that the gas can cool down before sound waves have time to restore pressure balance.

[8] This expression is obtained by linearizing the energy Equation (4.13) and then computing

$$\tau_c \equiv \frac{\delta e}{\left(\frac{d\,\delta e}{dt}\right)}$$

for an isobaric process.

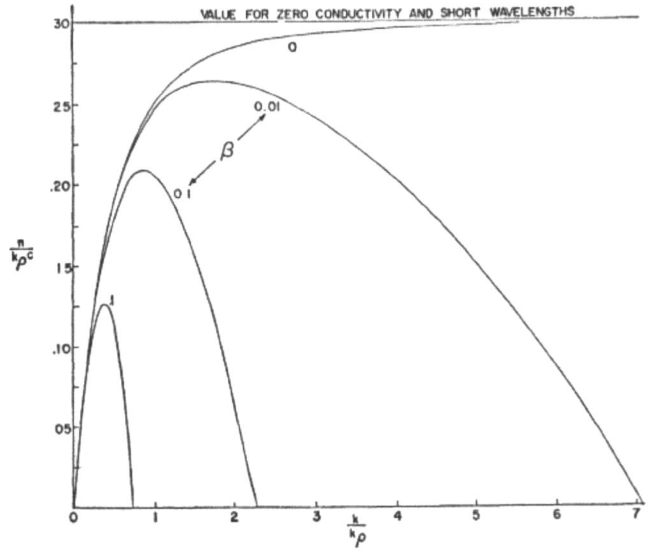

Figure 4.5. (*Solid line*): Analytical growth rate as a function of wavenumber as given by Equation (4.22) as computed by Field (1965) for various values of the thermal conductivity. The non-zero wavenumber at which the growth rate becomes zero corresponds to the so-called *Field length*. In the non-conducting case, the growth rate asymptotically approaches the cooling rate, and the fastest-growing scales are the smallest. Note that β in this plot is unrelated to the exponent in Equation (4.30), and rather refers to a parameter in Field (1965), from which this figure is taken. Reproduced from Field (1965).

These timescales then define two characteristic length scales. First, the so-called Field length (Begelman & McKee 1990) $\lambda_F = 2\pi/k_F$, with $k_F = (|\mathcal{L}_{T,P}|/\kappa_0')^{1/2}$, at which $\tau_\kappa \sim \tau_c$, and below which the growth of perturbations is inhibited by thermal conduction, because conduction restores thermal balance before the medium can cool radiatively. Second, the scale at which $\tau_s \sim \tau_c$, which we shall denote λ_{eq}. These two length scales then define three scale ranges, to which we refer as the "long" ($\lambda \gg \lambda_{eq}$), "intermediate" ($\lambda_{eq} \gg \lambda \gg \lambda_F$) and "short" ($\lambda \lesssim \lambda_F$) ranges. For the long-wavelength range, $\eta \ll 1$, while for the intermediate- and short-wavelength cases, $\eta \sim 1$ and $\eta \gg 1$, respectively.

In the case of short- and intermediate-wavelength perturbations,[9] the condensation mode evolves nearly isobarically, since $\tau_c > \tau_s$, and thus sound waves have ample time to restore pressure equilibrium before the gas cools. This also has the consequence that, in the case of zero diffusivity, the growth rate asymptotically approaches the cooling rate in the limit of large wavenumbers. That is, *the condensation can grow only as fast as the gas cools*. In the presence of diffusivity, the growth rate decreases again in the short-wavelength range, due to the increasing importance of thermal conduction, until the latter finally becomes faster than cooling at λ_F, and the instability is suppressed. These properties are illustrated in Figure 4.5.

[9] For convenience, in the remainder of this chapter, we will group intermediate- and small-scale perturbations into the small-wavelength ($\eta > 1$) category.

In the opposite limit of very long wavelengths ($k \to 0$), the condensation mode initially behaves isochorically, since $\eta \ll 1$, so that the cooling acts much more rapidly than the sound waves can travel across the perturbation to restore pressure balance. This implies that large pressure gradients can be set up, which in turn can trigger strong motions that can become locally supersonic in the cool gas (Field 1965; Meerson 1996). Moreover, in this regime, the thermal pressure is given by the condition of thermal equilibrium, because the rapid cooling always allows its establishment. In this case, the growth rate asymptotically approaches (Meerson 1996)

$$\omega_{\rm long} = \pm kc\left(-\frac{\mathcal{L}_{T,P}}{\mathcal{L}_{T,\rho}}\right)^{1/2}, \tag{4.35}$$

which is of the order of the inverse of the sound crossing time $\tau_s = 2\pi/kc$, explaining the vanishing of the growth rate as $k \to 0$ (Figure 4.5).

4.4.2 Adiabatic Perturbations

The case of adiabatic perturbations is perhaps the most relevant for a turbulent ISM because a perturbation that is set up by a dynamic compression (caused, for example, by supersonic turbulence) whose crossing time is much shorter than the cooling time, causing an approximately adiabatic perturbation, in which both the pressure and the density are enhanced by the *PdV* work without significant heat losses by cooling.

In this case, we define the bulk-velocity crossing time, $\tau_u = \lambda/u$, where u is the characteristic velocity of the perturbation. Furthermore, it is convenient to consider a generalized crossing (or *dynamical*) timescale as $\tau_{\rm dyn} = \min(\tau_s, \tau_u)$, so that the definition $\eta = \tau_c/\tau_{\rm dyn}$ can be preserved. Adiabatic perturbations with $\eta \gg 1$ (short wavelengths) are unstable when the adiabatic instability criterion (4.28), or $\beta < -3/2$, is satisfied, and excite the wave mode of the instability, which in this case consists of nearly adiabatic sound waves with growth rate $\omega = ikc + (1/2)(\mathcal{L}_{T,P}/c_P - \mathcal{L}_{T,\rho}/c_V)$, where the imaginary part gives the propagation speed and the real part gives a modulation that grows if the adiabatic criterion is satisfied. This is what was referred to by F65 as *overstability*. However, at $\eta < 1$ (long wavelengths), the complex conjugate roots become real, and the wave mode becomes a condensation mode, which is unstable whenever $\mathcal{L}_{T,P}/(\gamma\mathcal{L}_{T,\rho}) < 0$ (Shu 1992). For $\mathcal{L}_{T,\rho} > 0$, as is usually the case, this reduces to the isobaric criterion. Thus, in the atomic ISM, *wave-like (or "velocity") perturbations are linearly unstable only for $\eta < 1$, i.e., in the long-wavelength limit.* For nonlinear perturbations, however, the density increase in the waves can accelerate the cooling and locally cause η to become <1, allowing the instability to proceed even in cases of initial short-wavelength perturbations. A similar effect can occur even when the compression acts on the warm stable phase if it has a large (nonlinear) amplitude (Hennebelle & Pérault 1999; Koyama & Inutsuka 2000). We will discuss this further in Section 4.5.

4.4.3 An Illustrative Diagram

Figure 4.6 provides an illustration of how the various criteria for the condensation and wave modes operate. The figure shows the "standard" $P_{\rm eq}(\rho)$ curve for the atomic gas (Wolfire et al. 1995) as the sinusoidal, black solid curve, and corresponds to typical cooling functions Λ qualitatively similar to those shown in Figure 4.3 up to temperatures $T \approx 10^4$ K. The colors indicate the various types of perturbations that can be applied to a fluid parcel: blue corresponds to isobaric perturbations, red to isochoric, and green to adiabatic. The arrows depict the three types of perturbations considered here: isobaric, isochoric, and adiabatic, colored according to the color code. The black insets in the upper and lower parts of the $P_{\rm eq}(\rho)$ curve indicate the sign of the net cooling function $\mathcal{L}$ on the corresponding side of the curve, for perturbations of the type indicated by the color coding. The colored straight lines represent the critical (or limiting) effective logarithmic slopes (effective polytropic exponents, $\gamma_{\rm e,cr} = 1 - 1/\beta_{\rm cr}$, with $\beta_{\rm cr}$ being the critical exponent of the temperature in the cooling function) of the $P_{\rm eq}(\rho)$ curve required for instability of each type of perturbation (assuming that β satisfies its own requirement), and the colored insets

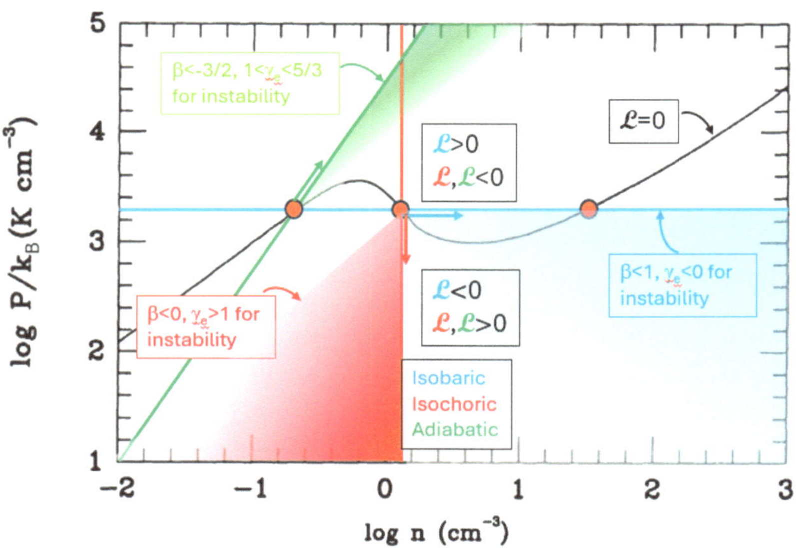

Figure 4.6. Illustration of the types of perturbations and respective instability criteria in the $\log(P/k_{\rm B})$-$\log n$ diagram. Isobaric (blue), isochoric (red), and adiabatic (green) perturbations (arrows) and criteria (colored lines, shaded areas, and colored insets) are shown. The black solid curve shows the "standard" thermal-equilibrium pressure curve $P_{\rm eq}(\rho)$ of Wolfire et al. (1995), for which the net cooling function satisfies $\mathcal{L} = 0$. The red filled circles show the dynamical equilibrium states in the warm (WNM; left), unstable (UNM; middle) and cold (CNM; right) phases at the mean thermal pressure of the ISM at the solar Galactocentric radius. The black insets above and below the $\mathcal{L} = 0$ curve indicate the sign of $\mathcal{L}$ in the corresponding region for the required values for instability of β, the exponent of the temperature in the cooling function Λ (cf. Equation (4.30)), for each type of perturbation. The colored insets indicate the critical values of the exponent β and of the effective polytropic exponent $\gamma_{\rm e}$ for each type of perturbation to be unstable. The colored lines indicate the critical slope of the $P_{\rm eq}(\rho)$ curve for the corresponding type of perturbation to be unstable. Adapted from Wolfire et al. (1995).

indicate the conditions on β and on γ_e for instability of the corresponding perturbations. The shaded areas qualitatively indicate the range of slopes γ_e of the $P_{eq}(\rho)$ curve where the mode of the corresponding color is unstable. The shading fades in the direction away from the critical value of γ_e.

The arrows originate from points (red circles) on the atomic ISM $P_{eq}(\rho)$ curve at the mean thermal pressure ($P/k_B \approx 2000$ K cm^{-3}) in the Galactic midplane at the Solar Galactocentric radius. These points represent the two stable equilibrium states, namely the WNM (left circle) and the CNM (right circle), as well as the unstable equilibrium known as the UNM (unstable neutral medium; middle circle). Note that the arrow corresponding to isobaric and isochoric perturbations originate from the unstable equilibrium, while the arrow corresponding to an adiabatic perturbation is drawn as originating from the WNM, since this is the phase where these types of perturbations are most typically applied by agents like shocks or expanding shells traveling through the ISM. Since this phase is linearly stable, typically strong, nonlinear adiabatic compressions are needed to induce instability and a transition to the CNM (Section 4.5).

Now let us consider the perturbations depicted as applied on the middle red circle. In this state, which corresponds to 0.5 cm$^{-3} \lesssim n \lesssim 5$ cm^{-3} and 300 K $\lesssim T \lesssim 3000$ K, it can be seen from Figure 4.3 that $\beta \lesssim 1$, implying that the local slope of the $P_{eq}(\rho)$ curve is $\gamma_e < 0$, as indeed it is seen to be the case in Figure 4.6. Thus, the isobaric instability criterion, $\beta < 1$, $\gamma_e < 0$, is satisfied. Let us then consider the isobaric perturbation, denoted by the blue arrow. This perturbation displaces a fluid parcel from the red circle to a position *above* the P_{eq} curve, in which the net cooling satisfies $\mathcal{L} > 0$ (as indicated by the inequality for the blue character "$\mathcal{L}$" in the black inset below the $P_{eq}(\rho)$ curve), since $n\Lambda$ has increased, implying that the parcel now begins to cool down, and to lose pressure. The parcel's environment, at the unperturbed pressure, will thus begin to compress it. If the perturbed parcel is small enough so that $\tau_s \ll \tau_c$ (the sound crossing time is much shorter than the cooling time), sound waves can quickly compress the parcel and restore pressure balance between it and its surroundings, and the parcel can remain nearly isobaric. But it is now at an even higher density due to the compression, and the process runs away, the perturbation thus being unstable.

Note, however, that the isobarically-perturbed parcel can only remain nearly isobaric if its size scale is small enough that sound waves are able to restore its pressure to the ambient value; i.e., if $\tau_s \ll \tau_c$. Otherwise, a larger-scale perturbation for which $\tau_s \gg \tau_c$ would continue to cool down and lose pressure, and would deviate toward becoming isochoric. Thus, intermediate-scale perturbations for which $\tau_s \sim \tau_c$ can evolve approximately along the P_{eq} curve. In any case, the growth of the perturbation's density can continue until the slope of the $P_{eq}(\rho)$ curve becomes positive again at $n \sim 5$ cm^{-3} ($\log(n/\text{cm}^{-3}) \sim 0.7$), so that the isobaric condition ceases to be satisfied. However, even though the fluid parcel is not thermally unstable anymore above this density, its pressure continues to be lower than the ambient's, and so compression continues until the parcel's pressure returns to the ambient value, at the position of the right red filled circle in Figure 4.6, corresponding to the CNM. Analogously, an isobaric perturbation in the opposite direction, toward lower density, will terminate its evolution at the locus of the left red circle, corresponding to the WNM.

Let us now consider an isochoric perturbation, as indicated by the red downward arrow. The future evolution of this parcel depends on β. If this exponent only satisfies the isobaric criterion but not the isochoric one (i.e., if $0 < \beta < 1$), then Λ decreases with the decrease in T, and the net cooling function satisfies $\mathcal{L} < 0$ below the P_{eq} curve, and so the parcel will be heated again, and returned to the equilibrium curve. So, this type of perturbation is *stable* when only the isobaric criterion is satisfied. In order for the parcel to continue cooling down, it is necessary to reverse the dependence of Λ with T. This occurs when $\beta < 0$; i.e., when the isochoric criterion is met. In this case, the cooling rate Λ *increases* with a decrease of the temperature and so does the net cooling $\mathcal{L}$. Thus, the isochoric perturbation puts the parcel in a region of net cooling ($\mathcal{L} > 0$, as indicated by the inequality with the red "$\mathcal{L}$" character in the black inset below the $P_{eq}(\rho)$ curve), and the parcel can continue cooling down, moving further away from the equilibrium curve, thus making the perturbation unstable. However, as already pointed out above, in general the cooling function Λ has $\beta \geqslant 0$ in the temperature range $T \lesssim 10^4$ K, and so the atomic medium is generally *stable* to isochoric perturbations. Instability ($\beta < 0$) only occurs at high temperatures, $T \gtrsim 10^5$ K, as seen from the plot of Λ in Figure 4.3.

Note, however, that similarly to the scale dependence discussed above for the isobaric perturbation, the isochorically-perturbed parcel can only remain nearly isochoric if its size scale is large enough that sound waves cannot restore its pressure to the equilibrium value; i.e., if $\tau_s \gg \tau_c$. Otherwise, an initially isochoric small-scale perturbation for which $\tau_s \ll \tau_c$ would have its pressure restored by sound waves, and would deviate toward becoming isobaric.

Finally, let us now consider the adiabatic perturbation, represented by the green arrow (which, for a monatomic gas, has a slope of $5/3$). This type of perturbation, as mentioned before, is perhaps the most natural perturbation mechanism in a supersonically turbulent ISM subject to a constant background heating, since rapid, small-scale compressions, for which $\tau_{dyn} \ll \tau_c$, will nearly adiabatically heat the fluid parcel.

For this type of perturbation, the adiabatic instability criterion ($\beta < -3/2$) is in general *not* satisfied in the atomic medium. Therefore, an adiabatic perturbation, which heats the gas by PdV work, puts the parcel in a net heating ($\mathcal{L} < 0$) region. This raises the temperature even more, and the parcel attempts to reexpand, returning to equilibrium. That is, similarly to the isochoric case, adiabatic perturbations are linearly *stable* in general for typical cooling functions in the atomic ISM, and instability only occurs for temperatures in the range 10^5 K $\lesssim T \lesssim 10^6$ K, where the cooling function Λ is seen to drop precipitously in Figure 4.3.

Finally, note that we have drawn this perturbation as being applied on the left red circle, corresponding to the WNM, as this phase is likely to occupy most of the ISM volume at the Galactic midplane, and therefore it is where turbulent compressions are most likely to be applied. As mentioned above, these are linearly stable, but they can be *nonlinearly* unstable, as discussed in Section 4.5.

4.5 Nonlinear Triggering of the Instability and Cloud Formation

4.5.1 Cloud Formation by Shock Compression and Phase Transition Fronts

As mentioned in Section 4.4.3, in the turbulent atomic ISM, the dominant perturbation mechanism is likely to be some form of compression acting on the *stable* WNM, since this phase occupies most of the volume of the Galactic disk outwards of ~5 kpc within a few hundred parsecs from the midplane (e.g., Ferrière 2001), although it is noteworthy that the total mean density at the midplane at the Solar Galactocentric radius is ~1 cm^{-3}. This suggests that the existence of both warm-diffuse and cold-dense components of the ISM at essentially the same pressure is already the result of condensation by thermal instability. We refer to this type of medium as *thermally bistable*. However, it is important to recognize that sizeable mass fractions of unstable gas have been reported both in observation and in simulations. This is discussed in Section 4.5.2.

Since the WNM is stable even with respect to the isobaric criterion, and adiabatic perturbations are stable even in the isobarically-unstable range, sufficiently strong (nonlinear) perturbations are necessary in order to trigger a phase transition. This process began to be investigated at the turn of the century (Hennebelle & Pérault 1999; Koyama & Inutsuka 2000, 2002)). In particular, Koyama & Inutsuka (2000, see also Kritsuk & Norman 2002) studied the instability of non-equilibrium states, into which a linearly stable WNM fluid parcel may be driven into instability by the passage of a strong shock wave. This is shown in Figure 4.7, which shows the evolution of the densest fluid parcel in a numerical simulation of a Mach-6 shock passing through the WNM (dashed lines). These authors also calculated the stability of an initially out-of-thermal-equilibrium configuration undergoing isobaric cooling, finding an unstable region in the T-ρ diagram depicted by the shaded area. This shows that a sufficiently strong compression applied in the WNM can nonlinearly trigger a phase transition to the CNM.

The one-dimensional structure of the shock and the subsequent condensation was studied by Koyama & Inutsuka (2000) and Vázquez-Semadeni et al. (2006), and is shown in Figure 4.8. The flow is modeled as a gas supersonically hitting a wall on the left-hand side (hatched area), which then triggers the formation of a shock that

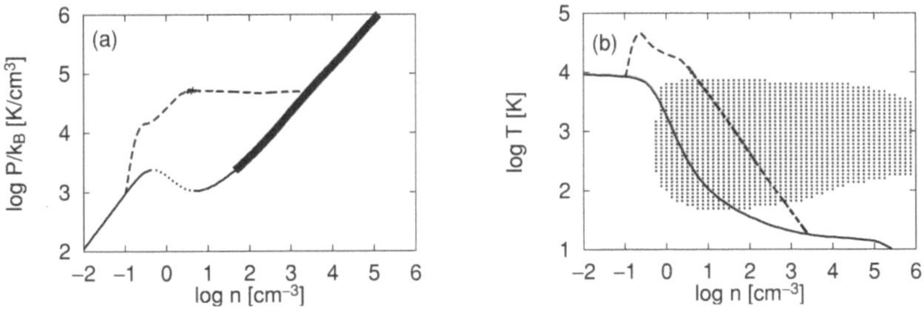

Figure 4.7. Nonlinear evolution of a shock-compressed fluid parcel initially in the WNM in the P-ρ (left panel) and the T-ρ (right panel) diagrams. The solid lines are the thermal equilibrium curve for atomic and molecular gas, and the dashed lines track the evolution of the densest fluid parcel of the simulation. The shaded region in the right panel corresponds to instability of a cooling initial state. Reprinted with permission from Koyama and Inutsuka (2000).

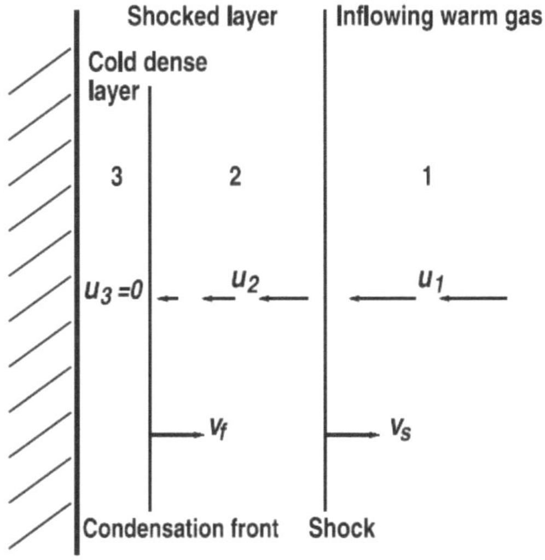

Figure 4.8. Structure of the post-shock region and condensed layer. The shock propagates to the right with speed v_s in the rest frame. The post-shock gas is thrown out ot thermal equilibrium, and cools on a cooling time. Therefore, roughly one cooling length behind the shock, a dense layer forms, bounded by a *condensation front*. Since this layer is roughly 100 times denser than the pre-shock gas, the fluid velocity is correspondingly slower, and the layer grows in thickness only very slowly, being bounded by a *condensation front* that propagates at speed $v_f \ll v_s$ (in the rest frame). Reprinted with permission from Vázquez-Semadeni et al. (2006). © 2006. The American Astronomical Society. All rights reserved. Printed in U.S.A.

propagates back into the incoming gas (right-hand vertical line). The shock initially propagates to the right with speed v_s until the total (thermal + ram) pressure on both sides of the shock becomes equal, at which time the shock becomes stationary. The post-shock gas is thrown out of thermal equilibrium, and cools on a cooling time τ_c (Equation (4.33)). Therefore, at a distance given by the *cooling length* $\ell_{\rm cool} = u_2 \tau_c$ behind the shock, where u_2 is the post-shock velocity, a dense layer appears. Moreover, since the cooling rate per unit volume $n^2\Lambda$ increases rapidly with density, the cooling accelerates as the density increases, and the dense layer appears quite suddenly, and is bounded by a thin *condensation (or phase-transition) front* (middle vertical line). Since this layer is roughly 100 times denser than the pre-shock gas, the condensation front advances with a speed v_f which is smaller than u_2 by nearly the same factor, so that the cold gas' boundary can be considered to be essentially stationary. The dense gas is at rest with respect to the wall.

4.5.2 Turbulent Mixing and the Existence of Unstable Gas

The fact that dynamic perturbations can cause a fluid parcel to move far away from the thermal equilibrium curve, and that this gas then evolves back to equilibrium but to the cold stable phase, suggests that at any moment in time, in a turbulent medium, there should exist a sizable fraction of gas in the unstable phase. Indeed, shortly after the thermal bistability of the atomic medium was proposed (Pikel'Ner 1968; Field et al. 1969), the existence of unstable gas, now known as the *unstable neutral medium*, or UNM, was reported observationally (e.g., Dickey et al. 1977; Kalberla et al. 1985; Spitzer et al. 1995; Heiles 2001; Kanekar et al. 2003; Kalberla & Kerp 2009) as well as numerically (e.g., Vazquez-Semadeni et al. 2000; Koyama & Inutsuka 2000; Gazol et al. 2001; Kritsuk & Norman 2002; Audit & Hennebelle 2005; for more recent results, see the review by McClure-Griffiths et al. 2023), with mass fractions ranging from 20 to 40%.

The actual fraction of unstable gas depends on the strength and characteristic scale (the energy injection scale) of the turbulence, because it is determined essentially by the ratio $\eta \equiv \tau_c/\tau_{\rm dyn}$, as shown in Figure 4.9, taken from Gazol et al. (2005). This figure shows the two-dimensional histogram (depicted by the contours) of points in a numerical simulation of the turbulent neutral atomic medium in the $\log P$-$\log n$ diagram, for several different scales of turbulent energy injection λ_d, ranging from $\lambda_d = 1/2$ (top left) to $\lambda_d = 1/16$ (bottom right) of the numerical box. The $P_{\rm eq}(\rho)$ curve used in that study, a piecewise power law, is indicated by the thick solid line.

The variation of the energy injection scale effectively changes the ratio η, since, for a given rms turbulent velocity dispersion, the turbulent crossing time $\tau_{\rm dyn}$ is proportional to λ_d. Thus, in Figure 4.9 we see that, as λ_d and $\tau_{\rm dyn}$ become smaller,

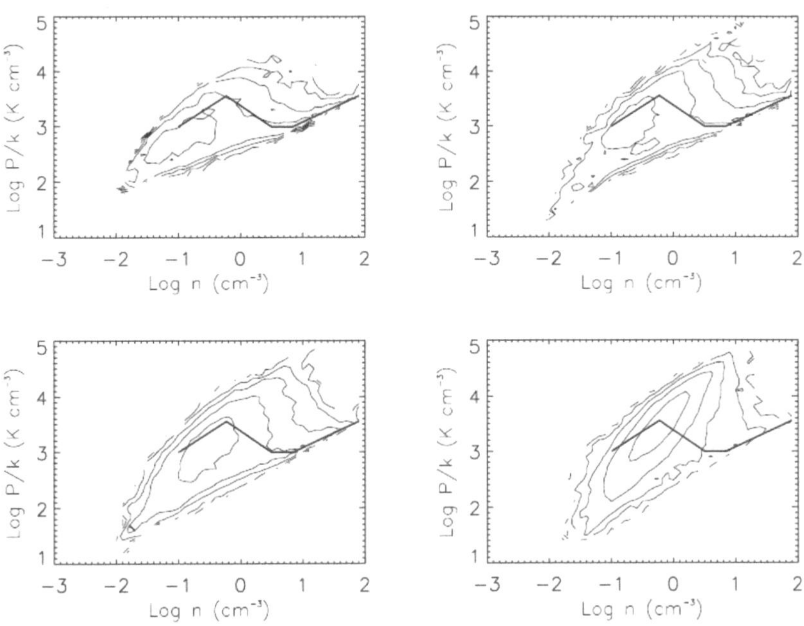

Figure 4.9. Two-dimensional histograms in the P-ρ plane of points in numerical simulations of driven turbulence in unstable gas with mean density 1 cm^{-3}, with the energy being injected at wavenumbers corresponding to 1/2 (top left), 1/4 (top right), 1/8 (bottom left) and 1/16 (bottom right) the size of the numerical box. The $P_{\rm eq}$-ρ curve for this system is shown by the broken solid line. It is seen that, as the injection scale, and thus $\tau_{\rm dyn}$, become smaller, the distribution of points shifts from enveloping the $P_{\rm eq}$-ρ curve (top left) to approximating an adiabatic process with slope $\gamma = 5/3$ (bottom right), although, at high densities, the cooling time is so short that most points lie on the thermal equilibrium curve always. Reprinted with permission from Gazol et al. (2005). © 2005. The American Astronomical Society. All rights reserved. Printed in U.S.A.

and η gets larger, the distribution of points in the diagram departs from the thermal equilibrium curve $P_{\rm eq}(\rho)$ and approaches a mean slope of 5/3, corresponding to adiabatic behavior (although at high densities there is always a significant fraction of gas in thermal equilibrium due to the rapid increase of the cooling rate with density).

We conclude this chapter by noting that thermal instability is one important intermediate step in the flow of gas from the diffuse warm phase to the molecular clouds, in which gravity dominates, and stars are formed, as will be discussed at length in Chapter 10.

References

Audit, E., & Hennebelle, P. 2005, A&A, 433, 1

Begelman, M. C., & McKee, C. F. 1990, ApJ, 358, 375

Dalgarno, A., & McCray, R. A. 1972, ARA&A, 10, 375

Dickey, J. M., Salpeter, E. E., & Terzian, Y. 1977, ApL, 211, L77

Dickey, J. M., Strasser, S., Gaensler, B. M., Haverkorn, M., Kavars, D., McClure-Griffiths, N. M., Stil, J., & Taylor, A. R. 2009, ApJ, 693, 1250

Elmegreen, B. G. 1991, The Physics of Star Formation and Early Stellar Evolution, ed. C. J. Lada, & N. D. Kylafis, (Dordrecht: Springer)

Ferrière, K. M. 2001, RvMP, 73, 1031

Field, G. B. 1965, ApJ, 142, 531

Field, G. B., Goldsmith, D. W., & Habing, H. J. 1969, ApL, 155, L149

Gazol, A., Vázquez-Semadeni, E., Sánchez-Salcedo, F. J., & Scalo, J. 2001, ApL, 557, L121

Gazol, A., Vázquez-Semadeni, E., & Kim, J. 2005, ApJ, 630, 911

Heiles, C. 2001, ApL, 551, L105

Hennebelle, P., & Pérault, M. 1999, A&A, 351, 309

Hunter, , & James, H. 1970, ApJ, 161, 451

Kalberla, P. M. W., & Kerp, J. 2009, ARA&A, 47, 27

Kalberla, P. M. W., Schwarz, U. J., & Goss, W. M. 1985, A&A, 144, 27

Kanekar, N., Subrahmanyan, R., Chengalur, J. N., & Safouris, V. 2003, MNRAS, 346, L57

Koyama, H., & Inutsuka, S.-I. 2000, ApJ, 532, 980

Koyama, H., & Inutsuka, S.-I. 2002, ApL, 564, L97

Kritsuk, A. G., & Norman, M. L. 2002, ApL, 569, L127

Krumholz, M. 2017, Star Formation (Singapore: World Scientific)

McClure-Griffiths, N. M., Stanimirović, S., & Rybarczyk, D. R. 2023, ARA&A, 61, 19

McKee, C. F., & Ostriker, J. P. 1977, ApJ, 218, 148

Meerson, B. 1996, RvMP, 68, 215

Myers, P. C. 1978, ApJ, 225, 380

Pikel'Ner, S. B. 1968, SvA, 11, 737

Shu, F. H. 1992, Physics of Astrophysics, Vol. II (Mill Valley, CA: Univ. Science Books)

Spitzer, , Lyman, J., & Fitzpatrick, E. L. 1995, ApJ, 445, 196

Sutherland, R. S., & Dopita, M. A. 1993, ApJS, 88, 253

Vazquez-Semadeni, E., Ostriker, E. C., Passot, T., Gammie, C. F., & Stone, J. M. 2000, Protostars and Planets IV, ed. V. Mannings, A. P. Boss, & S. S. Russell (Tucson, AZ: Univ. Arizona Press) 3

Vázquez-Semadeni, E., Gazol, A., Passot, T., et al. 2003, Turbulence and Magnetic Fields in Astrophysics, Lecture Notes in Physics Vol. 614, ed. E. Falgarone, & T. Passot (Berlin: Springer) 213

Vázquez-Semadeni, E., Ryu, D., Passot, T., González, R. F., & Gazol, A. 2006, ApJ, 643, 245

Wolfire, M. G., Hollenbach, D., McKee, C. F., Tielens, A. G. G. M., & Bakes, E. L. O. 1995, ApJ, 443, 152

Wolfire, M. G., McKee, C. F., Hollenbach, D., & Tielens, A. G. G. M. 2003, ApJ, 587, 278

AAS | IOP Astronomy

Interstellar Flow and Star Formation

Enrique Vázquez-Semadeni

Chapter 5

The Scalar Virial Theorem

The scalar virial theorem (VT), in the context of gas dynamics, is an equation that describes the energy balance within a certain volume V within the fluid, equating the work done by all the applied forces to the change in the mass distribution (the moment of inertia, I) and kinetic energy of the fluid in the volume. It is one of the fundamental estimators of the dynamical state of clouds in the ISM. In this chapter, we will discuss its derivation and interpretation, and then consider various applications, from highly idealized, to more realistic. We also discuss frequent misconceptions and caveats.

5.1 Derivation

The VT is obtained by taking the dot product of the momentum equation (Equation (2.75), which represents Newton's Second Law) with the position vector $\boldsymbol{x}$ and integrating over some volume V, which we will refer to as the "cloud" (although in practice, choosing its boundary is an ambiguous task; see Section 1.2.1). Recalling that the work done by a force along a trajectory is defined as $\mathcal{W} \equiv \int \mathbf{F} \cdot d\boldsymbol{x}$, we see that then the VT gives the work performed on a certain parcel of the flow by the forces acting on it. Traditionally, the volume V is taken in a *Lagrangian* frame of reference; that is, a frame moving with the fluid (Section 2.2.1). An *Eulerian* (i.e., in a fixed frame of reference) version of the VT was derived by McKee & Zweibel (1992). This may be a more convenient form for computation in practice, and we discuss it briefly in Section 5.1.1. Here we present the derivation of the Lagrangian form in detail, partly following the discussion in Vázquez-Semadeni (1999). Performing the operations mentioned above, one obtains

$$\int_V \rho \boldsymbol{x} \cdot \frac{d\boldsymbol{u}}{dt}\, dV = -\int_V \boldsymbol{x} \cdot \nabla P\, dV - \int_V \rho \boldsymbol{x} \cdot \nabla \varphi\, dV + \frac{1}{4\pi} \int_V \boldsymbol{x} \cdot (\nabla \times \mathbf{B}) \times \mathbf{B}\, dV, \tag{5.1}$$

doi:10.1088/2514-3433/adcb28ch5

where $d/dt \equiv \partial/\partial t + \boldsymbol{u} \cdot \nabla$ is the total or Lagrangian derivative operator. The left-hand side (LHS) of this equation can be rewritten as

$$\int_V \rho \boldsymbol{x} \cdot \frac{d\boldsymbol{u}}{dt} dV = \frac{1}{2} \int \frac{d^2 x^2}{dt^2} dm - \int_V \rho u^2 \, dV \equiv \frac{1}{2} \frac{d^2 I}{dt^2} - \int_V \rho u^2 \, dV, \tag{5.2}$$

where $x^2 = \boldsymbol{x} \cdot \boldsymbol{x}$, $\boldsymbol{u} \equiv d\boldsymbol{x}/dt$, $dm \equiv \rho \, dV$, $I \equiv \int_V x^2 dm$ is the *Lagrangian moment of inertia* of the fluid parcel, and we have used Reynolds' Transport Theorem (Section 2.2.2.2, Equation (2.15))

$$\frac{d}{dt} \int_V \alpha \, dV = \int_V \left[\frac{\partial \alpha}{\partial t} + \nabla \cdot (\alpha \boldsymbol{u}) \right] dV \tag{5.3}$$

to show that $\frac{d}{dt} \int \alpha \, dm = \int \frac{d\alpha}{dt} dm$ for any function $\alpha(\boldsymbol{x}, t)$.

For the right-hand side (RHS) of Equation (5.1), let us consider the various terms as follows. First, according to Newton's Law of Gravitation (or from the solution of Poisson's equation; see, e.g., Shu 1992), the gravitational term can be written as

$$W \equiv -\int_V \rho(\boldsymbol{x}) \boldsymbol{x} \cdot \nabla \varphi(\boldsymbol{x}) \, dV = -G \int_V \int_{\text{all space}} \frac{\rho(\boldsymbol{x}) \rho(\boldsymbol{x}') \boldsymbol{x} \cdot (\boldsymbol{x} - \boldsymbol{x}')}{|\boldsymbol{x} - \boldsymbol{x}'|^3} dV' dV. \tag{5.4}$$

If the contribution to the gravitational potential of the matter external to V is negligible, then the second integral can be performed also over volume V only. In this case, noting that the integrand is nearly symmetric with respect to the primed and unprimed variables, we can replace $\boldsymbol{x}$ by $(\boldsymbol{x} - \boldsymbol{x}')/2$ and write

$$W = -\frac{1}{2} G \int_V \int_V \frac{\rho(\boldsymbol{x}) \rho(\boldsymbol{x}')}{|\boldsymbol{x} - \boldsymbol{x}'|} dV' dV. \tag{5.5}$$

Thus, W is the gravitational energy of the mass distribution within volume V. It is important to stress, however, that, although performing both integrals over the same volume is a very common strategy, it is *not* justified in general when the mass distribution *around* the cloud is important in setting the full gravitational potential.

Second, the remaining thermal pressure and magnetic terms can be dealt with using Gauss' divergence theorem, and the facts that $\nabla \cdot \boldsymbol{x} = 3$ and that (in tensor notation) $\partial x_i / \partial x_j = \delta_{ij}$. Additionally, for the magnetic term, it is most convenient to work with Maxwell's magnetic stress tensor $\mathbf{T} \equiv T_{ij} \equiv B_i B_j / 4\pi - |\mathbf{B}|^2 \delta_{ij} / 8\pi$, so that we can write $(\nabla \times \mathbf{B}) \times \mathbf{B}/4\pi = \nabla \cdot T_{ij}$ in Equation (5.1). Thus, gathering together the LHS from Equation (5.2) and the RHS from the above considerations, Equation (5.1) becomes

$$\begin{aligned} \frac{1}{2} \frac{d^2 I}{dt^2} - \int_V \rho u^2 \, dV = & \left(3 \int_V P \, dV - \oint_S P \boldsymbol{x} \cdot d\boldsymbol{S} \right) \\ & + \left(\frac{1}{8\pi} \int_V B^2 \, dV + \oint_S \boldsymbol{x} \cdot \mathbf{T} \cdot d\boldsymbol{S} \right) + W. \end{aligned} \tag{5.6}$$

This equation can then be written as

$$\frac{1}{2}\frac{d^2I}{dt^2} - 2K = \left(2U - \oint_S P\boldsymbol{x}\cdot d\boldsymbol{S}\right) + \left(\mathcal{M} + \oint_S \boldsymbol{x}\cdot \mathbf{T}\cdot d\boldsymbol{S}\right) + W, \quad \text{(Lagrangian)} \tag{5.7}$$

where S is the surface enclosing V, and K, U, and $\mathcal{M}$ are, respectively, the kinetic, internal, and magnetic energies contained within volume V, as defined by the respective integrals in Equation (5.6). Expression (5.7) is the most common form of the *scalar* Virial Theorem, which will be discussed in the remainder of this chapter. A *tensor* form can also be derived (e.g., Zweibel 1990), which allows the treatment of multidirectional transport phenomena, but we will not discuss it here.

Several comments and warnings are in order. First, note that, contrary to most treatments of the VT, we have written the kinetic energy term, $2K$, on the LHS of the equation, rather than on the RHS. We have chosen to do this because the kinetic energy is *imparted* onto the cloud by the work performed by each of the forces, analogously to the standard notion from classical mechanics that the work performed on a particle equals the change in its kinetic energy. That is, the change in kinetic energy of a fluid parcel is an *effect* of the work performed on it. Indeed, it comes from the LHS of the momentum equation, which is the *response*—or *acceleration*—of the parcel that results from the forces applied on the parcel (on the RHS of the equation), as can be seen in Equation (5.2).[1] However, since we are considering a parcel with a finite volume, the work performed on it also causes a change of its internal mass distribution (described by the second derivative of the moment of inertia), in addition to changing its total kinetic energy. Therefore, the LHS of Equation (5.7) can be thought of as the *effect* on the cloud, while the various terms on the RHS can be interpreted as the *causes* of the effect. For example, if some energy-injection term is included in the momentum equation to drive turbulence (as is done, for example, in simulations of continuously-driven turbulence), then the driving term should be included in the RHS of the equation, and the corresponding work done on volume V will appear on the RHS of the VT. This has the effect of changing the mass distribution within a cloud, as well as imparting some kinetic energy to it. The same applies to a possible viscosity term, or an effective numerical viscosity, that dissipates kinetic energy. The kinetic energy, however, continues to be part of the *response* of the flow to the work done on it by the various forces.

Second, a cloud is customarily said to be in *virial equilibrium* if

$$\frac{d^2I}{dt^2} = 0. \tag{5.8}$$

However, it is important to note that this definition allows for the case of a constant rate of change in the cloud's moment of inertia. Thus, "virial equilibrium" does not

[1] I am indebted to my colleague Javier Ballesteros-Paredes for pointing out this important realization to me.

necessarily imply a static cloud configuration, although the expression is almost always interpreted to mean this.

Third, note that both the thermal pressure and the magnetic contributions consist of a *volumetric* and a *surface* term (the volume and surface integrals within the first and second parentheses in Equation (5.6), respectively). In fact, if either the thermal pressure or the magnetic field are uniform, then the corresponding surface and volumetric terms cancel each other, indicating balance between internal and external stresses (see Section 5.2.1). This corresponds to the fact that if the gradients of P, φ, and $\boldsymbol{B}$ are zero, then these physical agents apply no net force on the flow (see, e.g., Section 5.2.1). Equivalently, the surface terms can only be ignored if the corresponding fields can be neglected at the boundary surface. However, it is frequently encountered in the literature that virial balance is considered only among the volumetric terms.

Fourth, the thermal pressure surface term describes the "confining" effect of the external pressure on the cloud (volume V). A similar comment applies to the magnetic surface term, although in this case this term is not exclusively confining, as it includes the effects of both the external magnetic pressure as well as the magnetic tension along field lines. Moreover, in the Eulerian version of the theorem, an analogous surface term of the form $\int \boldsymbol{x} \cdot \rho \boldsymbol{u}\boldsymbol{u} \cdot d\boldsymbol{S}$ for the macroscopic velocity field also appears (see Section 5.1.1), where $\rho\boldsymbol{u}\boldsymbol{u}$ is a tensor describing momentum transfer due to the fluid motions, particularly important in turbulent flow. However, due to the strongly fluctuating nature of the turbulent velocity field, which may contain large-scale motions (comparable to the full system's size), this term can hardly be considered as producing a confining effect; instead, it produces transport of momentum across the cloud's boundary and a redistribution of the mass within the cloud, thus contributing to the change in its moment of inertia.

5.1.1 Eulerian Virial Theorem

For completeness, we reproduce here, without derivation, the Eulerian form of the VT obtained by McKee & Zweibel (1992), which applies to a volume V and surface S defined as fixed in some inertial reference frame. This is the relevant form to use when considering the energy balance in fixed regions of space. It reads

$$\begin{aligned}\frac{1}{2}\frac{d^2 I_{\rm E}}{dt^2} &- \left(\int_V \rho u^2\, dV + 2\oint_S \boldsymbol{x} \cdot \rho \boldsymbol{v}\boldsymbol{v} \cdot d\boldsymbol{S}\right) - \frac{d}{dt}\oint_S (\rho x^2 \boldsymbol{v}) \cdot d\boldsymbol{S} \\ &= \left(3\int_V P\, dV - \oint_S P\boldsymbol{x} \cdot d\boldsymbol{S}\right) + \left(\frac{1}{8\pi}\int_V B^2\, dV + \oint_S \boldsymbol{x} \cdot \mathbf{T} \cdot d\boldsymbol{S}\right) + W \text{ (Eulerian)},\end{aligned} \tag{5.9}$$

where $I_{\rm E} = \int_V \rho x^2\, dV$ is the *Eulerian moment of inertia*. We see that two additional terms appear in comparison to Equation (5.6). One is the second term within the parentheses on the left-hand side, which is nothing but the surface term corresponding to the volumetric kinetic energy term, analogous to the pressure and magnetic surface terms on the right-hand side. It represents the flux of momentum across the (fixed) surface S. The second additional term is the last term on the left-hand side, which can be interpreted as the rate of change of the moment of inertia flux across surface S (Ballesteros-Paredes et al. 1999). Both of these terms are absent in the Lagrangian version of the VT because the boundary moves with the flow, and so no material crosses it.

5.2 Applications in Ideal Cases

We now turn to the application of the VT to a variety of idealized configurations, often under the assumption of virial *equilibrium* (i.e., when $d^2I/dt^2 = 0$), and with no flux of mass nor energy across the bounding surface S of volume V in Equation (5.7). These constitute the most frequently considered applications of the VT, but it should be noted that neglecting mass and energy flux across S may render these solutions too unrealistic as models of interstellar clouds. More realistic applications are covered in Section 5.3.

5.2.1 Pressure Confinement

One of the simplest applications of the VT is to the case of a spherical region of radius R immersed in a non-magnetized medium of uniform density and pressure, with negligible self-gravity. In this case, taking $\mathbf{B} = \boldsymbol{u} = \nabla\varphi = 0$, and assuming $P = P_0$ cst. in Equation (5.7) gives

$$\frac{1}{2}\frac{d^2I}{dt^2} = 3\int_V P\,dV - \oint_S P\boldsymbol{x}\cdot d\boldsymbol{S}, \tag{5.10}$$

where S is the surface bounding volume V, and $d\boldsymbol{S} = \hat{\boldsymbol{n}}\,dS$, with $\hat{\boldsymbol{n}}$ being the unit vector normal to the surface at each point on S. Since the integration domain is the surface of the sphere, it is convenient to choose the origin of coordinates at the center of the sphere. With this choice, the position vector equals the radius vector at each point on the surface, $\boldsymbol{x} = \boldsymbol{R}$, and the normal unit vector $\hat{\boldsymbol{n}}$ is the unit vector in the radial direction, $\hat{\boldsymbol{n}} = \boldsymbol{R}/R$.

It is easy to see that the two terms on the RHS of this equation cancel each other out in this case. For constant pressure $P = P_0$, the volume integral is

$$2U = 3\int_V P_0\,dV = 3P_0V = 4\pi P_0R^3, \tag{5.11}$$

while the surface integral is

$$\oint_S P_0\boldsymbol{x}\cdot d\boldsymbol{S} = P_0\oint_S \boldsymbol{x}\cdot\hat{\boldsymbol{n}}\,dS = P_0R^3\oint_S d\Omega = 4\pi P_0R^3, \tag{5.12}$$

where Ω is the solid angle. Therefore, the right-hand side of Equation (5.10) vanishes, and thus $d^2I/dt^2 = 0$, and the cloud is in virial equilibrium. This is, for example, the case of any parcel of the air in a room with no drafts.

5.2.2 Thermal Support, Hydrostatic Equilibrium, and the Jeans Length

Now assume an isothermal sphere of volume V, in which $\mathbf{B} = \boldsymbol{u} = 0$, a certain mean pressure $\langle P\rangle$ inside, and $P = 0$ outside, but keeping self-gravity. In this case, the virial equilibrium condition $d^2I/dt^2 = 0$ implies $W = -2U$ in Equation (5.7). The gravitational energy can be computed as

$$W = -G\int \frac{M(r)}{r}\,dM(r) = -\frac{3GM^2}{5R}, \tag{5.13}$$

where $M(r)$ is the mass interior to radius r.

On the other hand, the internal energy term is again

$$2U = 3\int_V P\,dV = 4\pi\langle P\rangle R^3. \tag{5.14}$$

In virial equilibrium, we thus obtain

$$\frac{3GM^2}{5R} = 4\pi\langle P\rangle R^3, \tag{5.15}$$

so that the condition of *hydrostatic equilibrium* for a uniform sphere in the absence of external pressure requires the mean internal pressure to be

$$\langle P\rangle_{\rm eq} = \frac{3GM^2}{20\pi R^4}. \tag{5.16}$$

Expressing this condition in terms of the density, we can obtain the size of the sphere that can be in equilibrium at a given density and temperature. Recalling Equation (3.50), we write in general $P = c_{\rm s}^2\rho$, where the sound speed $c_{\rm s}$ may or may not be a constant depending on whether the flow is isothermal or not. Substituting this expression in Equation (5.15), and using $M = 4\pi\langle\rho\rangle R^3/3$, where $\langle\rho\rangle$ is the mean density in the sphere, we find

$$R_{\rm eq} = \left(\frac{15c_{\rm s}^2}{4\pi G\langle\rho\rangle}\right)^{1/2}, \tag{5.17}$$

the equilibrium radius of a uniform self-gravitating sphere in the absence of external pressure. This is the VT equivalent of the the *Jeans length*, which will be defined in Section 6.1. It differs from the Jeans length defined in that chapter only in the coefficient of the term $(c_{\rm s}^2/G\rho)^{1/2}$, due to the different configuration assumed here (a finite sphere with average density $\langle\rho\rangle$), compared to the one considered there (an infinite uniform medium with sinusoidal perturbations).

5.2.3 Stability of Hydrostatic Equilibrium, the Effective Polytropic Exponent, and Dimensionality of Contraction

An important question is whether this hydrostatic equilibrium is stable or unstable, and under which conditions stability can occur. This can be investigated in approximate form using the VT for the case of a polytropic gas for which $P = P_0(\rho/\rho_0)^{\gamma_{\rm e}}$, where $\gamma_{\rm e}$ is the effective polytropic exponent (cf. Section 4.2). In this case, we can write the ratio of the gravitational to the thermal energy as

$$\frac{|W|}{2U} = \left(\frac{4\pi}{3}\right)^{(\gamma_e - 1)} \frac{GM^{(2-\gamma_e)}\rho_0^{\gamma_e}}{5P_0} R^{(3\gamma_e - 4)}. \tag{5.18}$$

Under the assumption that M remains constant,[2] this result implies that, for $\gamma_e > 4/3$, the ratio $|W|/2U$ is an *increasing* function of R, so that, if initially $|W|/2U = 1$ and the sphere is slightly compressed (its radius is decreased), the ratio decreases, and the sphere becomes dominated by its internal energy, tending to re-expand and to return to the initial radius. Therefore, the equilibrium is stable for $\gamma_e > 4/3$. Conversely, if $\gamma_e < 4/3$, then the ratio $|W|/2U$ is a *decreasing* function of R, and, if the sphere is compressed, the ratio increases, so that now the gravitational energy dominates over the internal energy, and the sphere continues to contract. Therefore, the equilibrium is stable for $\gamma_e > 4/3$ and unstable for $\gamma_e < 4/3$.

Another way of interpreting this is that, *for* $\gamma_e > 4/3$, *the Jeans length increases with increasing density*, and vice versa. So, for $\gamma_e > 4/3$, the Jeans length *increases* during gravitational contraction, and so eventually it becomes larger than the contracting object's size, halting the collapse. This can happen, for example, when the collapsing gas becomes so dense that it becomes optically thick, trapping the radiation inside of the core, and therefore storing the heat produced by the PdV work by the collapse, rather than cooling by radiation. The gas thus effectively becomes nearly adiabatic, with $\gamma_e \to \gamma$ (i.e., the dependence of the pressure on the density ceases to be determined from radiative cooling (cf. Section 4.2) and returns to being determined by the internal degrees of freedom of the gas molecules).

From the above discussion, we conclude that $\gamma_{e,cr} = 4/3$ is thus a critical value separating the stable from the unstable regimes of hydrostatic equilibrium when the compressive perturbations are three-dimensional. However, it can be shown that, if the compressions are applied in two or one dimensions, then the critical values of the effective polytropic exponent are, respectively, $\gamma_{e,cr} = 1$ and $\gamma_{e,cr} = 0$. This can be accomplished by writing the volume as $V \propto R^{\eta} R_0^{3-\eta}$, where R_0 remains constant and R is varied, and $\eta = 2$ for two-dimensional compressions, while $\eta = 1$ for one-dimensional compressions (see Section 10.1.2.2). The latter are, for example, of the kind exerted by a moving piston in a closed chamber, or the passage of a planar shock through a given location. In particular, the fact that the critical value for 1D compressions is zero is part of the reason why, when magnetic support is strong enough, collapse cannot occur in molecular cloud cores (Section 9.2.1.2): if the magnetic field prevents the collapse in the two directions perpendicular to it, then the only possible direction of contraction is *along* the field lines, and thus it is effectively one-dimensional, so that the critical value of the effective polytropic exponent for instability is $\gamma_{e,cr} = 0$. This is significantly smaller than the actual value of the exponent for nearly isothermal (i.e., with $\gamma_e \approx 1$) molecular clouds, and thus the contraction along the field is stable, ultimately reaching a state of hydrostatic equilibrium in this direction.

[2] This often-made, but seldom-recognized assumption is, however, not valid in general, as discussed in Section 5.3.

It is also important to note that, when the equilibrium configuration is unstable ($\gamma_e < 4/3$), a compressive perturbation will induce collapse of the sphere, while an expansive perturbation will cause runaway expansion. This is in particular the case for nearly isothermal structures (in which $\gamma_e \approx 1$), such as molecular clouds and their substructures. Thus, a hydrostatic equilibrium of clumps and cores within molecular clouds is impossible unless they are truncated at some radius, and embedded in a more diffuse environment that provides pressure without adding significant mass (because if the environment does add mass, then it is equivalent to making the sphere's radius larger than the Jeans length). This configuration is known as a *truncated Bonnor–Ebert sphere*, and is discussed further in Section 5.2.4 in the context of the VT, and in Section 6.2 in the context of the full momentum equation.

5.2.4 Combined Thermal Support and Pressure Confinement

We can now consider the case of a self-gravitating uniform sphere, which is also subject to an external pressure, and investigate its hydrostatic configurations. This is the virial theorem equivalent of the so-called *Bonnor–Ebert* (BE) structures that will be discussed in Section 6.2. There, the detailed force balance between the pressure gradient and self-gravity will be discussed at every radius in the sphere. Here, the virial theorem gives instead a global, approximate balance between just the interior and the exterior of the sphere.

To better understand the physical situation, note that an isothermal sphere for which $|W| < 2U$ will simply undergo runaway expansion if it is not confined by an external pressure. We then consider the balance between the internal and external pressures, and self-gravity. In this case, the virial equilibrium condition reduces to

$$0 = 3\int P\,dV - \oint_S P\boldsymbol{x}\cdot d\boldsymbol{S} + W \approx 3V(P_{\rm int} - P_{\rm ext}) + W, \tag{5.19}$$

where $P_{\rm int} \equiv c_s^2\langle\rho_{\rm int}\rangle = 3Mc_s^2/4\pi R^3$ is the average internal pressure, $\langle\rho_{\rm int}\rangle$ is the mean internal density, M is the gas sphere's mass, and $P_{\rm ext}$ is the (constant) pressure applied at the surface. Then, the previous equation becomes

$$P_{\rm ext} = \frac{3M}{4\pi}\left(\frac{c_s^2}{R^3} - \frac{GM}{5R^4}\right) \equiv P_{\rm int} - P_{\rm grav}, \tag{5.20}$$

where we have defined the gravitational "pressure"[3] $P_{\rm grav}$ as the second term after the first equality in Equation (5.20). This equation gives the external pressure required in order to maintain a gaseous isothermal sphere of mass M and radius R in equilibrium, and is plotted in Figure 5.1 for a simplified case with arbitrary units for which $M = c_s = G = 1$. Several features are worth noting in this plot. First, it is important to recall that this plot represents *a sequence of hydrostatic equilibrium configurations*. It does *not* represent any kind of evolution of the gaseous sphere.

[3] Or, alternatively, the gravitational energy density.

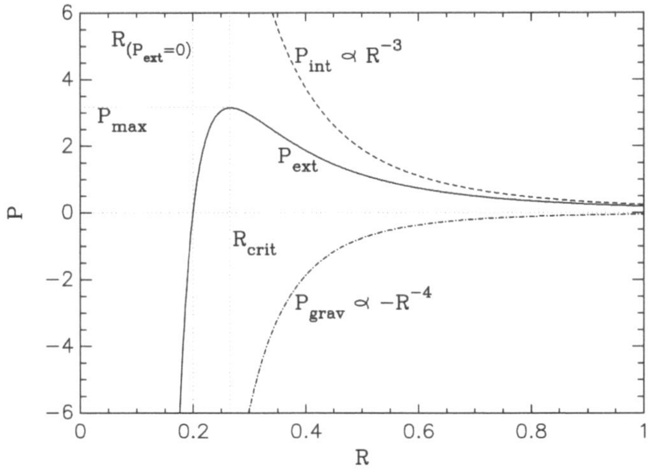

Figure 5.1. Internal pressure ($P_{\rm int}$; dashed line), gravitational "pressure" ($P_{\rm grav}$; dash-dotted line) and required external pressure for hydrostatic equilibrium ($P_{\rm ext}$; solid line) versus R for a spherical, uniform-density, isothermal cloud of size R and mass M, as given by Equation (5.20). Here, we have taken arbitrary units where $M = c_{\rm s} = G = 1$.

Figure 5.1 can be understood as follows. When the sphere in equilibrium has a large radius (i.e., when the mass M is spread over a large volume), the gravitational pressure $P_{\rm grav}$ is almost zero, meaning that the gravitational contribution to the confinement is negligible, and thus the external pressure required for equilibrium is almost equal to $P_{\rm int}$, as in the case of simple pressure confinement (see the region $R \to 1^-$ in Figure 5.1).

Obtaining smaller equilibrium radii for the sphere (moving from right to left in the plot) is seen to first require larger external pressures. However, a certain radius is reached ($R_{\rm crit} \approx 0.27$ in Figure 5.1; see Equation (5.21) below) at which $P_{\rm ext}$ reaches a maximum, $P_{\rm max}$. Equilibria with radius $R < R_{\rm crit}$ require *smaller* values of $P_{\rm ext}$. This is because, as the sphere is forced to be smaller (and denser), the gravitational contribution to the confinement increases, reducing the required contribution from $P_{\rm ext}$. The turnover is due to the faster dependence of $P_{\rm grav}$ on R ($P_{\rm grav} \propto -R^{-4}$) than that of $P_{\rm int}$ ($P_{\rm int} \propto R^{-3}$). Therefore, as R is decreased, $P_{\rm grav}$ begins to dominate, and the required $P_{\rm ext}$ for equilibrium begins to decrease, until it eventually becomes zero (at $R = R_{P_{ext}=0} = 0.2$ in the plot). This point corresponds to the case of precise hydrostatic balance between the internal pressure and self-gravity. At even smaller radii, negative values of $P_{\rm ext}$ would be required to maintain the sphere in equilibrium. Since this is impossible, this means that the equilibrium cannot be realized when the sphere has a radius smaller than the value at which $P_{\rm ext}$ becomes zero. Also, no equilibrium is possible for $P_{\rm ext} > P_{\rm max}$.

It is important to note that for radii $R_{P_{\rm ext}=0} \leqslant R \leqslant R_{\rm crit}$, the equilibria are unstable. This can be seen considering a point on the curve in this range, and applying a small increment to the external pressure. This reduces the radius of the sphere, but at the new radius, a *smaller* value of $P_{\rm ext}$ is required to maintain the equilibrium, while the actual external pressure is larger, therefore inducing further compression, and runaway collapse ensues. Instead, for $R > R_{\rm crit}$, the equilibria are seen to be stable, as now, for a point in this range, a small increment on the pressure brings the radius to a smaller value, for which the new higher pressure can maintain the equilibrium.

An important property of the critical radius is the value it implies for the mean density inside the sphere. This can be found by differentiating Equation (5.20) and setting the derivative to zero, to find the radius for which $P_{\rm ext}$ is maximum. The critical radius is found to be

$$R_{\rm crit} = \frac{4GM}{15c_{\rm s}^2} = \left(\frac{45c_{\rm s}^2}{16\pi G\rho_{\rm int}}\right)^{1/2}. \tag{5.21}$$

For the simple case of $c_{\rm s} = G = M = 1$ shown in Figure 5.1, $R_{\rm crit} = 4/15 \approx 0.27$. Substituting the critical radius back in Equation (5.20) we find that, at the maximum, the external pressure at the maximum satisfies

$$P_{\rm max} = \frac{1}{4}\langle P_{\rm int}\rangle. \tag{5.22}$$

Note that, in the equation above, we have written $\langle P_{\rm int}\rangle$ because $P_{\rm int}$ is actually the *mean* pressure inside the sphere, as is clear from the first term in Equation (5.19). Therefore, Equation (5.22) implies that the maximum possible mean internal density for a *stable* sphere is related to the density at the surface by

$$\langle \rho_{\rm int}\rangle_{\rm max} = 4\rho_{\rm ext}. \tag{5.23}$$

Values of the mean internal density larger than this correspond to *unstable* equilibria.

It is important to reiterate that the equilibrium configurations we have found using the VT involve only the *average* values of the pressure and density inside the sphere. In Section 6.2 we will discuss the *detailed* equilibrium configurations, in which the hydrostatic condition holds at every radius of the sphere, determining the dependence of the density and pressure with radius (the density and pressure *radial profiles*). Nevertheless, the treatment using the VT illustrates most of the fundamental properties of the equilibrium configurations, such as the existence of a range of stable and unstable solutions, and the fact that, for values of the mean internal density larger than a certain value, the equilibria are unstable.

5.2.5 Turbulent "Support" and Turbulent "Pressure"

Assume now that a small-scale turbulent velocity field is present within volume V. By small-scale we mean that the largest turbulent eddies are much smaller than the linear size of this volume. This requirement, usually referred to as *microturbulence*, is necessary so that the turbulent velocity dispersion can be considered to be isotropic, and therefore can act as an additional source of isotropic pressure. In this case, the condition of virial equilibrium becomes $W = -2(U + K)$, indicating that turbulence can provide, in principle, additional support against gravity.

Comparing the K and U terms in Equations (5.6) and (5.7), we note that, in this case, the quantity ρu^2 plays a role analogous to that of the thermal pressure, and therefore it is common to refer to this quantity as the turbulent “pressure”. However, note that the assumption that the turbulent motions are constrained to size scales much smaller than that of the system is highly questionable because, as discussed in Chapter 3, turbulence in general involves a large range of scales that extend from the energy-injection scale to the dissipation scale, and moreover, the largest velocities occur at the largest scales. Indeed, using the technique of principal component analysis, Brunt et al. (2009) have shown that the energy-injection scale in molecular clouds is of the order of the size of the clouds themselves. This implies that the role of turbulence in the clouds cannot be to provide support for maintaining hydrostatic equilibrium, but rather it consists in distorting (compressing or shearing) or even assembling the clouds (Ballesteros-Paredes et al. 1999). We will return to this in Chapter 10.

5.2.5.1 The Virial Parameter and the Larson Ratio: Equilibrium and Non-Equilibrium

A widely used measure of the gravitational binding of clouds and their substructures (clumps and cores) is the so-called *virial parameter*, defined as (Bertoldi & McKee 1992)

$$\alpha_{\rm vir} \equiv \frac{2(K+U)}{|W|} = \frac{5\sigma^2 R}{GM}, \tag{5.24}$$

where $\sigma \equiv \sqrt{c_s^2 + \sigma_{\rm 1D}^2}$ is the total (thermal+turbulent) one-dimensional velocity dispersion in the cloud, with $\sigma_{\rm 1D} \equiv \sqrt{\sigma_{\rm 3D}^2/3}$ being the one-dimensional turbulent velocity dispersion, and

$$\sigma_{\rm 3D}^2 \equiv \frac{1}{V}\int_V u^2\, dV$$

being the three-dimensional turbulent velocity dispersion.[4] The turbulent kinetic energy is then approximated as $K \approx 1/2\ M\sigma_{\rm 3D}^2 \equiv 3/2\ M\sigma_{\rm 1D}^2$. Also, in Equation (5.24), R is the cloud's measured radius, M is its mass, and we have assumed a uniform-density sphere, so that W is given by Equation (5.13). Note that for a strongly supersonic medium, $\sigma_{\rm 1D} \gg c_s$, and so $\sigma \approx \sigma_{\rm 1D}$.

Ideally, $\alpha_{\rm vir}$ would take the value of unity for a uniform-density sphere in hydrostatic equilibrium with no external pressure, although it must be kept in mind that real clouds are far from being uniform spheres, are immersed in an external medium, and need not be in hydrostatic equilibrium. Values of $\alpha_{\rm vir}$ differing from unity, both smaller and larger, are often observed in molecular clouds and their dense substructures (e.g., Kauffmann et al. 2013; Leroy et al. 2015). Such values can be interpreted in different ways, as we now discuss.

[4] Assuming $\langle u \rangle = 0$.

Clouds with values of $\alpha_{\rm vir}$ larger than unity are most commonly interpreted as being gravitationally unbound, and would require the presence of an external pressure in order to remain confined (e.g., Keto & Myers 1986; Bertoldi & McKee 1992; Field et al. 2011). Indeed, we can rewrite Equation (5.20) in terms of $\alpha_{\rm vir}$ as

$$\alpha_{\rm vir} = \frac{4\pi R^3 P_{\rm ext}}{|W|} + 1. \tag{5.25}$$

This equation shows that, if $P_{\rm ext} = 0$, virial equilibrium requires $\alpha_{\rm vir} = 1$, but, for $P_{\rm ext} > 0$, virial equilibrium implies $\alpha_{\rm vir} > 1$ in general. Alternatively, unbound clouds can also be in a state of dispersal (e.g., Colín et al. 2013).

Clouds with $\alpha_{\rm vir} \approx 2$, which implies $K \approx |W|$ are sometimes described as being "marginally bound". However, it is important to note that gravitationally collapsing clouds can also have $\alpha_{\rm vir} \approx 2$. Specifically, a cloud that has been undergoing unimpeded gravitational contraction from infinity also satisfies the condition $K \approx |W|$, and therefore has $\alpha_{\rm vir} \approx 2$. Note also that the condition of free-fall is actually a more strongly gravitationally bound situation than virial equilibrium, because in this case the kinetic energy is the *effect* of the gravitational force, rather than being an agent capable of opposing gravity. Which possibility is actually realized depends on whether the velocity field is mainly directed inwards as a consequence of the collapse, or it is randomly oriented with no dominant infall component, so that it can oppose self-gravity.

A related quantity is the ratio $\sigma/R^{1/2}$, often referred to as the *Larson ratio*, denoted $\mathcal{L}$. For virial equilibrium between self-gravity and turbulent kinetic energy, it is obtained by setting $\alpha_{\rm vir} = 1$ in Equation (5.24) and neglecting the thermal contribution U, to yield

$$\mathcal{L} \equiv \frac{\sigma_{\rm 1D}}{R^{1/2}} = \left(\frac{\pi G \Sigma}{5}\right)^{1/2}, \tag{5.26}$$

where $\Sigma \equiv M/\pi R^2$ is the cloud's column density.

5.2.6 Magnetic Support and the Mass-to-Flux Ratio

We now consider the role of the magnetic field in providing support against the self-gravity of a cloud. This was the preferred mechanism for cloud support until the first decade of the 21st century, when statistically significant evidence began to show that in general the magnetic field is too weak to support most molecular clouds and their substructures ("clumps" and "cores"; Crutcher 2012; for a more recent compilation, see Pattle et al. 2023). Nevertheless, it is important to understand the mechanism and the effects of the magnetic field even when it cannot support the structures.

Consider a uniform-density, *isolated* spherical cloud of fixed mass M, radius R, permeated by a uniform magnetic field $\mathbf{B} = B_0 \hat{\mathbf{e}}_x$ inside. The magnetic volumetric term $\mathcal{M}$ in the VT is

$$\mathcal{M} = \frac{B_0^2 R^3}{6} = \frac{\varphi^2}{6\pi^2 R}, \tag{5.27}$$

where

$$\varphi \equiv \int_A \boldsymbol{B} \cdot \hat{\boldsymbol{n}}\, dA = \pi R^2 B_0, \tag{5.28}$$

is the *magnetic flux*, in which the area A is a *cross section* of volume V. According to the flux-conservation equation, Equation (2.77), φ is conserved in the limit of *ideal MHD* (i.e., in the limit of zero electric resistivity; e.g., Shu 1992; see also the chapter "Molecular clouds" in Ballester et al. 2018).

For this setup, we can investigate the required magnetic field strength to support a cloud against self-gravity. Assuming hydrostatic equilibrium, so that $\boldsymbol{u} = 0 = d^2I/dt^2$, and neglecting the effects of thermal pressure, we obtain the relevant virial equilibrium condition:

$$\mathcal{M} + \oint_S \boldsymbol{x} \cdot \mathbf{T} \cdot d\boldsymbol{S} + W = 0. \tag{5.29}$$

For simplicity, let us assume now that the magnetic field vanishes on the surface of the volume, so that the surface term in the above equation can be neglected.[5] In this case, Equation (5.29) gives $\mathcal{M} = -W$ and, recalling Equations (5.13) and (5.27), this implies

$$\frac{\varphi^2}{6\pi^2 R} = \frac{3GM^2}{5R},$$

from which we find the condition

$$\left(\frac{M}{\varphi}\right)_{\rm cr} = \sqrt{\frac{5}{18\pi^2 G}}. \tag{5.30}$$

This expression gives the critical value of the *mass-to-flux ratio*, M/φ, (hereafter, MFR) in order for the magnetic and gravitational terms of the VT to be equal for a uniform sphere with no magnetic field outside. Note that this critical value of the MFR is strictly constant, so that, in particular, it is independent of the radius or density of the gaseous sphere, as long as its mass remains constant. Other geometric configurations, such as, for example, cylinders (e.g., Nakano & Nakamura 1978), give a slightly different numerical value of $(M/\varphi)_{\rm cr}$, but they nevertheless retain the property of being constant.

Moreover, it is seen from Equations (5.13), (5.27), and (5.30) that, for a spherical, uniform-density cloud (not necessarily in equilibrium), the ratio of the gravitational term to the volumetric magnetic term in the VT is in general related to the MFR by

$$\frac{|W|}{\mathcal{M}} = \left[\frac{M/\varphi}{(M/\varphi)_{\rm cr}}\right]^2 \equiv \mu^2 \tag{5.31}$$

[5] It is important to note that this condition violates Gauss's Law $\nabla \cdot \boldsymbol{B} = 0$, because it implies that the magnetic field lines terminate at the surface of the volume, and it is therefore unrealizable. So, it should be considered only as an illustrative example.

The identity in the above expression defines μ, the MFR normalized to the critical value, although this quantity is often referred to simply as the MFR of a cloud, implicitly considered to be normalized to the critical value. Clouds with $\mu < 1$ have $|W| < \mathcal{M}$, and thus can be supported by the magnetic field against their self-gravity. These are said to be *magnetically subcritical.* Instead, clouds with $\mu > 1$ cannot be supported, and are said to be *magnetically supercritical.*

A very useful property of the MFR is that it can be expressed in terms of directly observable quantities: the *column density* $\Sigma \equiv M/\pi R^2$ and the magnetic field strength, since

$$\frac{M}{\varphi} = \frac{M}{\pi R^2 B} = \frac{\Sigma}{B}, \tag{5.32}$$

and so the normalized MFR μ can also be written as

$$\mu = \frac{\Sigma/B}{(\Sigma/B)_{\rm cr}}, \tag{5.33}$$

where we also have

$$\left(\frac{\Sigma}{B}\right)_{\rm cr} = \sqrt{\frac{5}{18\pi^2 G}}, \tag{5.34}$$

again noting that the exact value on the right-hand side depends on the geometry.

It is also important to reiterate then that, *for a fixed-mass cloud, and under ideal MHD conditions,* in which the magnetic flux is conserved, *the MFR is conserved.* This is a consequence of the fact that both W and $\mathcal{M}$ have the same R^{-1} dependence. In particular, this implies that μ remains constant regardless of any compression the cloud may be subjected to.

However, if the MFR were really a robust invariant, then the important question arises as to why different clouds are observed to have different values of the MFR. That is, if the MFR were strictly conserved, how can clouds formed in the same environment (with a given value of the MFR) end up having different values of the MFR? The resolution of this apparent conundrum is related to the question of whether the constancy of a cloud's mass is a valid assumption. We return to this in Section 5.4.1 and later in Section 10.2.1, where we discuss the simultaneous evolution of a cloud's mass and the MFR.

5.2.7 Correction due to the Magnetic Surface Term

It is now time to forgo the unphysical assumption that the magnetic field vanishes outside the cloud. The simplest case to consider in this case is that of a uniform magnetic field everywhere. In this case, it can be easily shown that the magnetic surface term exactly cancels out the volume term, as was the case for the volume and surface pressure terms, and the cloud is again in virial equilibrium.

A more general, and interesting case, is when the magnetic field outside the cloud is smaller than the average field inside, but still nonzero. This is, for example, the

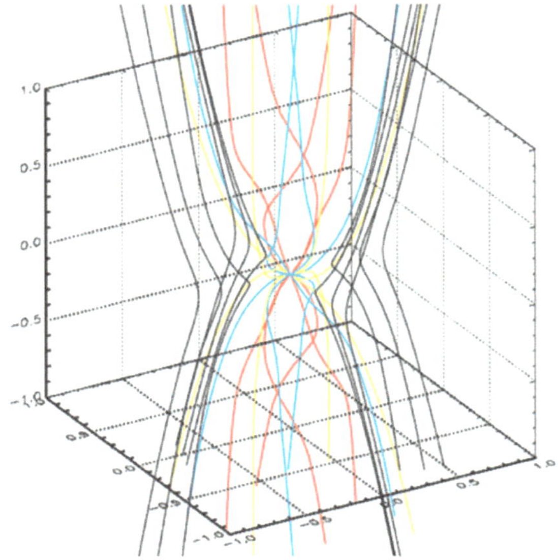

Figure 5.2. The "hourglass" magnetic field arising in a numerical simulation of the collapse of a magnetized core. Reproduced from Allen et al. (2003).

case of a core in which the magnetic field lines have been bent and pinched toward the core's center by the weight of the core, giving the classical "hourglass" shape (Mestel 1966), in either a magnetostatic or dynamically contracting state. An example is shown in Figure 5.2.

In this case, the precise value of the surface term may be difficult to calculate, but one can expect its effect to be intermediate between the cases of zero external field and that of a uniform field. Since in the latter case, the surface term completely cancels the volume term out, in the case of a weaker but nonzero external field we can expect that it will partially counteract the support provided by the volume term. Following Shu (1992), this can be described by writing

$$\mathcal{M} + \oint_S \boldsymbol{x} \cdot \mathbf{T} \cdot d\boldsymbol{S} = \beta \frac{\varphi^2}{6\pi^2 R}, \tag{5.35}$$

where $0 \leqslant \beta \leqslant 1$ is a correction factor to take into account the counterbalancing effect of the surface term in the expression for the magnetic term in terms of the magnetic flux, with $\beta = 0$ corresponding to the case when the field is uniform (i.e., is equal inside and outside the cloud), and $\beta \to 1$ when $\boldsymbol{B}_{\rm outside} \to 0$. However, since in general this factor is expected to be smaller than unity, this means that clouds or clumps in which subcritical values of the MFR are estimated from measurements of Σ/B may actually be supercritical.

5.2.8 Combined Thermal and Magnetic Support: The "Reduced Mass"

Let us now consider a *supercritical* isothermal cloud in which both thermal and magnetic support are active. In this case, the VT reads

$$\frac{1}{2}\frac{d^2I}{dt^2} - 2K = \left(2U - \oint_S P\boldsymbol{x} \cdot d\boldsymbol{S}\right) + \left(\mathcal{M} + \oint_S \boldsymbol{x} \cdot \mathbf{T} \cdot d\boldsymbol{S}\right) + W. \tag{5.36}$$

Note that, since the cloud is supercritical, we do not assume virial equilibrium, as the cloud is likely in the process of gravitational contraction. We therefore keep the time derivative and kinetic energy terms. For simplicity, however, we neglect the external pressure because, if the cloud is gravitationally contracting, then self-gravity is already overpowering the internal pressure, with no need for external confinement. Using Equations (5.13), (5.14) and (5.35), and writing $P = c_s^2 \rho$, Equation (5.36) becomes

$$\frac{1}{2}\frac{d^2I}{dt^2} - 2K = 4\pi c_s^2 \rho R^3 + \frac{1}{R}\left(\frac{\beta\varphi^2}{6\pi^2} - \frac{3GM^2}{5}\right). \tag{5.37}$$

We note that, if we define an effective *reduced mass* as

$$M_{\rm red} \equiv M\left(1 - \frac{\beta}{\mu^2}\right)^{1/2}, \tag{5.38}$$

then Equation (5.37) can be written as

$$\frac{1}{2}\frac{d^2I}{dt^2} - 2K = 4\pi c_s^2 \rho R^3 - \frac{3GM_{\rm red}^2}{5R}, \tag{5.39}$$

implying that the cloud can be qualitatively described as a system of mass $M_{\rm red} < M$ with only thermal support.

5.3 (Somewhat) More Realistic Configurations

Up to this point, we have considered the extremely idealized case of a cloud with a *fixed* mass, and therefore, with clearly defined boundaries. Although very useful for illustrative purposes, such a system is actually highly unrealistic. First and foremost, somewhere along the way, we have lost the fundamental notion that the clouds are just regions of enhanced density in a compressible *continuum*. This means that, in general, the clouds do not have sharp, well-defined boundaries and, of utmost importance, depending on how they are defined, *they can either move with the gas, or exchange mass, momentum, and energy with their surroundings.*

5.3.1 The Arbitrariness of Cloud Boundaries and Its Implications

As we discussed in Section 1.2.1, the "boundaries" of a cloud are completely arbitrary, and are usually defined under a variety of criteria, such as the presence of a flow feature (e.g., a shock, a condensation front, etc.), a chemical feature (e.g., a transition in chemical composition), or a threshold intensity level.

Formally, in the context of the VT, the clouds are gaseous parcels occupying volume V and delimited by its enclosing surface S, according to the notation of Equation (5.6). However, in the Lagrangian version of the VT, the surface S moves with the gas velocity, and so it is continuously distorted in time. Thus, even if S was initially defined by the location of a shock or a condensation front, at later times it will *not* coincide with the front, since in general the gas *flows* (passes) through these fronts. Therefore, the gas parcel initially contained within volume V will later occupy a different volume, delimited by the new location of surface S at that time.

Conversely, in the Eulerian version of the VT, surface S can be defined as fixed in a reference frame moving with the front (as long as it is not accelerating). But then there will be a net mass, momentum and energy flux across the surface, since the gas is passing through it. In this case, it is essential to consider such fluxes across the boundaries by means of the surface terms in the VT, and also the time dependence of the configurations. In the remainder of this chapter we will attempt to take a few more steps in this direction beyond the standard cases.

5.3.2 The VT in the Presence of Turbulence Driving and Dissipation

A very common assumption is that the kinetic energy term $2K$ in the VT corresponding to turbulence in a self-gravitating gaseous cloud can function as a counterbalance for the gravitational term to maintain virial equilibrium, in analogy with the situation for N-body systems such as globular clusters and elliptical galaxies. However, in the case of a fluid this is an incorrect interpretation, because fluid turbulence dissipates rapidly, in roughly one crossing time in the absence of continuous driving (cf. Section 3.5.1). Therefore, the complete description must include the driving and dissipation terms in the VT. To this end, let us write the momentum equation, Equation (2.75), including generic terms for an additional force (per unit mass) driving the turbulence, $\mathbf{f}_{\rm driv}$, and the viscous force, $\mathbf{f}_{\rm visc}$. We obtain

$$\frac{d\boldsymbol{u}}{dt} = -\frac{\nabla P}{\rho} - \nabla\varphi + \frac{1}{4\pi\rho}(\nabla \times \boldsymbol{B}) \times \boldsymbol{B} + \mathbf{f}_{\rm driv} + \mathbf{f}_{\rm visc}. \tag{5.40}$$

For illustrative purposes, let us only consider the part of the viscous force applicable to incompressible flows. The other viscous terms in Equation (2.75) including $\nabla \cdot \boldsymbol{u}$ are qualitatively similar, since they all consist of second spatial derivatives of the velocity. Thus, writing $\mathbf{f}_{\rm visc} = \nu\, \nabla^2 \boldsymbol{u}$, the VT becomes

$$\frac{1}{2}\frac{d^2 I}{dt^2} - 2K = \left(2U - \oint_S P\boldsymbol{x} \cdot d\boldsymbol{S}\right) + \left(\mathcal{M} + \oint_S \boldsymbol{x} \cdot \mathbf{T} \cdot d\boldsymbol{S}\right) + W + \mathcal{W}_{\rm driv} + \mathcal{W}_{\rm visc}, \tag{5.41}$$

where

$$\mathcal{W}_{\rm driv} = \int_V \boldsymbol{x} \cdot \mathbf{f}_{\rm driv}\, dV \tag{5.42}$$

and

$$\mathcal{W}_{\rm visc} = \int_V \nu \boldsymbol{x} \cdot \nabla^2 \boldsymbol{u}\, dV \tag{5.43}$$

are the works done, respectively, by $\mathbf{f}_{\rm driv}$ and $\mathbf{f}_{\rm visc}$. Here we do not further evaluate $\mathcal{W}_{\rm driv}$ and $\mathcal{W}_{\rm visc}$, but instead we only make a number of important remarks:

- The viscous force in general does not need to cancel the driving force, because the former depends on the *total* velocity field, which results from the combined action of all forces. For example, it can contain an infall component generated by the gravitational force, and a random, "turbulent" component generated by $\mathbf{f}_{\rm driv}$.
- The kinetic energy term, $2K$, is determined by this total velocity field, which has contributions from the velocities generated by the work done by each of the forces.

5.3.3 Dissipation of Turbulent Support and Gravitational Collapse

Let us now consider the important case of a non-magnetized, self-gravitating cold cloud, so that we can neglect the magnetic and thermal pressure terms. In this case we are left with

$$\frac{1}{2}\frac{d^2I}{dt^2} - 2K = W + \mathcal{W}_{\rm driv} + \mathcal{W}_{\rm visc}. \tag{5.44}$$

The condition of virial equilibrium, $d^2I/dt^2 = 0$, is then

$$-2K = W + \mathcal{W}_{\rm driv} + \mathcal{W}_{\rm visc}, \tag{5.45}$$

rather than simply $-2K = W$, which is the case for an undriven, non-dissipative system of particles. This implies that a continuous source of kinetic energy injection, $\mathcal{W}_{\rm driv}$, is necessary in order to maintain virial equilibrium in a dissipative gas.

Alternatively, we can consider the case of a cloud undergoing gravitational collapse, in which the work done by gravitational force is the source of the kinetic energy. Because of the nonlinear transfer of kinetic energy among modes in the turbulent velocity field (cf. Chapter 3), even minute turbulent motions can be amplified by transfer from the infall motions. Robertson & Goldreich (2012) have shown that the energy in the turbulent modes approaches equipartition with the energy in the infall motions. Note, however, that, in this case, the system is *not* in equilibrium, but rather in a dynamical state, and so the VT reads

$$\frac{1}{2}\frac{d^2I}{dt^2} - 2K = W + \mathcal{W}_{\rm visc}, \tag{5.46}$$

meaning that, for a collapsing cloud, the change in the spatial mass distribution (i.e., in the moment of inertia, I) and the work done by the viscous force must be taken into account. Guerrero-Gamboa & Vázquez-Semadeni (2020) have found in numerical simulations that a pseudo-virial equilibrium regime arises during the collapse, in which $2K \approx -W$. However, in this case this condition cannot be taken as evidence of virial equilibrium, because the system is dynamic, and at least two more terms need to be considered in the virial balance of the system. In particular, if $2K \approx -W$, then this implies that

$$\frac{1}{2}\frac{d^2I}{dt^2} \approx \mathcal{W}_{\rm visc}. \tag{5.47}$$

That is, the energy dissipated by the viscosity accounts for the contraction rate of the cloud.

5.4 Accretion: Clouds with Time-Variable Mass

We now consider clouds with feature-tracking boundaries (cf. Section 1.2.1). A particularly relevant case is that of clouds defined by a phase transition, such as that from the warm to the cold ISM (cf. Section 4.3).

5.4.1 Caveats on the Mass-to-Flux Ratio

5.4.1.1 The Fundamentally Distinct Nature of the Mass and the Magnetic Flux
In Section 9.2.1 we showed that, *for a fixed mass cloud*, the conservation of magnetic flux implies that the MFR also remains constant. Failure to conserve the mass-to-flux ratio is most often considered to be the consequence of non-ideal MHD (Mestel & Spitzer 1956), in which case the magnetic flux is not perfectly conserved due to diffusive or dissipative effects. In classical treatments (e.g., Mouschovias & Spitzer 1976), it was customary to consider a cloud of finite extension, confined by the pressure of a hotter intercloud medium, so that the cloud's mass was well defined and constant in time. However, as discussed in Section 4.5.1, not even in this case is the cloud's mass fixed, since the boundary between the cold dense and warm diffuse gas is not a rigid boundary, but rather a phase-transition front, across which mass readily flows, in particular allowing the cloud's mass to grow. When one recognizes that the cloud is part of a continuum, with boundaries that are only operationally defined, and not true physical boundaries, then the notion of the "mass-to-flux ratio of the cloud" becomes blurry.

This can be seen in particular by noting that the mass of the cloud is defined over a *three-dimensional* domain: it involves the integration of the density field over the *volume* V of the cloud:

$$M = \int_V \rho\, dV.$$

Instead, the magnetic flux is defined over a *two-dimensional* domain—a cross section of V, so that it is computed as a *surface* integral:

$$\varphi = \int_A \boldsymbol{B} \cdot d\boldsymbol{A},$$

where $\boldsymbol{A}$ is a cross-sectional area of volume V. Therefore, M and φ *are defined over different domains in space.* This implies that the MFR in general depends on the choice of V and $\boldsymbol{A}$, and is thus not uniquely defined. In particular, it is possible to change M while keeping the magnetic flux fixed simply by changing the extension of volume V along a flux tube, since, by construction, the magnetic flux is constant along the tube. Instead, the mass depends on the extension along the tube. Even in the case of a cloud immersed in a warm medium the *dense* gas mass need not be constant, since mass can be flowing in across a phase transition front from the diffuse phase. In conclusion, the MFR in general is *not* constant for clouds in the ISM, but rather a property of the clouds that evolves in time. We discuss this further in Section 10.2.1.

5.4.1.2 The Range of MFR Values for a Cloud Fragment

As discussed in the previous subsection, the MFR will depend on the volume chosen to define the cloud. To illustrate this point, consider the following examples (Vázquez-Semadeni et al. 2005). First, consider a spherical region ("cloud") of radius R within a medium of uniform density ρ_0, permeated by a uniform magnetic field of intensity B_0. In this case, the cloud's mass is $M = 4\pi\rho_0 R^3/3$, while the magnetic flux through the cloud's midplane perpendicular to the magnetic field is $\varphi = \pi R^2 B_0$. Therefore, the MFR measured for the cloud is seen to depend linearly on R:

$$\frac{M}{\varphi} = \frac{4\rho_0}{3B_0}R. \tag{5.48}$$

Of course, this example is a purely academic exercise, since in practice it is not possible to observationally discriminate between subregions of different sizes in a medium of uniform density and magnetic field. But it illustrates the dependence of the MFR on the choice of the boundary when only a fragment (a piece) of the cloud is considered.

The opposite extreme is the case where an isolated spherical cloud contracts or expands from radius R_1 to radius R_2, assuming ideal (i.e., non-resistive) MHD, for which field freezing holds exactly. This case is considered in Section 9.2.1, where the *magnetic support* model for clouds is discussed. In that case, the magnetic flux is conserved by virtue of Equation (2.77), while the mass is conserved by construction (the entire cloud's mass contracts from R_1 to R_2, corresponding to choosing a Lagrangian boundary). Therefore, $M/\varphi = \text{cst.}$ (independent of radius) in this case.

The two cases above bracket the range of possible values of the MFR of a cloud's fragment of radius $R_f < R_c$, density $\rho_f > \rho_c$ and mass $M_f < M_c$, so that

$$\frac{R_f}{R_c}\mu_c < \mu_f < \mu_c, \tag{5.49}$$

where the quantities with subindex "c" correspond to the parent cloud.

5.4.1.3 A Centrally Concentrated Cloud

Another interesting example is the case of a spherical cloud with a magnetic field strength given approximately[6] by $B = B_0(r/r_0)^{-m}$, and a density profile, $\rho(r) = \rho_0(r/r_0)^{-n}$, where $\mathbf{B}_0$ and ρ_0 are the field strength and density at radius r_0,

[6] This scaling may hold *approximately* at sufficiently distant positions from the cloud's center, if the cloud has contracted gravitationally, producing an "hourglass" morphology for the magnetic field. For $m > 0$, it cannot be valid at the center, since it would violate the condition $\nabla \cdot \boldsymbol{B} = 0$.

respectively, and $0 \leqslant m < 2$ and $0 < n < 3$ are constants. In this case, the mass out to radius R is given by

$$M(R) = \int_V \rho\, dV = \frac{4\pi\rho_0 r_0^n}{(3 - n)} R^{3-n},$$

while the magnetic flux out to the same radius is

$$\varphi(R) \approx \frac{2\pi B_0 r_0^m}{2 - m} R^{2-m}.$$

Therefore, the MFR measured out to radius R in this case is

$$\frac{M}{\varphi}(R) \approx \frac{2(2 - m)\rho_0 r_0^{n-m}}{(3 - n)B_0} R^{1+m-n}, \tag{5.50}$$

and we see that again, in general, the MFR depends on the choice of the boundary (given by the radius R). In particular, for density profiles with $n < m + 1$, Equation (5.50) implies that the measured MFR should *increase* with increasing radius of the measurement region. Such an increase has indeed been qualitatively observed between dense regions and their immediate environments (Crutcher et al. 2009; Arzoumanian et al. 2021).

5.4.1.4 The MFR as a Boundary Condition

The cases discussed above clearly illustrate that the MFR of a cloud or clump is as elusive and arbitrary as the object's own boundary, and therefore, the most appropriate way of thinking about it is as a *boundary condition* of a particular problem (an analytic model, a numerical simulation or an observation). Moreover, the MFR should only be expected to be conserved, even under ideal MHD conditions, when Lagrangian boundaries are used to define the cloud or clump.

However, as mentioned in Section 1.2.1, in practice clouds are generally not defined in a Lagrangian way, neither in observations nor in numerical simulations. In observations, a Lagrangian definition is plain impossible, since the flow is not known in detail. In simulations, defining a Lagrangian cloud is possible in principle, but in practice, a cloud's boundaries are generally defined mimicking observational definitions, which are not fixed on the flow. Therefore, the MFR is expected in general to depend on the chosen location of the boundary in both observations and simulations.

5.4.1.5 The Accumulation Length

A corollary of the discussion in Section 5.4.1.1 is that, given a medium of uniform density n_0 and magnetic field strength B_0, there exists a length, known as the *accumulation length*, along a flux tube beyond which the mass contained in the tube is magnetically supercritical (e.g., Mestel 1985; Hartmann et al. 2001; Vázquez-Semadeni et al. 2011). This length is obtained by noting that the ratio of mass-to-flux is equivalent to the ratio of column density to magnetic field strength, since $M/\varphi \approx M/BA = \Sigma/B$, where $\Sigma = M/A$ is the surface density, A being the cross-sectional area of the cloud. In turn,

$$\Sigma = \int_L \rho \, d\ell \approx \langle \rho \rangle L,$$

where L is the length of the cloud and $\langle \rho \rangle$ is its mean density. Therefore, the critical mass-to-flux ratio, Equation (5.30), is equivalent to

$$L_c \approx 470 \left(\frac{B_0}{5 \mu G} \right) \left(\frac{n_0}{1 \text{ cm}^{-3}} \right)^{-1} \text{pc}, \tag{5.51}$$

thus giving the length beyond which gas of number density n_0 permeated by a uniform field of strength B_0 is magnetically supercritical.

Because magnetic field lines do not end,[7] and because the mean magnetic field in the Galactic disk is preferentially azimuthal, it is likely that all flux tubes in the Galaxy are longer than the accumulation length, and so it could be said in principle that the ISM at large is always magnetically supercritical. Local detections of subcritical regions can be interpreted, as discussed in Section 5.4.1.1, as simply a consequence of having performed the measurement over too small a region.

However, what is locally relevant for the the magnetic support of self-gravitating clouds is the MFR *of the mass that generates the local gravitational potential*, which the magnetic field opposes, as discussed in Section 9.2.1. This is the case, in particular, of clouds that have formed via a phase transition from the warm/diffuse phase to the cold dense phase of the atomic component of the ISM, or, more generally, undergone rapid cooling. For such clouds, a natural feature-tracking boundary is the phase transition front where the gas transits from the warm/diffuse phase to the cold/dense one (Section 4.5). This front is often observed in simulations to be very sharp, although it is not a shock (e.g., Vázquez-Semadeni et al. 2006). Therefore, it is the mass of cold, dense gas that should be considered in the calculation of the MFR, to determine whether this mass is locally supported by the magnetic field.

However, because this boundary is of the feature-tracking kind, the gas is free to cross it, increasing the cloud's mass and column density, until eventually it becomes supercritical and begins to collapse. According to Equation (5.51) above, if the diffuse gas is crossing the boundary at speed u_f, then the time necessary for the cloud to become supercritical is

$$\tau_{\text{sup}} \approx 47 \left(\frac{B_0}{5 \mu G} \right) \left(\frac{n_0}{1 \text{ cm}^{-3}} \right)^{-1} \left(\frac{u_f}{10 \text{ km s}^{-1}} \right)^{-1} \text{Myr}. \tag{5.52}$$

It is important to note that Hartmann et al. (2001) showed that cold atomic clouds accreting diffuse material in the Solar neighborhood are expected to become magnetically supercritical, Jeans-unstable, and molecular at roughly the same time (cf. Section 10.2.1).

[7] Due to the condition $\nabla \cdot \boldsymbol{B} = 0$, magnetic field lines either are closed or extend to infinity.

References

Allen, A., Li, Z.-Y., & Shu, F. H. 2003, ApJ, 599, 363
Arzoumanian, D., Furuya, R. S., Hasegawa, T., et al. 2021, A&A, 647, A78
Ballester, J. L., Alexeev, I., Collados, M., et al. 2018, SSRv, 214, 58
Ballesteros-Paredes, J., Vázquez-Semadeni, E., & Scalo, J. 1999, ApJ, 515, 286
Bertoldi, F., & McKee, C. F. 1992, ApJ, 395, 140
Brunt, C. M., Heyer, M. H., & Mac Low, M. M. 2009, A&A, 504, 883
Colín, P., Vázquez-Semadeni, E., & Gómez, G. C. 2013, MNRAS, 435, 1701
Crutcher, R. M. 2012, ARA&A, 50, 29
Crutcher, R. M., Hakobian, N., & Troland, T. H. 2009, ApJ, 692, 844
Field, G. B., Blackman, E. G., & Keto, E. R. 2011, MNRAS, 416, 710
Guerrero-Gamboa, R., & Vázquez-Semadeni, E. 2020, ApJ, 903, 136
Hartmann, L., Ballesteros-Paredes, J., & Bergin, E. A. 2001, ApJ, 562, 852
Kauffmann, J., Pillai, T., & Goldsmith, P. F. 2013, ApJ, 779, 185
Keto, E. R., & Myers, P. C. 1986, ApJ, 304, 466
Leroy, A. K., Bolatto, A. D., Ostriker, E. C., et al. 2015, ApJ, 801, 25
McKee, C. F., & Zweibel, E. G. 1992, ApJ, 399, 551
Mestel, L. 1966, MNRAS, 133, 265
Mestel, L. 1985, Protostars and Planets II, ed. D. C. Black, & M. S. Matthews, 320
Mestel, L., & Spitzer, L. 1956, MNRAS, 116, 503
Mouschovias, T. C., & Spitzer, L. 1976, ApJ, 210, 326
Nakano, T., & Nakamura, T. 1978, PASJ, 30, 671
Pattle, K., Fissel, L., Tahani, M., Liu, T., & Ntormousi, E. 2023, Protostars and Planets VII; ASP Conf. Ser. Vol. 534, ed. S. Inutsuka, Y. Aikawa, T. Muto, K. Tomida, & M. Tamura (San Francisco: ASP) 193
Robertson, B., & Goldreich, P. 2012, ApL, 750, L31
Shu, F. H. 1992, Physics of Astrophysics (Vol. II), (Mill Valley, CA: Univ. Science Books)
Vázquez-Semadeni, E. 1999, Millimeter-Wave Astronomy: Molecular Chemistry & Physics in Space; Proc.of the 1996 INAOE Summer School of Millimeter-Wave Astronomy Vol. 241, ed. W. F. Wall, A. Carramiñana, & L. Carrasco; Dordrecht: Kluwer Academic Publishers) 143
Vázquez-Semadeni, E., Kim, J., Shadmehri, M., & Ballesteros-Paredes, J. 2005, ApJ, 618, 344
Vázquez-Semadeni, E., Ryu, D., Passot, T., González, R. F., & Gazol, A. 2006, ApJ, 643, 245
Vázquez-Semadeni, E., Banerjee, R., Gómez, G. C., Hennebelle, P., Duffin, D., & Klessen, R. S. 2011, MNRAS, 414, 2511
Zweibel, E. G. 1990, ApJ, 348, 186

Enrique Vázquez-Semadeni

Chapter 6

Self-Gravity I: Equilibrium and Stability

In this chapter, we will study hydrostatic configurations of self-gravitating structures, with a special emphasis on their stability and the situations where stable configurations can and cannot arise.

Equilibrium at all locations in a given body of gas in the presence of self-gravity can only occur when the latter is balanced everywhere by some other force in the momentum equation, Equation (2.75). But moreover, in order for a strictly hydrostatic configuration to arise in nature, it is also necessary for the configuration to be *stable*; i.e., that slight perturbations of this equilibrium evolve toward returning to the equilibrium state. If they evolve away from the equilibrium, then the equilibrium is unstable and is not expected to be reached. One of the most widely used stability analyses in astronomy is that due to Jeans (1902), which applies (with some caveats) to uniform, infinite media, and is discussed in Section 6.1. After that, we consider the equilibria of spherical configurations, and discuss their stability using the virial theorem.

6.1 Jeans Analysis: The Jeans Length and Mass

Jeans (1902) considered the stability of the equilibrium between the thermal pressure gradient and self-gravity of both a spherical cloud and of an infinite medium. Here we consider the latter, for a polytropic gas (cf. Section 4.2), described by a polytropic exponent γ. In this case, the relevant equations are the continuity and momentum equations, Equations (2.74) and (2.75), which we rewrite here as

$$\frac{\partial \rho}{\partial t} + \nabla \cdot (\rho \boldsymbol{u}) = 0 \tag{6.1}$$

$$\frac{\partial \boldsymbol{u}}{\partial t} + (\boldsymbol{u} \cdot \nabla)\boldsymbol{u} + \frac{1}{\rho}\nabla P + \nabla \phi = 0 \tag{6.2}$$

doi:10.1088/2514-3433/adcb28ch6

supplemented by Poisson's (self-gravity) equation,

$$\nabla^2\phi = 4\pi G\rho, \tag{6.3}$$

and the equation of state

$$P = \frac{\rho k_B T}{\mu m_H}, \tag{6.4}$$

where ρ is the density, $\boldsymbol{u}$ is the flow velocity, P is the thermal pressure, ϕ is the gravitational potential, k_B is Boltzmann's constant, m_H is the hydrogen mass, and μ is the mean particle (atom or molecule) mass in units of m_H.

Jeans considered the equilibrium of a medium at rest with uniform density and pressure, so that $\rho = \rho_0 =$ cst., $P = P_0 =$ cst., $\boldsymbol{u} = 0$. At this point, the (in)famous *Jeans' swindle* should be noted, which consists in the fact that this equilibrium state is formally only consistent with zero density. Indeed, substituting the equilibrium state in Equation (6.2) we see that, since $P_0 =$ cst., then $\nabla P_0 = 0$, implying that $\nabla\phi_0 = 0$, which contradicts Equation (6.3), except when $\rho_0 = 0$. Nevertheless, the approach works for determining the size scale of the unstable regions, and several vindications have been offered in the literature (e.g., Kiessling 1999). The fact that the approach produces physically correct results may arise because, after all, an infinite homogeneous medium is equally unrealistic, and in particular, if the density were perfectly uniform, the gravitational force on such a medium would cancel out in all directions and everywhere, so in this case Poisson's equation might be written as $\nabla^2\phi = 4\pi G(\rho - \rho_0)$, meaning that only the density *fluctuations* around the mean density in such an infinite medium produce a nonzero gravitational acceleration.

Let us then consider small (linear)[1] perturbations (denoted by a subindex '1') on the equilibrium values of the variables, such that, for any of the variables, generically written A, we have $|A_1| \ll |A_0|$. Therefore, we write

$$\begin{aligned} \rho &= \rho_0 + \rho_1 \quad \boldsymbol{u} = \boldsymbol{u}_1 \\ P &= P_0 + P_1 \quad \phi = \phi_0 + \phi_1. \end{aligned} \tag{6.5}$$

In addition, to first order, the adiabatic sound speed is given by

$$c_s^2 \equiv \left(\frac{\partial P}{\partial \rho}\right)_s \approx \frac{\gamma P_0}{\rho_0} \approx \frac{P_1}{\rho_1}, \tag{6.6}$$

as can be verified by performing a Taylor expansion to first order of the adiabatic equation $P/P_0 = (\rho/\rho_0)^\gamma$, where γ is the heat capacity ratio (cf. Section 2.5.1). Substituting Equations (6.5) in Equations (6.1) and (6.2), we obtain the evolution equations for the perturbations:

$$\frac{\partial \rho_1}{\partial t} + \rho_0 \boldsymbol{\nabla} \cdot \boldsymbol{u}_1 = 0 \tag{6.7}$$

[1] The nomenclature "linear" for small-amplitude fluctuations comes from the fact that, upon substitution in the gas dynamics equations, terms of second order and higher in the perturbation can be neglected, leaving only first order, or *linear* terms.

$$\frac{\partial \boldsymbol{u}_1}{\partial t} + \frac{c_s^2}{\rho_0}\boldsymbol{\nabla}\rho_1 + \boldsymbol{\nabla}\phi_1 = 0 \tag{6.8}$$

$$\nabla^2\phi_1 = 4\pi G\rho_1. \tag{6.9}$$

Differentiating Equation (6.7) with respect to time, and taking the divergence of Equation (6.8) we can eliminate the velocity, to obtain

$$\frac{\partial^2\rho_1}{\partial t^2} - c_s^2\,\nabla^2\,\rho_1 = 4\pi G\rho_0\rho_1, \tag{6.10}$$

which can be identified as a wave equation augmented by the term on the right-hand side. Introducing traveling wave solutions of the form

$$\rho_1(\mathbf{r},\, t) = B\rho_0 \exp\left[-i\boldsymbol{k}\cdot\mathbf{r} + i\omega t\right], \tag{6.11}$$

in Equation (6.10) for all variables, where $\omega = 2\pi/\tau$ is the angular frequency, τ is the *period* of the wave, $\boldsymbol{k}$ is the wavevector, and $B \ll 1$ is a constant, we obtain the *dispersion relation*

$$\omega^2 = c_s^2 k^2 - 4\pi G\rho_0, \tag{6.12}$$

where $k \equiv |\boldsymbol{k}| = 2\pi/\lambda$ is the *wavenumber*, and λ is the wavelength of the traveling wave solution.[2]

6.1.1 The Jeans Length

Equation (6.12) implies that, for

$$k > \left(\frac{4\pi G\rho_0}{c_s^2}\right)^{1/2} \equiv k_J, \tag{6.13}$$

or, equivalently, for

$$\lambda = \frac{2\pi}{k} < \left(\frac{\pi c_s^2}{G\rho_0}\right)^{1/2} \equiv L_J, \qquad \text{(Jeans length)} \tag{6.14}$$

ω is real, and the solutions are indeed ondulatory, and remain bounded. Instead, for $\lambda > L_J$, ω is imaginary, and the perturbations evolve exponentially in time as $e^{|\omega|t}$. Since Equation (6.12) is a polynomial with real coefficients, given any complex solution, its complex conjugate is also a solution, and therefore there will always exist a solution such that $i\omega > 0$, which will grow in time. Thus, ω is known as the *growth rate.* Therefore, perturbations with wavelength $\lambda < L_J$ are *stable*, while perturbations with $\lambda > L_J$ are *unstable*. L_J is known as the *Jeans length.*

[2] Note that the wavenumber k literally is the number of waves of wavelength λ that fit in the unit of length, and therefore it can be thought of as the *spatial* frequency of the wave.

The physical meaning of Equation (6.12) is that the evolution of a perturbation in the medium depends on the competition between the propagation of sound waves (which tend to restore pressure balance in the medium) and the tendency to collapse gravitationally due to the self-gravity of the medium. For perturbations with wavelength $\lambda < L_J$, the *sound crossing time*, $\tau_s \equiv \lambda / c_s$ is shorter than the *collapse time* [3]

$$\tau_{\text{coll}} \equiv \sqrt{\frac{\pi}{G\rho_0}}, \tag{6.15}$$

and so a sound wave is able to restore pressure balance before the region collapses. The opposite occurs when $\lambda > L_J$. In this case, the collapse time is shorter than the sound crossing time, and so a perturbation is able to collapse before sound waves are able to restore equilibrium.

6.1.2 The Jeans Mass

It is convenient to define the *Jeans mass* as the mass contained in a sphere with radius $R = L_J/2$, at the uniform density ρ_0:

$$M_J \equiv \frac{\pi^{5/2}}{6} \frac{c_s^3}{G^{3/2}\rho_0^{1/2}}. \qquad \text{(Jeans mass)} \tag{6.16}$$

This is approximately the maximum mass that can be supported against self-gravity by the thermal pressure gradient. It is only approximate because it considers a spherical region, while the Jeans length gives the critical length in each spatial direction, and so there is no precise correspondence between the two definitions.

It is also convenient to compare the above definition of the Jeans length (Equation (6.14)) with the estimate obtained using the virial theorem (VT; Section 5.2.2, Equation (5.17)). The functional dependence on the physical variables, the sound speed (or, equivalently, the temperature) and the density, is the same, but the VT estimate is a factor $\sqrt{15/4\pi^2} \approx 0.62$ smaller. The numerical difference again arises because the spatial mass distribution is different in two cases, and also because the VT estimate is only an average over the spherical volume, while the Jeans analysis considers detailed pressure balance at all points.

6.1.3 The Dependence of the Growth Rate on the Length Scale

An important point to note is that, according to the dispersion relation, Equation (6.12), unstable perturbations of different wavelengths $\lambda = 2\pi/k$ grow at different rates ω. Indeed, the growth rate $|\omega|$ increases as $k \to 0$ in Equation (6.12); i.e., as $\lambda \to \infty$. Physically, this occurs because the larger the scale, the less relevant the pressure gradient is compared to the self-gravity of the perturbation, and so the less it opposes the gravitational contraction. Therefore, larger scales grow faster, approaching the free-fall rate at very large wavelengths.

[3] Which is similar, but not identical, to the *free-fall time*, to be discussed in Section 7.2.1.

The above result implies that, if perturbations of different wavelengths but the same amplitude coexist in a medium, the larger-scale perturbations will grow faster than the smaller-scale ones, overwhelming them. This was the reason why *gravitational fragmentation* (Section 7.5) was discarded as a real possibility for a while (e.g., Tohline 1980). However, it is important to note that this objection is only valid for *linear* (i.e., low-amplitude) perturbations, which can be assumed to all have nearly the same density as the background, ρ_0, and therefore the same collapse time. For *nonlinear* perturbations, such as those produced by supersonic turbulence, the local density in the fluctuations can be much larger than ρ_0. Moreover, in a hierarchical, supersonically-turbulent medium, the density fluctuations are expected to occur as a consequence of a succession of shock passages (Vazquez-Semadeni 1994), and so the highest densities are expected to occur *inside* larger regions with previously enhanced densities. Thus, the highest densities (and therefore, the shortest collapse times) are expected to occur at the smallest scales. This allows small-scale perturbations to grow faster than larger-scale ones, and therefore the objection to gravitational fragmentation does not apply in the nonlinear case.

Furthermore, this objection is only applicable in three dimensions. As will be discussed in Section 6.4.2, for perturbations in a thin sheet, the maximum growth rate occurs for perturbations of size $2L_J$, and so in this case structures much larger than the Jeans length will tend to fragment into substructures of size $2L_J$, even in the case of linear perturbations. Therefore, early objections to gravitational fragmentation were most likely unfounded.

6.2 Hydrostatic Equilibrium of Isothermal Self-Gravitating Spheres: Bonnor–Ebert Spheres and the Singular Isothermal Sphere

We now turn to the equilibrium states of an isothermal gas sphere. Again, we note here that the equilibrium we will discuss here is *detailed*, meaning that it occurs at every radius in the sphere. That is, at each radius, the pressure gradient must balance the weight of the material in a thin shell at that radius. This allows us to determine the *density profile*, i.e., the density as a function of radius. In contrast, the calculation done using the VT (Section 5.2.2) involved only the competition between the pressure gradient and self-gravity for the bulk volume of the cloud; that is, the VT allows only for a *global*, volume-averaged equilibrium calculation.

Under the condition of hydrostatic equilibrium, considering only self-gravity and thermal pressure, the momentum equation, Equation (2.75), reduces to

$$\nabla P = -\rho \nabla \phi. \tag{6.17}$$

Furthermore, assuming an isothermal gas and spherical symmetry, the above equation becomes

$$\frac{c_s^2}{\rho}\frac{d\rho}{dr} = -\frac{d\phi}{dr}, \tag{6.18}$$

whose solution is

$$\rho = \rho_0 e^{-\phi/c_s^2}, \tag{6.19}$$

where ρ_0 is the density at a reference radius where the gravitational potential ϕ is set to zero. The potential, in turn, satisfies Poisson's equation (Equation (2.78)), which in spherical coordinates reads

$$\frac{1}{r^2}\frac{d}{dr}\left(r^2\frac{d\phi}{dr}\right) = 4\pi G\rho. \tag{6.20}$$

This equation admits both *regular* and *singular* solutions, which, respectively, have finite and diverging density at the origin.

6.2.1 Solutions with Central Boundary Conditions

To derive the regular solutions, we introduce the hydrostatic solution, Equation (6.19), in Equation (6.20) and set $\rho_0 = \rho_c$, where ρ_c is the (finite) central density, to get

$$\frac{1}{r^2}\frac{d}{dr}\left(r^2\frac{d\phi}{dr}\right) = 4\pi G\rho_c e^{-\phi/c_s^2}. \tag{6.21}$$

We next change to nondimensional variables, defined in terms of the central density as[4]

$$\psi \equiv \frac{\phi}{c_s^2} = -\ln\left(\frac{\rho}{\rho_c}\right); \qquad \xi \equiv \left(\frac{4\pi G\rho_c}{c_s^2}\right)^{1/2} r. \tag{6.22}$$

In terms of these, Equation (6.21) becomes

$$\frac{1}{\xi^2}\frac{d}{d\xi}\left(\xi^2\frac{d\psi}{d\xi}\right) = e^{-\psi}. \tag{6.23}$$

Equation (6.23) is known as the *Emden–Chandrasekhar* equation, and is a special case of the more general *Lane–Emden* equation. The latter (not discussed here) applies in the general case of a polytropic gas, and reduces to the Emden–Chandrasekhar equation in the case of isothermal flow, as considered here.

The Emden–Chandrasekhar equation can be solved numerically for the nondimensional gravitational potential as a function of radius, integrating outwards from the center, with boundary conditions at the center

$$\psi(\xi = 0) = 0, \qquad \psi'(\xi = 0) = 0. \tag{6.24}$$

[4] Note that, defined in this manner, ψ represents the ratio of the gravitational potential (i.e., the gravitational energy per unit mass) to the thermal energy per unit mass, while ξ represents the radius in units of the inverse Jeans wavenumber for the central density ($k_J^{-1}(\rho_c)$), which is of the order of the Jeans length (cf. Equations (6.13) and (6.14)).

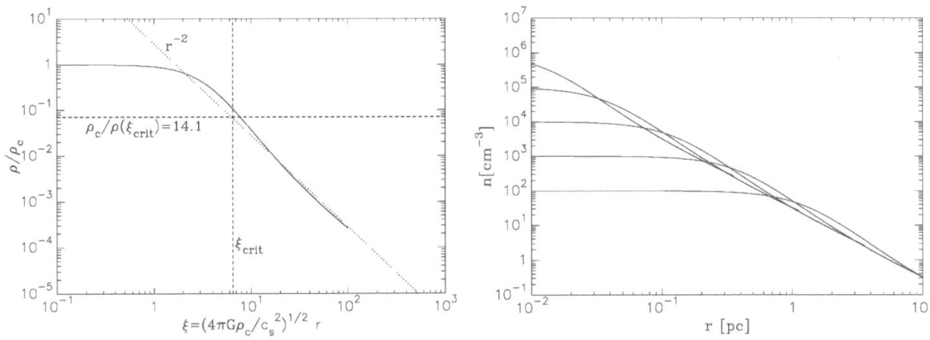

Figure 6.1. *Left:* Nondimensional regular solution (solid line) of the Emden–Chandrasekhar equation, Equation (6.23), and the singular solution (the SIS; dotted line). The vertical dashed line shows the critical nondimensional radius for stability, and the horizontal dashed line shows the value of the central-to-external density ratio at that point. *Right:* Dimensional regular solutions for central densities $n_c = \rho_c/\mu m_H = 10^2, 10^3, 10^4, 10^5$, and 10^6 cm − 3.

These correspond to requiring that $\rho(r = 0) = \rho_c$ and $\rho'(r = 0) = 0$, the latter indicating that ρ has a maximum at the center.

The left panel of Figure 6.1 shows the nondimensional regular solution for these boundary conditions. The right panel shows a family of dimensional solutions for several different values of the central density. These regular solutions are known as the *Bonnor–Ebert* (BE) spheres (Ebert 1955; Bonnor 1956) and, at small distances from the center ($r \ll cs/\sqrt{4\pi G\rho_c} = L_J(\rho_c)/2\pi$), they are seen to have a nearly flat density profile. On the other hand, at large distances from the center ($r \gg L_J(\rho_c)/2\pi$) they asymptotically approach the singular solution

$$\rho(r) = \frac{c_s^2}{2\pi G r^2}, \qquad \frac{d\phi}{dr} = \frac{2c_s^2}{r}, \tag{6.25}$$

which is known as the *singular isothermal sphere* (SIS). This solution is the limit when the central density approaches infinity, in which case the central flat part reduces to zero radius and the entire sphere has an r^{-2} density profile.

It is important to note that, in principle, the integration outwards from the center can be carried out all the way to infinity, in which case the sphere's mass diverges. However, the integration can also be stopped at some arbitrary radius $R = R_{ext}$, or equivalently, $\xi = \xi_{ext}$. In this case, in order to maintain the equilibrium, it is necessary to exert an external pressure $P_{ext} = c_s^2\rho(R_{ext})$ at R_{ext}. This external pressure is said to *confine* the BE sphere, which would otherwise expand freely. In turn, such a sphere is known as a *truncated* BE sphere.

6.2.2 Stability of the BE Sphere

Similarly to the stability of a uniform-density sphere investigated using the virial theorem (Section 5.2.4), truncated regular BE spheres (i.e., with finite central densities) can be either stable or unstable. Here we do not repeat the stability analysis for the BE sphere, but, qualitatively, it can be understood in the same terms as those for the uniform sphere.

As discussed in Section 5.2.4, the stability of a uniform sphere of mass M depends on the sign of dP_{ext}/dR (i.e., the slope of the $P_{ext}(R)$ curve; see Figure 5.1). In turn, the sign of dP_{ext}/dR arises from the different dependences of the thermal and "gravitational" pressure gradients on the radius of the sphere. When $dP_{ext}/dR < 0$, an *increase* in P_{ext} causes a reduction of the equilibrium sphere's radius, and therefore an increase in its density. As a consequence, the internal pressure also increases, and so it can eventually reach the value required to balance the increased P_{ext}. This happens because, in this range of radii, the sphere is dominated by its internal thermal pressure gradient, and so the balance between the internal and external pressures determines the equilibrium of the sphere.

However, when $dP_{ext}/dR > 0$ (but still $P_{ext} > 0$), an increase in P_{ext}, and the consequent decrease in radius, bring the sphere to a configuration in which a *smaller* P_{ext} is required to maintain the equilibrium. This is because the negative gravitational "pressure" is not negligible anymore, and moreover increases faster (in absolute value) than the thermal pressure. Therefore, the reduction in radius causes an increase in the gravitational binding of the sphere, reducing the external pressure necessary to confine it. However, since the external pressure had increased in the first place, the sphere is now out of equilibrium, and proceeds to collapse.

Finally, when the radius is small enough that the gravitational pressure is larger (in absolute value) than the internal pressure, no equilibrium is possible, and the sphere collapses even without perturbing the external pressure. This corresponds to the $P_{ext} < 0$ region in Figure 5.1, in which a negative value of P_{ext} would be necessary in order to maintain equilibrium.

The stability of the BE spheres was worked out by Bonnor (1956), who found that the critical nondimensional radius for stability (the BE equivalent of R_{crit} in Figure 5.1) is $\xi_{crit} \approx 6.5$, at which $\rho_c/\rho(\xi_{crit}) \approx 14.1$. These values are shown by the dashed lines in the left panel of Figure 6.1.

6.2.3 Impossibility of Hydrostatic BE Spheres and the SIS in Single-Phase Media

Several points are important to consider in relation with the stability of BE spheres:

- A corollary of the discussion in the previous section is that, in order to maintain the stability of a BE sphere truncated at $\xi < \xi_{crit}$, *the external pressure must be applied without adding weight (mass)* to the sphere on its outer shells. Otherwise, the confining medium would also have to satisfy the detailed balance between its pressure gradient and self-gravity at every radius, thus constituting a continuous extension of the sphere, rendering it equivalent to an *unstable* sphere. That is, the system consisting of the "truncated" sphere plus the also self-gravitating confining medium would be equivalent to a larger, unstable, non-truncated sphere.
- As a consequence, *the common notion that a hydrostatic dense core can be confined (in equilibrium) by the pressure exerted by the weight of its surrounding molecular cloud is erroneous.* Since the surrounding molecular material is in general at roughly the same temperature as the core, the core-

cloud system constitute a single-phase medium in which no abrupt, static density discontinuity can occur, and is thus equivalent to an unstable system.

- Abrupt discontinuities can occur in a single-phase medium in the presence of shocks, but in this case the system is no longer hydrostatic, but rather evolves in time, accreting mass and eventually going into collapse, or else re-expanding.
- In the multi-phase ISM, the only way to accomplish a truncation is by means of a phase transition to a warm, diffuse phase, which can exert pressure without adding significant mass. That is, *a discontinuity of density and temperature (a phase transition) must occur at the boundary* in order for a BE sphere to be effectively truncated.
- However, as discussed in Section 4.5.1, phase transitions in the ISM are generally *not* impenetrable contact discontinuities. Quite the opposite, phase transitions are in general *fronts*, across which the gas can flow freely. Therefore, the mass of a BE sphere bounded by a phase transition does not have to remain constant in general.
- The SIS is the most extreme case of an *unstable* BE sphere, because it corresponds to a diverging value of the ratio ρ_c/ρ_{ext}. Therefore, although it is a *formal* solution of the Emden–Chandrasekhar equation, it is not expected to occur in nature: it does not persist in equilibrium given the slightest perturbation, nor will a sphere that is already contracting from a less concentrated stage settle onto that unstable equilibrium.
- A noteworthy feature is that *the SIS contains one BE mass* (see Sections 6.2.4 and 7.4.3) *at every radius*. However, this also implies that it is at the verge of instability *at every radius*.

6.2.4 Solutions with External Boundary Conditions

In practice, it is more convenient to specify *outer* boundary conditions for the sphere, since they may be easier to determine observationally than the central density. The relevant external conditions are the total sphere's mass and the pressure at its boundary. A formulation of the problem in terms of these variables can be obtained by considering the (dimensional) mass internal to radius R_{ext}:

$$M(R_{ext}) = \int_0^{R_{ext}} 4\pi\rho(r)r^2 dr, \tag{6.26}$$

and expressing it in terms of the nondimensional variables defined in Equation (6.22) to obtain

$$M(\xi_{ext}) = 4\pi\rho_c\left(\frac{c_s^2}{4\pi G\rho_c}\right)^{3/2}\int_0^{\xi_{ext}} e^{-\psi}\xi^2 d\xi. \tag{6.27}$$

We next note that, integrating the Emden–Chandrasekhar equation (Equation (6.23)), we have

$$\int_0^{\xi_{ext}} e^{-\psi}\xi^2 d\xi = \left(\xi^2\frac{d\psi}{d\xi}\right)_{\xi_{ext}}, \tag{6.28}$$

so that the mass within radius $R_{\rm ext}$ for the BE sphere, in terms of the nondimensional variables, is

$$M(\xi_{\rm ext}) = 4\pi\rho_{\rm c}\left(\frac{c_{\rm s}^2}{4\pi G\rho_{\rm c}}\right)^{3/2}\left(\xi^2\frac{d\psi}{d\xi}\right)_{\xi_{\rm ext}}. \tag{6.29}$$

We can now define a corresponding nondimensional mass as

$$m(\xi_{\rm ext}) \equiv \frac{P_{\rm ext}^{1/2}G^{3/2}M(\xi_{\rm ext})}{c_{\rm s}^4} = \left(4\pi\frac{\rho_{\rm c}}{\rho_{\rm ext}}\right)^{-1/2}\left(\xi^2\frac{d\psi}{d\xi}\right)_{\xi_{\rm ext}}, \tag{6.30}$$

so that the mass of the truncated BE sphere can be computed from the numerical solution of the Emden–Chandrasekhar equation. In particular, it is found that the critical nondimensional mass ($m_{\rm crit}$) that corresponds to the critical ratio, $(\rho_{\rm c}/\rho)_{\rm crit} \approx 14.1$, is $m_{\rm crit} \approx 1.18$. The corresponding dimensional value of the mass is known as the *Bonnor–Ebert mass*, and is given by

$$M_{\rm BE} \equiv \frac{m_{\rm crit}c_{\rm s}^4}{P_{\rm ext}^{1/2}G^{3/2}} = \frac{m_{\rm crit}c_{\rm s}^3}{G^{3/2}\rho_{\rm ext}^{1/2}}. \tag{6.31}$$

The BE mass can be compared to the Jeans mass, Equation (6.16), which is similar, but larger by a factor ~2.5 when evaluated at the same density and temperature. Both expressions give a critical mass for stability, but the Jeans mass is larger because it corresponds to a uniform-density region at the same density as its surroundings. Instead, the BE mass corresponds to a centrally-peaked region, for which the gravitational potential is deeper than for a uniform-density configuration of the same mean density and radius (e.g., Bertoldi & McKee 1992). Therefore, a smaller mass is sufficient to destabilize the structure. Since most realistic situations will generally involve centrally-peaked regions, the BE mass is probably a more accurate criterion for stability, although the Jeans mass nevertheless provides a useful reference estimate and is used widely.

6.2.5 Understanding the Structure of the BE Sphere

6.2.5.1 From Pressure Confinement to Self-Gravitating Confinement and Beyond: The Need for a Diffuse *Confining Medium*

Equations (6.30) and (6.31) show that the BE mass depends both on the external pressure and on the central density. This means that the transition from stability to instability of a BE sphere can be triggered either by increasing the external pressure on a sphere of constant mass M (thus decreasing its radius and its corresponding value of $M_{\rm BE}$ until $M_{\rm BE} < M$, as depicted in Figure 5.1), or else by increasing its mass (that is, increasing its central density) at constant radius until $M > M_{\rm BE}$. The latter process is a convenient way to understand the behavior of BE spheres.

Let us imagine a rigid container with a spherical central cavity of radius R_0, into which gas can be pumped to any pressure, as illustrated in Figure 6.2. The rigid wall can exert any pressure required to confine the gas within the cavity.

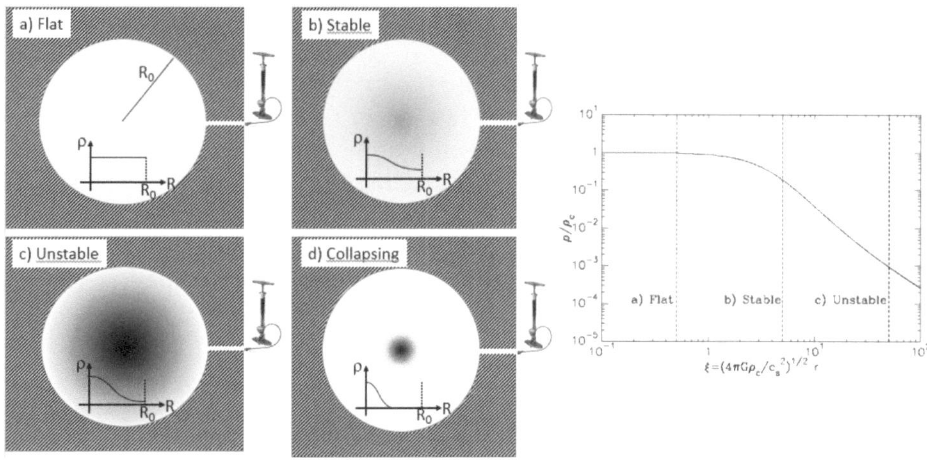

Figure 6.2. *(a)–(d):* Schematic illustration of the progression from stable to unstable BE spheres presented as a laboratory experiment in which mass is pumped into a rigid cavity of fixed radius that can always exert the necessary pressure to keep the gas from expanding. See text for a description of the various states. The state in panel *(d)* is not hydrostatic, but rather collapsing. *Right:* Plot of the radial density profile for the BE sphere in nondimensional units.

When the gas mass and density in the cavity are low, the self-gravity of the gas is negligible, and therefore it is confined exclusively by the pressure exerted by the cavity wall. In this case, the density is uniform throughout the cavity (equivalently, the density profile from the center of the cavity is flat), as is the case, for example, of any spherical parcel of air in the room you are now. This is illustrated in Figure 6.2(a). The external pressure needed to confine this sphere equals the (uniform) internal pressure of the sphere.

If we keep pumping gas into the cavity, its mass and mean density increase, and the self-gravity of the gas begins to be noticeable. In this case, the central density and pressure of the gas sphere must begin to increase, in order to compensate for the weight of the outer parts (Figure 6.2(b)). Nevertheless, as long as the density contrast between the central density and that at the boundary is less than the critical value, $[\rho_c/\rho(R_0)]_{\rm crit} \approx 14.1$, the sphere remains stable, in the sense discussed in Section 5.2.4, that if $P_{\rm ext}$ were increased at constant mass and the radius were allowed to vary, the sphere would settle onto a new, smaller radius, for which the new, larger $P_{\rm ext}$ is the necessary value for the sphere to remain in equilibrium. This corresponds to the negative-slope part of the $P_{\rm ext}$ *vs.* R curve in Figure 5.1.

However, further pumping of gas into the cavity will cause the mass and mean density of the sphere to continue increasing, until the ratio $\rho_c/\rho(R_0)$ exceeds the critical value of 14.1. At this point, the self-gravity of the gas becomes dominant in confining the sphere, which then becomes unstable (Figure 6.2(c)). Now, if $P_{\rm ext}$ is increased at constant mass, the radius will decrease, but the new external pressure required to maintain the sphere in equilibrium is *lower*, because the self-gravity is even more dominant over $P_{\rm ext}$. Since the actual $P_{\rm ext}$ is larger, the sphere will then undergo runaway contraction. This regime corresponds to the positive-slope, positive-$P_{\rm ext}$ part of the curve in Figure 5.1.

Finally, if we continue pumping gas into the cavity, we will reach the point where the external pressure required to confine the sphere is zero; that is, the sphere is now in hydrostatic equilibrium between its self-gravity and internal pressure gradient exclusively. This corresponds to a sphere of radius $R(P_{\rm ext} = 0)$ in Figure 5.1. Adding any more mass to the cavity will render its internal pressure incapable of supporting the cloud's weight, and so the sphere will proceed to collapse (Figure 6.2(d)).

Note that this latter point reiterates the impossibility of confining a BE sphere with material at the same temperature as the sphere, because in this case the "confining" material adds mass to the BE sphere and brings it to the unstable (or fully collapsing) regime. The only possible way to confine a BE sphere and maintain it in stable configuration is *to apply pressure without adding weight*, either by placing the sphere in a solid container as in the idealized laboratory experiment described above (but impossible in practice in the ISM because there are no rigid containers for molecular clouds), or by providing the pressure via a much hotter and rarefied gas (a warmer gas phase) that is indeed capable of exerting the pressure without adding any significant mass.

6.2.5.2 The Physical Origin of the Different Slopes in the Inner and Outer Parts of the BE Sphere

Another way to view the transition from pressure confinement to gravitational confinement, and the physical origin of the different slopes of the density profile in the inner and outer regions of the BE sphere, is by considering the force balance at increasing radii (ξ) from the center. This can be compared to the discussion in the previous section, in terms of an increasing mass within a fixed-radius, truncated BE sphere. To this end, we can refer to Figure 6.2, and consider the equivalence between moving outwards from the center in the right panel, and increasing the mass at a fixed radius. This is possible because of the nondimensional nature of the nondimensional variable $\xi \equiv (4\pi G\rho_c/c_s^2)^{1/2}r$, which allows interpreting it as the radial coordinate at fixed central density, or as the mass (measured by the central density) within a fixed radius.

The inner, uniform-density region labeled *(a)* of the density profile in the right panel of Figure 6.2, is equivalent to the low-mass BE sphere depicted in panel *(a)* of Figure 6.2. As discussed in Section 6.2.5.1, the self-gravity of such a low-mass sphere is unimportant. Therefore, this region is confined only by the thermal pressure gradient. Since no other force is active in this region, the only way for the sphere to remain in equilibrium is for the thermal pressure gradient to vanish, implying a uniform thermal pressure. Since, for our isothermal sphere, the thermal pressure depends only on density, the density must be uniform as well.

As we move outwards, into the region labeled *(b)* in the density profile, the self-gravity of the material internal to radius ξ begins to become important in confining the sphere, and therefore, the external pressure required to maintain the sphere in equilibrium begins to decrease. As a consequence, the density profile begins to bend downwards. In this region, thus, both the sphere's self-gravity and the external pressure contribute to the confinement.

Moving further outwards, into region *(c)*, the self-gravity of the sphere out to that radius becomes dominant over the external pressure, meaning that the density must now decrease with radius as required to maintain balance between the internal pressure gradient and the self-gravity at every radius. That is, at large radii, where the density profile approaches a power-law, the BE sphere can be thought of as having *one Jeans mass at each radius.*

The necessary slope of the density profile for this to happen can be derived by considering Equation (6.16) for the Jeans mass. This can be compared to the total mass of the sphere as a function of radius. At large radii, we can approximate the BE sphere as simply having a power-law radial density distribution of the form

$$\rho(r) = \rho_0 \left(\frac{r}{r_0} \right)^{-p}, \tag{6.32}$$

where ρ_0 and r_0 are normalization constants, and p is the *logarithmic slope* of the density profile. Then, the mean density of the sphere out to radius r is

$$\langle \rho \rangle_r = \frac{1}{V} \int_0^V \rho dV' = \frac{3\rho_0}{3-p} \left(\frac{r}{r_0} \right)^{-p}. \tag{6.33}$$

It is noteworthy that the mean density up to radius r for a power-law density profile scales with the same power of the radius as the density profile itself. The total mass within radius r is then

$$M(r) = \frac{4\pi}{3} \langle \rho \rangle_r r^3, \tag{6.34}$$

We wish to compare this mass with the Jeans mass for the mean density for the same density profile up to that radius, which, from Equation (6.16), we write as

$$M_J(r) = \frac{\pi^{5/2}}{6} \frac{c_s^3}{G^{3/2} \langle \rho \rangle_r^{1/2}}. \tag{6.35}$$

Equating Equations (6.34) and (6.35), we find the condition

$$\frac{3}{4\pi} \frac{\pi^{5/2} c_s^3}{6 G^{3/2}} r^{-3} = \langle \rho \rangle_r^{3/2} = \left(\frac{3\rho_0 r_0^p}{3-p} \right)^{3/2} r^{-3p/2}. \tag{6.36}$$

Equating the exponents and coefficients of r in the above equation, we finally find

$$p = 2; \qquad \rho_0 = \frac{\pi c_s^2}{4G r_0^{-2}}, \tag{6.37}$$

which implies that the required slope for $M(r)$ to equal M_J at all radii is $p = 2$. This is indeed the slope approached by the BE sphere at large radii. That is, the profile slope $p = 2$ causes the Jeans mass to increase with radius at the same rate as the total mass inside that radius, so that equilibrium between internal pressure and self-gravity can be maintained at all (large) radii for the BE sphere. The balance between self-gravity and the thermal pressure gradient at each radius is often referred to as a *detailed mechanical balance* condition (e.g., Shu 1977).

It should be noted, however, that, as shown in Section 5.2.3, the pure equilibrium between self-gravity and the internal pressure gradient is *unstable* for a polytropic sphere with polytropic exponent $\gamma_e < 4/3$. Therefore, in particular, the isothermal sphere (with $\gamma_e = 1$) we are considering here is unstable when it is confined by its self-gravity only (i.e., when the external pressure required for equilibrium has become negligible compared to the self-gravity). This is the reason why a very extended BE sphere (with $\rho_c/\rho(R_0) > [\rho_c/\rho(R_0)]_{\rm crit} \approx 14.1$) becomes unstable, albeit it is still an equilibrium configuration.

6.3 Hydrostatic Equilibrium of Infinite Isothermal Cylinders

During the second decade of the 21st century it became widely recognized that molecular clouds contain a complex network of interconnected filaments, motivating a rebirth of the interest in both hydrostatic and dynamical cylindrical configurations studied previously. In this section we briefly recall (without derivation) the main results concerning equilibrium in cylindrical symmetry.

Employing a treatment analogous to that in Section 6.2.1 but using cylindrical coordinates and assuming an isothermal cylinder of infinite length, Ostriker (1964) found that, in this case, the equilibrium radial density profile has the form

$$\rho = \frac{\rho_c}{\left(1 + \frac{1}{8}\xi^2\right)^2}, \tag{6.38}$$

where ξ is the nondimensional radius, defined as in the second of Equations (6.22), and ρ_c is the density of the filament at its central axis. Note that the density is assumed to be uniform in the longitudinal direction (i.e., along the cylinder's axis).

The following features are noteworthy about solution (6.38):

1. At large radii from the cylinder's axis, the density profile falls off as r^{-4}, while at small radii, the density is nearly flat. Thus, the density profile is qualitatively similar to that of a BE sphere, but with a large-radius falloff as r^{-4}, instead of r^{-2}.
2. Since the cylinder is infinite in the axial direction, its total mass is also infinite. However, the mass *per unit length* (the *linear mass density*) remains finite at all radii, as can be readily seen by integrating Equation (6.38) to $\xi \to \infty$ over a unit-length segment of the cylinder. The linear mass density of the equilibrium isothermal cylinder is

$$M_{\rm line,eq} = \frac{2c_s^2}{G} \approx 16\ M_\odot {\rm pc}^{-1}, \tag{6.39}$$

where the second approximate equality applies for molecular gas at a temperature ~10 K. A cylinder with a linear mass density larger than the value given by Equation (6.39) is therefore unstable to gravitational collapse in the radial direction.

3. The assumption of infinite length in the axial direction is clearly impossible to fulfill in nature, and therefore, this solution is also not expected to occur in nature, similarly to the case of the SIS. In practice, if the cylinder is finite, cancellation of the gravitational pull from the positive- and negative-direction parts of the filament can only occur at its middle point, and the filament will also contract along the longitudinal direction, unless it is accreting material from its environment, as will be discussed in Section 11.4.

6.4 Fragmentation, Growth Rates, and Geometry

The ability of a structure to fragment into smaller units once it has become unstable is a fundamental requirement in order to form ever smaller structures, allowing for the possibility of eventually forming objects as small as stars out of initially much larger and massive structures such as molecular clouds.

The ability of an unstable structure to fragment results from the different speeds at which perturbations of different wavelengths can grow. When the perturbations are linear (i.e., of small amplitude around the unperturbed state), the growth rate for each wavelength is determined by the dispersion relation. On the other hand, when the perturbations are of large amplitude (i.e., nonlinear), the growth rate must be inferred directly from the new physical conditions of the perturbation, and may be independent of its wavelength (i.e., size scale).

An important property of the dispersion relation is that it is modified when different geometries are considered, or when the contraction is restricted to occur in certain directions only. In this section we first summarize a number of cases of linear growth relevant for the study of the structures within molecular clouds. Larson (1985) has provided a comprehensive summary of the structure and stability of a variety of geometrical shapes. Afterwards, we briefly discuss the case of nonlinear perturbations.

6.4.1 The Uniform-Density Sphere

A perhaps somewhat surprising property of the standard Jeans analysis for a uniform medium is that its dispersion relation, Equation (6.12), implies that the growth rate $|\omega|$ of unstable perturbations (those for which ω is imaginary) approaches its maximum as $k \to 0$, or equivalently, for $\lambda \to \infty$. That is, for a finite-size object, perturbations at the scale of the object itself grow fastest. This has often been interpreted as implying that a collapsing sphere will not fragment, since the growth of the largest scales overwhelms that of the smaller scales, which then cannot ever "protrude" from the large ones (e.g., Layzer 1963; Tohline 1980).

6.4.2 Infinitely Thin Isothermal Sheet

A first case of interest is that of an infinitely thin sheet, characterized by its surface density Σ. This is relevant for molecular clouds because there exists significant evidence that they may be flattened in general (cf. Section 8.2.1). In this case, the stability analysis yields a dispersion relation of the form (Larson 1985)

$$\omega^2 = c_s^2 k^2 - 2\pi G \Sigma k. \tag{6.40}$$

As in the case of the standard Jeans analysis, there exists a critical wavenumber,

$$k\mathrm{c} = \frac{2\pi G \Sigma}{c_s^2}, \tag{6.41}$$

below which ω is imaginary, and the perturbation grows exponentially. The corresponding length scale is

$$L_{\mathrm{J,2D}} = \frac{c_s^2}{G\Sigma}. \tag{6.42}$$

However, an important distinction with the Jeans case is that for the thin sheet the maximum growth rate occurs at the nonzero wavenumber $k_{\max} = k\mathrm{c}/2$, as can be easily checked by differentiating Equation (6.40) and setting $d\omega/dk = 0$. That is, for the thin-sheet case, the scale $\lambda = 2L_{\mathrm{J,2D}}$ grows fastest, and thus the original unstable region will tend to fragment into objects of half the original size. Once the fragments become nonlinear, they will in turn fragment into new fragments of half their size, and so on.

6.4.3 Dependence of the Critical Polytropic Exponent for Stability of Hydrostatic Equilibrium on the Dimensionality of Contraction

In Section 5.2.3, using the VT, we showed that the hydrostatic equilibrium of a polytropic uniform-density sphere, for which $P = P_0(\rho/\rho_0)^{\gamma_e}$, is stable if $\gamma_e > 4/3$ and unstable otherwise. It is important to note, however, that this result implicitly assumed that the contraction occurs isotropically in three dimensions. Instead, if the contraction is assumed to occur in just one or two dimensions (to form a dense sheet or filament, respectively), then the critical value of the polytropic exponent is different. A simple way to address this problem (Vazquez-Semadeni et al. 1996) is to consider the Jeans length, given by Equation (6.14), noting that in the polytropic case the sound speed is given by

$$c_s^2 = \frac{dP}{d\rho} = \frac{P_0}{\rho_0^{\gamma_e}} \gamma_e \rho^{\gamma_e - 1} = \frac{\gamma_e P}{\rho}. \tag{6.43}$$

Then, substituting this in Equation (6.14), we find

$$L_{\mathrm{J}}(\rho_0) = \left(\frac{\pi \gamma_e}{G} \frac{P_0}{\rho_0^{\gamma_e}} \rho^{\gamma_e - 2} \right)^{1/2} \propto \left(\frac{M}{V} \right)^{\frac{\gamma_e - 2}{2}}, \tag{6.44}$$

where we have written $\rho = M/V$, with M and V being respectively the mass (assumed fixed) and volume of the cloud.

Let us now assume that the cloud's volume is given by $V \propto L_0^{3-n} L^n$, where L_0 is a length that remains fixed, while L is a length that may vary, so that the cloud contracts in n dimensions only. Therefore, the proportionality in expression (6.44) becomes

$$L_J(\rho_0) \propto L^{-\frac{n(\gamma_e - 2)}{2}}, \tag{6.45}$$

which gives the dependence of the Jeans length on the size of the cloud as it contracts in n dimensions. If the exponent of L is negative, then $dL_J/dL < 0$, and L_J increases as L decreases during a contraction. Thus, if a cloud starts in equilibrium, with $L_J = L$, then a reduction in the cloud's size implies that L_J becomes larger than L, and so the equilibrium is stable. The converse is true when $-n(\gamma_e - 2)/2 > 0$, and the equilibrium is then unstable.

The critical value of the polytropic exponent is obtained by requiring that $L_J(\rho_0)$ remains stationary upon perturbations in L—i.e., by setting the exponent in Equation (6.45) equal to zero. We thus find

$$\gamma_{e,cr} = 2\left(1 - \frac{1}{n}\right). \tag{6.46}$$

In particular, we obtain the following values for $n = 1$, 2, and 3:

$$\gamma_{e,cr} = \begin{cases} \frac{4}{3} & \text{if } n = 3; \\ 1 & \text{if } n = 2; \\ 0 & \text{if } n = 1. \end{cases} \tag{6.47}$$

A number of points can be noted in regards to these results:

- For a contraction in three dimensions we recover the value $\gamma_{e,cr} = 4/3$ we obtained applying the VT to a uniform sphere in Section 5.2.3.
- For contractions in two dimensions (i.e., in the radial direction in cylindrical geometry, which may be relevant for the formation of filamentary structures), the gas must be isothermal or "softer" (i.e., with $\gamma_e < 1$) in order for the equilibrium to be unstable.
- For one-dimensional compressions (such as those producing thin sheets), only thermally unstable gas, with $\gamma_e \leqslant 0$, can undergo gravitational instability. This is, for example, the reason why a magnetically subcritical cloud, which cannot contract significantly in the direction perpendicular to the magnetic field, eventually reaches a hydrostatic configuration in the direction along the field and cannot collapse (e.g., Shu et al. 1987), unless it accretes mass along the field to eventually become supercritical (cf. Section 12.3.2.2).
- As in Section 5.2.3, the treatment is only approximate, since it assumes a structure of fixed mass and a well-defined boundary. As discussed in Section 12.3.2.2, these assumptions are in general not satisfied in real clouds. Nevertheless, Equation (6.47) provides useful guidelines for understanding the ability of compressions in n dimensions to induce gravitational collapse.

6.5 Conclusions

In this chapter we have covered a series of topics related to gravitational equilibrium and stability in dense clouds. The main often-neglected conclusions to be drawn are:

- The dispersion relations for gravitational instability for 3D and 2D (planar) configurations differ fundamentally in that the maximum growth rate in 3D occurs at the largest scale of the system, while in 2D it occurs at twice the Jeans length. This implies that gravitational fragmentation can readily occur in flattened clouds.
- Isothermal BE spheres can only be stable when the ratio of their central to peripheral density satisfies $\rho_c/\rho_{per} \lesssim \rho_c/\rho(\xi_{crit}) \approx 14.1$, where $\xi_{crit} \approx 6.5$ is the critical nondimensional radius for stability. Hydrostatic BE spheres with larger density contrasts than this are *not expected to be observed, as they are unstable.*
- In order for a BE sphere to be stable, *it must be truncated at a nondimensional radius smaller than* ξ_{crit}. Here, by "truncated" it is meant that *the mass beyond that radius must be negligible*, since otherwise the sphere is equivalent to a BE sphere with a larger radius, and thus unstable. The truncation can only accomplished in reality by a phase transition to a warmer, diluted medium.
- Therefore, a BE sphere cannot be bounded and stabilized by the weight of the molecular cloud around it, as the system of the sphere plus the cloud is equivalent to a larger, unstable BE sphere, which must be collapsing.
- The SIS is the most unstable member of the family of BE spheres, and thus an already-contracting sphere has a vanishing probability of settling onto it. Therefore, the SIS must be considered an idealized configuration of pedagogical interest, but which cannot occur in the real ISM.
- Radially hydrostatic cylinders are also idealized configurations, as longitudinal collapse can only be prevented in cylinders with an infinite longitudinal extension, or in numerical simulations with periodic boundary conditions. Otherwise, the cylinder must be contracting longitudinally.
- The critical value of the polytropic exponent for the stability of thermal support against self-gravity depends on the dimensionality of the allowed gravitational contraction, and is given by Equation (6.47).

References

Bertoldi, F., & McKee, C. F. 1992, ApJ, 395, 140

Bonnor, W. B. 1956, MNRAS, 116, 351

Ebert, R. 1955, ZA, 37, 217

Jeans, J. H. 1902, RSPTA, 199, 1

Kiessling, M. K. H. 1999, Mathematical Vindications of the "Jeans Swindle", *arXiv e-prints*, pages astro–ph/9910247

Larson, R. B. 1985, MNRAS, 214, 379

Layzer, D. 1963, ApJ, 137, 351

Ostriker, J. 1964, ApJ, 140, 1056

Shu, F. H. 1977, ApJ, 214, 488
Shu, F. H., Adams, F. C., & Lizano, S. 1987, ARA&A, 25, 23
Tohline, J. E. 1980, ApJ, 239, 417
Vazquez-Semadeni, E. 1994, ApJ, 423, 681
Vazquez-Semadeni, E., Passot, T., & Pouquet, A. 1996, ApJ, 473, 881

AAS | IOP Astronomy

Interstellar Flow and Star Formation

Enrique Vázquez-Semadeni

Chapter 7

Self-Gravity II: Gravitationally Driven Collapse Flow and Gravitational Fragmentation

7.1 What We Mean by "Collapse"

In this chapter we now discuss many facets of gravitational collapse. However, it is important to note that the term "collapse" can be interpreted in at least two significantly different ways. One of them is its more traditional interpretation as the contraction of a well-defined "object" to a singularity (a point of diverging density, corresponding in our case to the formation of a star). This interpretation can be referred to as an "event", and is often thought of as lasting a relatively short time, known as the *free-fall time*, as in the statement "a dense molecular cloud core collapses to form stars in a free-fall time."

The other interpretation refers to a certain (possibly long) stretch of time during which a *flow* is set in motion by the action of gravity, sliding down a gravitational potential well, and that may last as long as there is a supply of mass at low densities. This interpretation can be referred to as a "process", as in the flow of water downhill from the mountains and along a river all the way to a lake or the ocean, which may last as long as an entire rain season. In this chapter, we will first discuss some classical concepts that pertain to the "event" view, but, as we advance, we will gradually turn to the "process" view.

With this preamble in mind, in this chapter we will then first discuss some basic concepts relating to spherical free-fall (i.e., unimpeded) collapse, and then we will proceed to more involved settings, such as the collapse in the presence of an opposing pressure gradient, and the collapse of non-spherical objects. We will also discuss the growth of cores in the presence of a continuous mass supply, gravitational fragmentation, and the asynchronicity of the onset of collapse of different levels of the fragmentation hierarchy.

doi:10.1088/2514-3433/adcb28ch7

7.2 Pressureless Gravitational Collapse and Collapse Timescales

In this section we will consider the pressureless approximation, to estimate the collapse timescale for finite objects of various shapes and dimensionalities. It is important to note that, contrary to what might be thought at first sight, the assumption of pressureless collapse can be a useful first-order approximation for the collapse of gaseous parcels in which thermal pressure *is* present, as long as the parcel masses are much larger than their mean Jeans mass. This is because a cloud containing precisely one Jeans mass is, by construction, in balance between the pressure gradient and its self-gravity. On the contrary, in a cloud of mass much larger than its Jeans mass (equivalently, of diameter much larger than the Jeans length), self-gravity overwhelms the pressure gradient at all scales larger than the Jeans length. Therefore, the collapse of multi-Jeans mass clouds can be approximated as pressureless down to the Jeans length.

7.2.1 Spherical, Homologous Collapse and the Free-Fall Time

The simplest case of collapse is that of a uniform, spherical mass distribution, although, as is often the case, the mathematically simplest case is not one that arises in reality. So, it is important to realize that this mode of collapse should *not* be expected to occur in actual interstellar clouds. On the other hand, it provides a useful starting point for the study of more realistic configurations.

Under the assumption of spherical symmetry, we can take advantage of the well-known fact that a point mass located at a certain radius R in the sphere is subject only to the gravitational pull of the mass interior to R, and the force is directed toward the center of the sphere. Therefore, a first consequence is that the spherical geometry is conserved during the collapse (although any perturbations will cause the collapsing object to deviate from the spherical geometry, as will be discussed in Section 7.2.2).

We can now compute the infall speed and the time to reach the collapse center of a fluid parcel located at some radius r in the sphere. We consider this radius to be a Lagrangian coordinate, so that it follows the collapse flow. In this case, the infall speed can be written as a function of r as

$$u(r(t)) = \frac{dr}{dt} < 0, \tag{7.1}$$

where the inequality denotes the facts that the radius is decreasing and the velocity is directed inwards.

In addition, using Newton's law of gravity, the acceleration is

$$\frac{d^2r}{dt^2} = \frac{du}{dt} = -\frac{GM}{r^2}, \tag{7.2}$$

where M is the mass of the sphere, which remains constant through the collapse because r is a Lagrangian coordinate. Also, ρ is the instantaneous density of the sphere which, as we will see, remains uniform within the sphere, albeit not constant in time.

We now note that

$$\frac{d^2r}{dt^2} = \frac{d}{dt}\frac{dr}{dt} = \frac{d}{dt}(u(r(t))) = \frac{du}{dr}\frac{dr}{dt} = u\frac{du}{dr} = \frac{1}{2}\frac{du^2}{dr}, \tag{7.3}$$

where we have applied the chain rule for derivatives in the fourth equality. Thus, the equation of motion, Equation (7.2), can be written as

$$\frac{1}{2}du^2 = -\frac{GM}{r^2}dr.$$

Integrating this equation with the initial conditions $r(t = 0) = r_0$, $u(r) = 0$, and $M = 4\pi\rho_0 r_0^3/3$, where $\rho_0 \equiv \rho(t = 0)$, we obtain

$$u = -\left[2GM\left(\frac{1}{r} - \frac{1}{r_0}\right)\right]^{1/2}, \tag{7.4}$$

where we have taken the negative square root because $u \leqslant 0$. Then, using Equation (7.1), we can write

$$dt = -\frac{dr}{\left[\frac{2GM}{r_0}\left(\frac{r_0}{r} - 1\right)\right]^{1/2}}. \tag{7.5}$$

Introducing the variable $\xi = r/r_0$, the above equation becomes

$$dt = -\left(\frac{2GM}{r_0}\right)^{-1/2}\frac{r_0 d\xi}{\left(\frac{1}{\xi} - 1\right)^{1/2}} = -\left(\frac{8\pi G\rho_0}{3}\right)^{-1/2}\left(\frac{\xi}{1 - \xi}\right)^{1/2}d\xi. \tag{7.6}$$

This equation can now be integrated from $r = r_0$ to $r = 0$ to give the total time it takes for the sphere to complete its collapse (reach zero radius), known as the *free-fall time*:

$$\tau_{\rm ff} = -\left(\frac{8\pi G\rho_0}{3}\right)^{-1/2}\int_{\xi=1}^{0}\left(\frac{\xi}{1 - \xi}\right)^{1/2}d\xi. \tag{7.7}$$

Using the substitution $\xi = \sin^2\theta$, the integral is found to equal $-\pi/2$, so that we finally write the free-fall time as

$$\tau_{\rm ff} = \left(\frac{3\pi}{32G\rho_0}\right)^{1/2} \qquad \text{(Free–fall time)}. \tag{7.8}$$

7.2.1.1 Caveats

The free-fall time derived above is seen to depend only on the density, and not on the size of the sphere. Therefore, *for a spherical cloud of uniform density*, the material at all radii must reach the center at the same time. This form of collapse is known as *homologous* or *monolithic*.

Note that, although this derivation is standard in astrophysics, the free-fall time has been calculated assuming a spherical and uniform distribution. This is an extremely idealized configuration that is not expected to occur in clouds, and therefore *homologous collapse is not expected to occur in real clouds, either*. In fact, the collapse of arbitrarily large uniform clouds would imply that they would have to reach arbitrarily large infall speeds (e.g., larger than the speed of light!) in order to reach the center in one free-fall time. In reality this does not occur for two main reasons:

1. An equilibrium self-gravitating structure—a likely initial condition for the collapse—is not expected to have a uniform density, but rather is expected to be centrally peaked (cf. Section 6.2). The collapse of such a structure is therefore expected to occur *over an extended period of time*, because the outer, lower-density material has a longer free-fall time than the inner, denser material, and therefore arrives to the collapse center (the singularity) at later times. That is, a realistic collapse is expected to develop a gravity-driven *accretion (or collapse) flow*, with gas continuously arriving to the singularity over an extended period of time, rather than occurring as a single collapse *event*, in which all the material would reach the singularity at the same time. The duration of the collapse flow may be estimated to first order as the range of free-fall times implied by the range of densities present in the initial system. For example, if a clump has a central density a hundred times larger than the density at its periphery, the outer parts are expected to reach the center at a time ten times larger than the inner parts.
2. When a cloud begins to collapse, it is generally only marginally out of equilibrium, and so the collapse is not expected to proceed unimpeded. Instead, especially during the early stages of the collapse, the forces opposing gravity almost cancel it, and the collapse proceeds rather slowly. For example, in his pioneering simulations of spherical collapse, Larson (1969, Appendix C) reported that the actual duration of the collapse timescale was $\sim$1.6 times the free-fall time.

In view of the above, the free-fall time should be considered as a *lower bound* to the actual *stretch of time* over which realistic clouds collapse.

7.2.2 Non-Spherical Collapse: Anisotropic Collapse of a Pressureless Triaxial Ellipsoid: Sheet and Filament Formation

The next level of approximation after relaxing the spherical geometry assumption is to assume that the clouds are triaxial ellipsoids and that they are pressureless. This mode of collapse is a direct extension to the case of the pressureless sphere (Section 7.2.1), but for a triaxial ellipsoid, allowing the cloud to maintain an ellipsoidal shape and a uniform internal density, with both the uniform density and the eccentricity of the ellipsoid increasing over time, as we now discuss, following Lin et al. (1965).

For simplicity, we consider an ellipsoid whose axes are parallel to the coordinate axes. In order for the density to remain uniform and the shape to remain triaxial throughout the evolution, it is necessary that the coordinates of any point mass on the spheroid initially at position (x_0, y_0, z_0) evolve through *homologous* transformations, so that at time t the position of the point mass is

$$x = x_0X(t), \qquad y = y_0Y(t), \qquad z = z_0Z(t), \tag{7.9}$$

where $X(0) = Y(0) = Z(0) = 1$, and the uniform density inside the ellipsoid is

$$\rho(t) = \frac{\rho_0}{X(t)\ Y(t)\ Z(t)}, \tag{7.10}$$

ρ_0 being the density at $t = 0$.

Now, the gravitational potential inside a uniform ellipsoid is (Lyttleton 1953)

$$\varphi = -\pi G\rho(\alpha x^2 + \beta y^2 + \gamma z^2 - \delta), \tag{7.11}$$

where

$$\begin{aligned}
\alpha &= abc\int_0^\infty \frac{d\lambda}{(a^2+\lambda)\Delta},\\
\beta &= abc\int_0^\infty \frac{d\lambda}{(b^2+\lambda)\Delta},\\
\gamma &= abc\int_0^\infty \frac{d\lambda}{(c^2+\lambda)\Delta},\\
\delta &= abc\int_0^\infty \frac{d\lambda}{\Delta},\\
\Delta^2 &= (a^2+\lambda)(b^2+\lambda)(c^2+\lambda),
\end{aligned} \tag{7.12}$$

and a, b, and c are constants such that $a\colon b\colon c = 1\colon X/Y\colon Y/Z$ are the axis ratios. Therefore, the equations of motion of the point mass are

$$\frac{d^2x}{dt^2} = -2\pi G\rho\alpha x, \qquad \frac{d^2y}{dt^2} = -2\pi G\rho\beta y, \qquad \frac{d^2z}{dt^2} = -2\pi G\rho\gamma z, \tag{7.13}$$

and can be written in terms of the nondimensional variables X, Y, Z as

$$\frac{d^2X}{dt^2} = -\left(\frac{2\pi G\rho_0}{XYZ}\right)\alpha X, \qquad \frac{d^2Y}{dt^2} = -\left(\frac{2\pi G\rho_0}{XYZ}\right)\beta Y, \qquad \frac{d^2Z}{dt^2} = -\left(\frac{2\pi G\rho_0}{XYZ}\right)\gamma Z, \tag{7.14}$$

with the initial conditions

$$X(0) = Y(0) = Z(0) = 1, \qquad \left.\frac{dX}{dt}\right|_{t=0} = \left.\frac{dY}{dt}\right|_{t=0} = \left.\frac{dZ}{dt}\right|_{t=0} = 0. \tag{7.15}$$

Without loss of generality, we can choose the ellipsoid axes so that $a > b > c$, and so, since the integrands in Equations (7.12) are positive definite, we see that

$$\alpha < \beta < \gamma,$$

implying that

$$-\frac{1}{X}\frac{d^2X}{dt^2} < -\frac{1}{Y}\frac{d^2Y}{dt^2} < -\frac{1}{Z}\frac{d^2Z}{dt^2}. \tag{7.16}$$

This, in turn, implies that the ratios a: b: c steadily increase in time. Therefore, oblate spheroids evolve into thin sheets, while prolate spheroids and elliptical sheets evolve into filaments. That is, *pressureless gravitational collapse amplifies anisotropies*. As a consequence, it also amplifies the *aspect ratio* (the ratio of the largest to the shortest axis) of triaxial ellipsoids. This is the mechanism that gives rise, for example, to the so-called *cosmic web*, the network of dark matter filaments observed in numerical simulations of structure formation in the Universe (e.g., Bond et al. 1996).

Note, however, that a gaseous mass that is in equilibrium between thermal pressure support and self-gravity is expected to approach a spherical shape, due to the isotropic nature of thermal pressure (Pascal's Principle). Therefore, we conclude that anisotropic, aspect ratio-enhancing collapse should be expected for clouds *much more massive than their thermal Jeans mass*, for which the thermal pressure can be neglected, except at the (comparatively small) scale of the Jeans length. This is precisely the case of molecular clouds. Moreover, in Section 7.5 we discuss the increase in the number of Jeans masses contained in an isothermal (or more generally, a polytropic cloud with $\gamma_e < 4/3$) during its gravitational collapse, which implies that isothermal contracting clouds should evolve toward becoming increasingly anisotropic (filamentary).

7.2.3 Pressureless Collapse Timescale of Sheets and Filaments

The collapse time for uniform-density, pressureless spheres, the free-fall time (Equation (7.8)), has the remarkable property that it only depends on the size and mass of the structure through the combination M/R^3—i.e., through the structure's volume density. This is actually a peculiar property, as it implies that two uniform spheres of the same density should take the same time to collapse regardless of whether one of them contains barely over one Jeans mass and the other is orders of magnitude much larger than the Jeans mass.

In practice, however, clouds have neither a spherical shape nor a uniform density and, as shown in Section 7.2.2, any deviation from perfect sphericity in clouds strongly dominated by self-gravity should drive them to become increasingly anisotropic. For such clouds, the gravitational contraction timescale is in general longer than the spherical free-fall time for clouds of the same volume density (Pon et al. 2011, 2012; Toalá et al. 2012, 2015). The physical reason for this is simple: a strongly anisotropic cloud (a sheet or a filament) contains significantly less mass than a spherical one with the same density and with a diameter equal to the largest dimension of the anisotropic cloud. Therefore, the gravitational potential well generated by the anisotropic cloud is much shallower than that of the spherical one, thus generating lower gravitational velocities and longer collapse times.

7.2.3.1 Edge-On and Homologous Collapse Modes

The timescale for the radial collapse of a finite sheet or for the longitudinal collapse of a filament can be calculated under the assumption of either an edge-on or a homologous contraction regime, which we now discuss. As first noted by Burkert & Hartmann (2004, hereafter BH04), a *finite*, thin sheet is prone to *edge-on collapse*,[1] a mode in which the edge of the sheet falls in the fastest, effectively "sweeping" the rest of the sheet from the outside in. This happens because, for a finite-extent, infinitely-thin sheet, the gravitational acceleration is largest at the edge. This can be seen by considering the gravitational potential at a point r from the center of an infinitely thin sheet of radius R and surface density Σ, which is given by (Wyse & Mayall 1942)

$$\varphi = -4G\Sigma R E(r/R), \tag{7.17}$$

where $E(x)$ is the complete elliptic integral of the second kind. The radial acceleration is therefore

$$a(r) \equiv -\frac{\partial\varphi}{\partial r} = 4G\Sigma\frac{R}{r}\left[K\left(\frac{r}{R}\right) - E\left(\frac{r}{R}\right)\right], \tag{7.18}$$

where $K(x)$ is the complete elliptic integral of the first kind. BH04 also provide the following series expansion for the above expression:

$$a(r) = \pi G\Sigma\left[\frac{r}{R} + \frac{3}{8}\left(\frac{r}{R}\right)^3 + \frac{45}{192}\left(\frac{r}{R}\right)^5 + \dots\right]. \tag{7.19}$$

The *solid line* in Figure 7.1 shows the gravitational acceleration for a thin sheet as a function of r/R. It can be seen that the acceleration increases almost linearly with radius for $r \lesssim 1/2\ R$ but then it begins to increase nonlinearly, diverging at $r \to R$. BH04 note, however, that the divergence disappears for a finite thickness of the sheet.

For a thin filament of length l and radius h, the acceleration as a function of the longitudinal position z along the filament's axis, with $z = 0$ corresponding to the middle of the filament, is given by (BH04)

$$a(z) = -2\pi G\rho\{2z - [h^2 + (l + z)^2]^{1/2} + [h^2 + (l - z)^2]^{1/2}\}, \tag{7.20}$$

which, under the assumptions that $l \gg h$ and $l - z \gg h$, can be approximated by

$$a(z) \approx -\pi G\rho h^2\left(\frac{1}{l - z} - \frac{1}{l + z}\right). \tag{7.21}$$

The longitudinal acceleration for a thin cylinder as a function of position along its axis is also shown in Figure 7.1 by the *dotted line*, and is also seen to diverge at the edge, although, as in the case of the sheet, BH04 point out that the divergence disappears for a finite thickness h of the filament.

We thus see that, in both the cases of the sheet and the filament, the edge falls in most rapidly, and so the collapse of finite-extent, thin, sheets and filaments is an

[1] Also called *edge effect*.

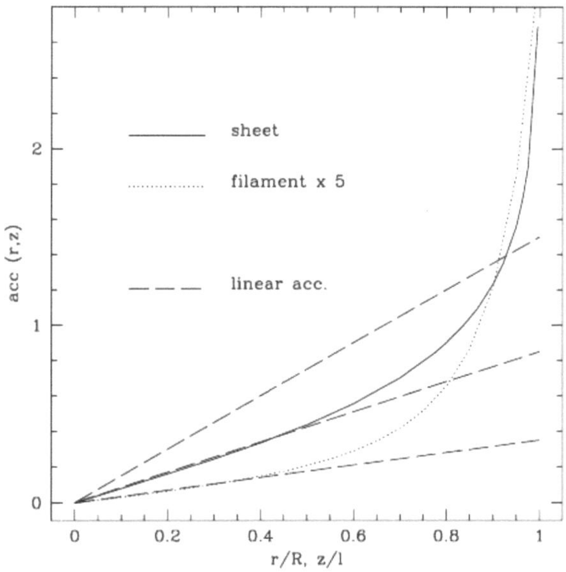

Figure 7.1. Gravitational acceleration in units of $4G\Sigma$ as a function of radius r within a circular, uniform thin sheet of external radius R and surface density Σ (solid line), and as a function of distance z from the midpoint of a finite, uniform filament of total length l and density ρ (dotted line), in units of $G\rho/5$. Reprinted with permission from Burkert and Hartmann (2004).

extreme example of outside-in collapse. Note, however, that the edge-on collapse mode requires the sheet or filament to be isolated and uniform. When the filament or sheet accrete (longitudinally or radially, respectively) from their environment, this behavior is generally not observed (e.g., Heitsch 2013; Gómez & Vázquez-Semadeni 2014; Heigl et al. 2022).

Figure 7.1 also shows that, far from the edge (for radii (for sheets) or longitudinal distances (for filaments) less than roughly half the sheet radius or the cylinder half length), the acceleration is almost linear, and so in these regions the contraction is approximately *homologous*. In this case, the density increases with time, albeit uniformly over the entire sheet or filament.

7.2.3.2 Collapse Timescale of a Circular, Pressureless, Isolated, Thin Sheet

We now calculate the free-fall timescale for a circular, thin sheet of radius $R(t)$, whose initial value is $R_0 \equiv R(t=0)$, following Toalá et al. (2015). We first consider the edge-on assumption. In this case, we can assume that the sheet's interior remains static until the edge sweeps through it. Thus, we assume that the interior surface density Σ remains constant.

As a first-order approximation, we consider Equation (7.19) keeping only the linear term, in the inner regions $r < R$. Noting that, in general, $a(r(t)) = dv/dt = (dv/dr)(dr/dt) = 1/2\ dv^2/dr$, we can then write the equation of motion *for the edge* as

$$\frac{1}{2}\frac{dv^2}{dR(t)} = -\pi G\Sigma.$$

Integrating, we obtain

$$\int_0^{v(R(t))} dv^2 = -2\pi G\Sigma \int_{R_0}^{R(t)} dR,$$

so that

$$v(R(t)) = \sqrt{2\pi G\Sigma R_0\left(1 - \frac{R(t)}{R_0}\right)}. \tag{7.22}$$

We can now introduce the *aspect ratio*, the ratio of the largest to the smallest dimension of the structure (the sheet in this case), given by $A \equiv R_0/h$, where h is the sheet thickness, which moreover provides the relation between the (constant) surface and volume densities, $\Sigma = \rho h$. We can then rewrite Equation (7.22) as

$$v(R(t)) = \sqrt{\frac{2\pi G\rho R_0^2}{A}\left(1 - \frac{R(t)}{R_0}\right)}. \tag{7.23}$$

To obtain an expression for the radius as a function of time, we integrate the previous equation, noting that $v(R(t)) = dR/dt$:

$$-\int_{R_0}^{R(t)} \frac{dR'}{\sqrt{1 - \frac{R'}{R_0}}} = \sqrt{\frac{2\pi G\rho R_0^2}{A}} \int_0^t dt',$$

so that

$$2R_0\sqrt{1 - \frac{R(t)}{R_0}} = \sqrt{\frac{2\pi G\rho R_0^2}{A}}\, t. \tag{7.24}$$

From the previous equation, the collapse time can be calculated as $\tau_{\mathrm{ff,2D}} = t(R = 0)$, obtaining

$$\tau_{\mathrm{ff,2D}} = \sqrt{\frac{2A}{\pi G\rho}} = \sqrt{\frac{64A}{3\pi^2}}\,\tau_{\mathrm{ff}}, \qquad \text{(Edge–on)} \tag{7.25}$$

where τ_{ff} is the standard spherical free-fall time, given by Equation (7.8).

This result implies that *the collapse time of a circular, thin sheet of density ρ depends not only on its density, but also on its aspect ratio A*. This time is thus proportional to $A^{1/2}$. This result can be understood because the mass of a full spherical structure of radius R is much larger than the mass of a circular sheet of the same radius and density, but with thickness $h \ll R$. As a consequence, the gravitational acceleration and infall speed are much smaller as well, and the collapse takes a longer timescale. The difference between the collapse time of a spherical structure and that of a flattened one becomes larger for larger aspect ratios.

A similar calculation, but *assuming homologous collapse*, and therefore a uniformly increasing density, gives a very similar collapse timescale (Toalá et al. 2015):

$$\tau_{\rm ff,2D} = \sqrt{\frac{\pi A}{8G\rho_0}} = \sqrt{\frac{4A}{3}}\,\tau_{\rm ff}. \qquad \text{(Homologous)} \tag{7.26}$$

7.2.3.3 Collapse Timescale of a Finite, Pressureless, Isolated, Uniform Filament
Similar calculations under the same assumptions (edge-on or homologous collapse), using the acceleration corresponding to a filament (Equation (7.20)) give the following longitudinal collapse timescales (Pon et al. 2012):

$$\tau_{\rm ff,1D} = \sqrt{\frac{32A}{\pi^2}}\,\tau_{\rm ff}, \qquad \text{(Edge-on)} \tag{7.27}$$

and

$$\tau_{\rm ff,1D} = \sqrt{\frac{2}{3}}\,A\tau_{\rm ff}, \qquad \text{(Homologous)} \tag{7.28}$$

Here we see that an important difference arises, since the collapse timescale for a filament under homologous collapse is *linearly* proportional to the aspect ratio A, while it is proportional to the *square root* of A under edge-on collapse. Since the edge-on collapse timescale is shorter, and in this mode the filament edges sweep over the interior parts, then we can expect this mode to be dominant *when it develops.* However, as mentioned above, this mode of collapse does not seem to appear in numerical simulations in which the filaments develop self-consistently, since in this case, the filament is just a part of a continuing flow from the cloud scale to the central core scale (Heitsch 2013; Gómez & Vázquez-Semadeni 2014; Heigl et al. 2022). We discuss the results from numerical simulations in Section 11.4.

From the results of this section, we conclude that the collapse timescale of anisotropic structures such as thin sheets and filaments is in general significantly longer than the standard, spherical free-fall time for objects of the same density.

7.3 Non-Homologous, Pressurized Collapse of Isothermal Spheres: Similarity Treatment

In this section we will now forgo the assumption of negligible pressure, and address the gravitational collapse of a spherical, pressurized, isothermal gas cloud, of arbitrary size. This is a problem that, when considered far away from its initial and boundary conditions (i.e., as it approaches an *asymptotic solution*), can be described by means of a *similarity* treatment, as discussed in Section 2.5.4. The self-similarity means that the flow at a certain spatial position and a certain moment in time resembles itself at another position and another time. Because of this, it is convenient to combine the spatial and temporal independent variables into a single variable known as the *similarity variable*. In the case of our collapsing isothermal sphere, the only dimensional physical parameter is the sound speed c_s, and so we can combine the radial and temporal variables into the similarity variable

$$\xi \equiv \frac{r}{c_s t}. \tag{7.29}$$

In addition, the only physical constant relevant to the problem is the gravitational constant G. Therefore, we expect to nondimensionalize the equations using only c_s and G.

We start by writing the continuity and momentum equations, Equations (2.74) and (2.75), in spherical coordinates, and assuming spherical symmetry, so that the only spatial dependence is on the radial coordinate. We obtain

$$\frac{\partial \rho}{\partial t} + \frac{1}{r^2}\frac{\partial}{\partial r}(r^2 \rho u) = 0, \tag{7.30}$$

$$\frac{\partial u}{\partial t} + u\frac{\partial u}{\partial r} = -\frac{c_s^2}{\rho}\frac{\partial \rho}{\partial r} - \frac{GM}{r^2}. \tag{7.31}$$

Given the spherical symmetry, it is convenient to write Equation (7.30) in terms of the mass in spherical shells. This mass is defined by

$$\frac{\partial M}{\partial r} = 4\pi r^2 \rho, \tag{7.32}$$

and thus the continuity equation in terms of the mass is

$$\frac{\partial M}{\partial t} + u\frac{\partial M}{\partial r} = 0. \tag{7.33}$$

We proceed now to the nondimensionalization of Equations (7.31) and (7.33). To do this, as in Section 2.5.2, we write all variables in the form

$$A(r, t) = A_0(t) \quad a(\xi(r, t)), \tag{7.34}$$

meaning that we write the variables as the product of a nondimensional variable a, which depends on radius and time only through the similarity variable ξ, times a characteristic value A_0, which may depend on time. The possible time dependence of A_0 is a manifestation of the fact that the properties of the flow may remain stationary, but at a radial position that varies with time. Noting that the units of the gravitational constant G are $[G]. = L^3 M^{-1} T^{-2} = [\rho]^{-1}[t]^{-2}$, we choose the following nondimensionalization (or *normalization*):

$$\rho(r, t) = \frac{\bar{\rho}(\xi)}{4\pi G t^2}, \qquad M(r, t) = \frac{c_s^3 t}{G} m(\xi), \qquad u(r, t) = c_s v(\xi), \tag{7.35}$$

where $\bar{\rho}$, m, and v are the nondimensional density, mass, and velocity, respectively. It is worth noting that the unit for velocity is, naturally, the sound speed, while for the density, we see that

$$\rho_0(t) = \frac{1}{4\pi G t^2}; \tag{7.36}$$

that is, $\rho_0(t)$ is the instantaneous value of the density for which the current time t is the *collapse time*, given by Equation (6.15), which is of the order of, but not identical to, the free-fall time.

In order to write the equations in terms of the nondimensional variables, we note that, for the generic notation of Equation (7.34) and recalling that A_0 depends only on time, the chain rule for derivatives implies that

$$\frac{\partial A}{\partial t} = a\frac{dA_0}{dt} + A_0\frac{da}{d\xi}\frac{\partial \xi}{\partial t} \tag{7.37}$$

$$\frac{\partial A}{\partial r} = A_0\frac{da}{d\xi}\frac{\partial \xi}{\partial r}. \tag{7.38}$$

Therefore, Equations (7.32) and (7.33) become

$$\frac{dm}{d\xi} = \xi^2\bar{\rho}, \tag{7.39}$$

$$m + (v - \xi)\frac{dm}{d\xi} = 0. \tag{7.40}$$

Combining the last two equations, we obtain an expression for the nondimensional mass within a certain similarity radius ξ:

$$m = \xi^2\bar{\rho}(\xi - v). \tag{7.41}$$

Using Equations (7.38), (7.37), and (7.41), we can rewrite the continuity (7.30) and momentum (7.31) equations as

$$\frac{dv}{d\xi} = \frac{\left[(\xi - v)\bar{\rho} - \dfrac{2}{\xi}\right](\xi - v)}{(\xi - v)^2 - 1}, \tag{7.42}$$

$$\frac{d\bar{\rho}}{d\xi} = \frac{\left[\bar{\rho} - \dfrac{2}{\xi}(\xi - v)\right](\xi - v)\bar{\rho}}{(\xi - v)^2 - 1}. \tag{7.43}$$

That is, we have managed to convert the original *partial* differential equations into *ordinary* ones, in which the only independent variable is the similarity variable ξ.

It is important to note that these equations have *singular points* at $\xi - v = \pm 1$, at which the denominator vanishes. For the solution to remain regular (i.e., to not diverge), it is therefore necessary that at these points the nondimensional density must satisfy $\bar{\rho} = \pm\, 2/\xi$. When this condition is satisfied, no singularity appears, and the flow can remain smooth, and in this case these points are referred to just as *critical* points.

7.3.1 Analytic Solutions

7.3.1.1 The SIS Solution

Equations (7.42) and (7.43) admit both *static* and *dynamic* solutions. The static ones coincide with those obtained in Section 6.2.1 as solutions of the Emden–Chandrasekhar equation, which is equivalent to the equations obtained here, in the special case of $v = 0$. In the following subsections, we review the most relevant solutions for the problem of star formation.

In particular, it is easy to check that the singular isothermal sphere (SIS), already discussed in Section 6.2.1 and written as

$$v = 0, \qquad \bar{\rho} = \frac{2}{\xi^2}, \qquad m = 2\xi, \tag{7.44}$$

is a static solution of Equations (7.42) and (7.43). In dimensional form, it corresponds to

$$u = 0, \qquad \rho = \frac{2c_s^2}{4\pi G r^2}, \qquad M = \frac{2c_s^2 r}{G}. \tag{7.45}$$

The SIS static solution is important as an academic exercise because it has been considered as the "initial condition" for the collapse in some cases (cf. Section 7.3.3), in spite of its being unrealizable in practice (Section 6.2.3).

7.3.1.2 The Infinite, Uniform-Density, Homologously Collapsing or Expanding Sphere

The other known analytical solution is the infinite, uniform-density, homologously collapsing or expanding sphere, given by

$$\bar{\rho} = \frac{2}{3}; \qquad v = \frac{2\xi}{3}. \tag{7.46}$$

Note that, in this case, negative times (i.e., $\xi < 0$) correspond to collapse ($v < 0$), while positive times ($\xi > 0$) correspond to expansion ($v > 0$). That is, the collapsing solution is restricted to the intervals t, $\xi < 0$. Moreover, from Equations (7.29) and (7.35), this solution implies that the physical speed is given by

$$u = \frac{2r}{3t}, \tag{7.47}$$

so that, at any given time, the speed increases linearly with radius, while the physical density is

$$\rho = \rho(t) = \frac{1}{6\pi G t^2}, \tag{7.48}$$

everywhere, so that it only depends on time. This solution highlights the fact that the equations apply for t either positive or negative. Specifically, for this solution, as the time varies in the range $-\infty < t < 0$ (corresponding to collapse), the (spatially uniform) density and the (negative) infall speed increase in magnitude with time, while for $0 < t < +\infty$ (corresponding to expansion), the density and (positive) expansion velocity decrease as time proceeds.

Of course, strictly speaking, this solution is physically impossible, because it extends to infinite size, has an infinite mass, and implies arbitrarily large velocities at arbitrarily large radii or small times (in absolute value). It can only be approached asymptotically as $\xi \to 0^-$, i.e., at small radii during the collapse stage, and in fact we will see in Section 7.3.2 that the collapsing part of this solution is approached by the inner regions of the general solution during the *prestellar* stage of collapse (i.e., before the formation of a protostar). Nevertheless, in order to remain physically plausible, the uniform density profile must end at some radius, and there connect with a decreasing density profile beyond that radius.

7.3.2 The General Non-Homologous Asymptotic Solutions: Inner and Outer Regions, Pre- and Protostellar Stages, and Global vs. Local Collapse

In general, the spherical collapse equations obtained from similarity analysis, Equations (7.42) and (7.43), admit a variety of solutions that depend on the chosen *boundary* conditions—which, given the combined spatial and temporal nature of the similarity variable ξ, can be interpreted as *initial* conditions as well. These solutions must in general be obtained numerically, as the only known analytical solutions are those discussed in Section 7.3.1.

Numerous researchers have investigated the general solutions of Equations (7.42) and (7.43) by means of asymptotic and numerical techniques. A non-exhaustive list of some prominent studies includes Larson (1969) and Penston (1969a, b), who performed full numerical studies (i.e., solving the full hydrodynamic equations in spherical symmetry, Equations (7.30) and (7.31)) of gravitational collapse and investigated the prestellar stage using similarity analysis; Shu (1977), who performed a similarity study of the *protostellar* (i.e., *after* the formation of a protostar) stage of the collapse under rather special initial conditions; Hunter (1977), who performed a similarity study describing *both* the prestellar and protostellar stages (therefore referred to as a *complete* solution), under a few restricted conditions; and Whitworth & Summers (1985, hereafter WS85), who performed a thorough similarity study of the entire set of physically meaningful solutions, obtaining both asymptotic and numerical solutions for the similarity equations. In the remainder of this subsection, we briefly discuss some of the latter authors' results, which encompass those of all the previous studies.

Specifically, WS85 discussed the asymptotic solutions of Equations (7.42) and (7.43), which can be obtained by writing v and $\bar{\rho}$ as Taylor series in either ξ or $1/\xi$ using these equations, and taking the limits $\xi \to |0|$ and $\xi \to |\infty|$, respectively (see Hunter, 1977). They found three asymptotic solutions which, to leading order in ξ, are:

1. A *settling* solution, relevant in the limit $\xi \to 0^-$, (i.e., $t < 0$), having the form

$$v(\xi) = \frac{2\xi}{3} + O(\xi^3), \tag{7.49}$$

$$\bar{\rho}(\xi) = \bar{\rho}_0 + O(\xi^2); \tag{7.50}$$

2. A *freefall* solution, relevant in the limit $\xi \to 0^+$, (i.e., $t > 0$), of the form

$$v(\xi) = -\left(\frac{2m_0}{\xi}\right)^{1/2} + O(\xi^{1/2}), \tag{7.51}$$

$$\bar{\rho}(\xi) = \left(\frac{m_0}{2\xi^3}\right)^{1/2} + O(\xi^{-1/2}), \tag{7.52}$$

and

3. A *cruising* solution, relevant in the limit $\xi \to |\infty|$, of the form

$$v(\xi) = v_\infty - \frac{\bar{\rho}_\infty - 2}{\xi} + O(\xi^{-2}), \tag{7.53}$$

$$\bar{\rho}(\xi) = \frac{\bar{\rho}_\infty}{\xi^2} + O(\xi^{-4}), \tag{7.54}$$

where $\bar{\rho}_0$ is the dimensionless central density, v_∞ is the dimensionless inflow velocity far from the center, and $\bar{\rho}_\infty$ is a dimensionless "far-field density",[2] representing the value of the density at some large distance from the center. The parameter m_0 is the mass of the central protostar in nondimensional units. These are, in effect, boundary condition values.

In their analysis, all times correspond to collapse, with the convention that $t = 0$ corresponds to the moment of formation of a singularity (a *protostar*) at the center of the sphere, so that the interval $-\infty < t < 0$ corresponds to the *prestellar* stage (referred to as the *early path* by WS85), while the interval $0 \leqslant t < +\infty$ corresponds to the *protostellar* stage (referred to as the *late path* by WS85), in which a collapsed object (the protostar) of finite mass M_0 already exists at the center of the core. During this stage, in addition to the gaseous mass, the protostar's mass must be taken into account when computing the total gravitational potential, and in fact, at the latest stages, or, equivalently, *very near the center*, it may be the dominant source of gravity. This is the case of the *freefall* solution given by Equations (7.51) and (7.52) as $\xi \to 0^+$ since, in this case, the self-gravity of the gas is negligible, and its velocity is simply the freefall speed due to the central mass. Indeed, using Equations (7.29) and (7.35) in the freefall solution, we see that the leading-order terms imply

[2] I am indebted to Prof. Anthony Whitworth for suggesting this name.

$$u(t) = \left(\frac{2GM_0}{r}\right)^{1/2}; \qquad \rho(r, t) = \frac{1}{4\pi t}\left(\frac{M_0}{2Gr^3}\right)^{1/2}, \tag{7.55}$$

where $M_0 = c_s^3 m_0 t/G$. These equations show that, very near the center, the infall velocity is simply the free-fall speed onto the central object, since the gas mass is negligible there, and that the density maintains an $r^{-3/2}$ profile. The density, however, decreases as t^{-1} at all radii, because the core's mass is exhausted as it is transferred to the central object.

We emphasize that both the settling and the freefall solutions apply in the limit $|\xi| \to 0$; that is, at very small radii or very large absolute values of the time. Therefore, they describe the flow near the center of the core (in what WS85 called the *inner* region), the settling solution corresponding to the prestellar stage ($t < 0$, before the formation of the protostar) and the freefall solution corresponding to the protostellar stage ($t > 0$). The cruising solution, on the other hand, applies in the limit $|\xi| \to \infty$; that is, very far from the center (in what WS85 called the *outer* region), and applies in both the pre- and the protostellar stages.

The three asymptotic solutions are shown schematically in Figure 7.2. The *left panel* of this figure shows the general case where the various constants $\bar{\rho}_0$, m_0, v_∞, and $\bar{\rho}_\infty$ are all different from zero. In this case, the collapse extends to infinity, and is thus "global". The *right panel* shows the special case of $v_\infty = 0$, so that the infall speed drops as ξ^{-1} at large $|\xi|$, according to the first non-vanishing term in Equation (7.53), and so the collapse can be considered to be "local". Note that, during the

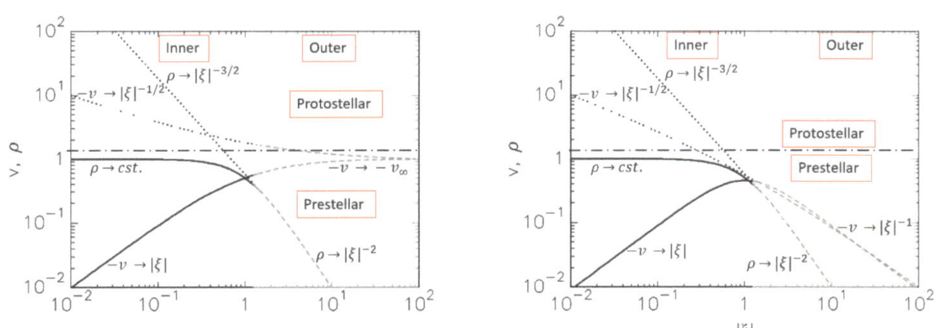

Figure 7.2. Asymptotic solutions of the collapse equations for the nondimensional velocity $v(\xi)$ and density $\bar{\rho}(\xi)$ defined in Equations (7.35). The white area represents the *inner* region (according to the nomenclature of WS85), where the limit $|\xi| \to 0$ applies. The shaded area represents the *outer* region, in which the limit $|\xi| \to \infty$ is applicable. The *dash-dotted* horizontal line approximately separates the *prestellar* (or "early", in WS85's nomenclature) stage from the *protostellar* (or "late") stage. The *solid* lines indicate the *settling* solution, applicable in the inner region during the prestellar stage of the collapse. The *dotted* lines indicate the *freefall* solution, applicable in the inner region during the protostellar stage. Finally, the dashed lines indicate the *cruising* solution, applicable in the outer region during both the pre- and protostellar stages. *Left panel:* A case with a finite infall speed at infinity, $v_\infty < 0$, corresponding to a global collapse. *Right panel:* A case with vanishing infall speed at large distances, $v_\infty = 0$, corresponding to local collapse, for example, from an initially marginally unstable BE sphere.

prestellar stage for the $v_\infty = 0$ case, the infall speed maximum occurs at a finite distance from the center, and so the collapse occurs from the "outside-in", having started at a radius of the order of the initial Jeans length. The solution with $v_\infty = 0$ is in fact the most realistic one, describing a real collapse that started from an initial zero infall speed, and confined to a region of finite size within a cloud.

7.3.2.1 The Moment of Protostar (Singularity) Formation

A crucial moment in the evolution of a collapsing core occurs when $t = 0$. Mathematically, this corresponds to the time at which a singularity forms—i.e., at which the density diverges. Physically, it corresponds to the time of formation of the protostar which, although still of finite density, is roughly 20 orders of magnitude denser than the molecular cloud in which it formed and so, compared to the initial conditions, it is akin to a singularity.

The moment of protostar formation can be investigated by considering the *cruising* asymptotic solution, given by Equations (7.53) and (7.54), in the limit $t \to 0^-$. In this limit, the central, flat-density region shrinks to zero physical radius, since it is bounded by $|\xi| = 1$, which in the dimensional variables corresponds to $r = c_s|t|$. So, as $|t|$ approaches zero, the physical radius r of the $|\xi| \leqslant 1$ region also shrinks to zero. The density profile of the entire core then approaches a single power law $\rho \propto r^{-2}$ all the way to the center (where the singularity appears), and the velocity approaches the constant v_∞. In the special case when $v_\infty = 0$, the solution corresponds to the hydrostatic but unstable SIS. When $v_\infty \neq 0$, the density profile resembles that of an SIS, although the core is not hydrostatic, but rather it is collapsing.

7.3.2.2 The Stationary Outer Density Profile of the Protostellar Stage and the Continuous Inflow Toward the Collapse Center

An important implication of the cruising asymptotic solution given by Equation (7.54) is that, at large distances from the center, the density profile becomes stationary—i.e., it does not depend on time, but only on radius. This can be seen by substituting the definitions of the nondimensional variables, Equations (7.29) and (7.35), into the asymptotic solution, Equation (7.54), from which we see that

$$\rho(r, t) = \rho(r) = \frac{\bar{\rho}_\infty}{4\pi G}\left(\frac{c_s}{r}\right)^2, \tag{7.56}$$

where $\bar{\rho}_\infty$ is a dimensionless constant. This indicates that the density only depends on the radius, but not on time—i.e., it becomes *stationary*. However, the gas is nevertheless continuously infalling toward the collapse center. This means that the asymptotic cruising solution constitutes a *stationary accretion inflow*, in which the gas continuously falls down deeper and deeper into a nearly stationary potential well, given by the nearly stationary density profile, in a manner analogous to water continuously running down a funnel. Note, however, that, since this solution applies for $|\xi| \to \infty$, it is valid only either very far from the center, or at $|t| \to 0$, where the mass of a possible central collapsed object is negligible compared to the total gas mass in the sphere. Otherwise, as the central mass increases, so does the depth of the potential well, and the density profile must also vary in time, changing from the r^{-2} slope corresponding to gas mass domination to the $r^{-3/2}$ corresponding to domination of the central mass. The break point between the two slopes, determined by $\xi = 1$ in nondimensional form, moves outwards in the dimensional variables.

7.3.2.3 The Similitude between the Prestellar Collapse Stage and the BE Sphere Density Profiles

A remarkable feature of the prestellar collapse stage is that its settling density profile, with a central flat part and an outer $\rho \propto r^{-2}$ power-law envelope, is surprisingly similar to that of the hydrostatic, truncated BE sphere (Section 6.2).

This similitude[3] is surprising because, as discussed in Section 6.2.5.2, the r^{-2} profile in the BE sphere is due to a *detailed mechanical balance* between self-gravity and the thermal pressure gradient at all radii in the sphere.[4] A necessary condition for this to happen is that sound waves within the core can establish pressure balance among different parts of the core. This condition is trivially satisfied in a hydrostatic core configuration, such as a BE sphere. However, it is *not* expected to be satisfied in a collapsing core, in which supersonic infall speeds develop, as already shown by the pioneering simulations of Larson (1969) and Penston (1969a, b). In the following subsections we discuss how this apparent dilemma led Shu (1977) to propose the well-known inside-out collapse solution, and how this profile may be established in general from generic, near-hydrostatic initial conditions.

7.3.3 Collapse from Singular Initial Conditions: The Inside-Out Collapse

The most widely known (although, unfortunately, perhaps the most unrealistic) solution of Equations (7.42) and (7.43) is the *inside-out* solution discussed by Shu (1977, hereafter S77). This is actually just one (extreme) case in the range of collapse solutions discussed by S77, but, because it was given special attention by S77 as the only case with a self-similar initial condition, the inside-out solution has become the assumed standard for collapse, even though it cannot occur in reality. For this reason, in what follows we discuss it in detail, in terms of the more general analysis of WS85, together with the reasons why it should be considered unrealistic in general.

As discussed in Section 7.3.2.3, the similar shape of the prestellar collapse to that of the hydrostatic BE sphere led S77 to note that a fundamental condition for detailed mechanical balance to occur in a core is that the motions within it remain

---

[3] We refer to the similar shape of the density profiles in the two cases as "similitude", to avoid confusion with the concept of "similarity" as the technique for solving the gas dynamics equations in the self-similar regime (Section 7.3).

[4] That detailed balance is an essential condition for the establishment of a hydrostatic configuration can be seen from the fact that the solution of the Emden–Chandrasekhar equation, Equation (6.23), is obtained by integrating outwards from the center, or inwards from the boundary, ensuring that the hydrostatic balance condition is satisfied at every new radius (cf. Section 6.2).

subsonic, so that detailed pressure balance can be established throughout the core, leading to the characteristic r^{-2} density profile, which precisely arises from this condition (Section 6.2). However, already in the early numerical simulations by Larson (1969), supersonic infall speeds were observed to arise during the prestellar collapse stage, seemingly making it impossible for sound waves to maintain the required detailed mechanical balance for the r^{-2} density profile. In view of this apparent contradiction, S77 wrote

> "How, then, can we understand the observed establishment of an r^{-2} law under a variety of initial conditions? The physical reason is simple, even if the mathematics is not: the convergence to an r^{-2} law represents the tendency of an isothermal self-gravitating gas to approach, as closely as possible, detailed mechanical balance. This tendency will be satisfied as long as different parts of the cloud can communicate with one another, i.e., as long as the initial conditions allow the early phases of the flow to occur subsonically. Once an r^{-2} law has been (nearly) established without the generation of considerable fluid velocities, the subsequent collapse flow occurs in a self-similar manner."

That is, Shu proposed that the establishment of the detailed balance occurred during an early quasistatic stage. This led him to consider the collapse of spheres *initially close to hydrostatic equilibrium*; that is, close to unstable BE spheres, including the extreme case of an SIS. Such near-equilibrium configurations could be thought of as small perturbations of an initial equilibrium one. In particular, S77 devoted special attention to the SIS as an "initial condition", since it is the only fully self-similar (i.e., with a single power-law density profile over its entire radial extension) hydrostatic configuration, thus constituting an exact solution of the similarity collapse problem at "all times before the presence of the outer boundary begins to play a dynamic role".

To find solutions that could arise from an initially hydrostatic condition, S77 looked for solutions satisfying the requirement that $v \to 0$ when $\xi \to \infty$; i.e., when $t \to 0$. We can readily recognize that this limit corresponds to the *outer cruising* solution of WS85, given by Equations (7.53) and (7.54) of Section 7.3.2, with the condition $v_\infty = 0$, giving

$$\bar{\rho} \to \frac{A}{\xi^2}, \qquad v \to -\frac{A-2}{\xi}, \qquad m \to A\xi, \tag{7.57}$$

where m is given by Equation (7.41) and, for consistency with S77's notation, we have replaced $\bar{\rho}_\infty$ in Equations (7.53) and (7.54) by A. We note that the case $A = 2$ corresponds precisely to the SIS (cf. Equations (7.44) and (7.45)), while cases with $A > 2$ correspond to dynamic *infall* (i.e., $v < 0$).

On the other hand, noting that the outer asymptotic solution must break down near the center, S77 also identified the *protostellar, freefall* solution (Equations (7.51) and (7.52)) as the appropriate asymptotic behavior for the evolution of the innermost parts ($\xi \to 0$) *after the formation of the singularity*, writing

$$\bar{\rho} \to \left(\frac{m_0}{2\xi^3}\right)^{1/2}, \qquad v \to -\left(\frac{2m_0}{\xi}\right)^{1/2}, \qquad m \to m_0. \tag{7.58}$$

In summary, S77 concluded that the asymptotic profiles of the density and infall speed for $A > 2$ *in the inner region* ($\xi \to 0^+$) should be

$$\begin{aligned} &\bar{\rho} \propto \xi^{-3/2}, \\ &v \propto \xi^{-1/2} \end{aligned} \tag{7.59}$$

(*dashed lines* in the *left half* of the top panel of Figure 7.3), corresponding to infall onto the central mass m_0, with the gravitational potential dominated by this mass (neglecting the self-gravity of the gas and temporal variation of m_0), while, *for the outer region,*

$$\begin{aligned} &\bar{\rho} \propto \xi^{-2}, \\ &v \propto \xi^{-1} \end{aligned} \tag{7.60}$$

(*dashed lines* in the *right half* of the top panel of Figure 7.3), corresponding to gravitational contraction dominated by the self-gravity of the gas (neglecting the gravity of the central mass).

It is important to note that, in the intermediate-ξ range, the flow regime transitions from one asymptotic regime to the other. The top panel of Figure 7.3 shows numerical solutions of the similarity Equations (7.42) and (7.43), obtained by integrating inwards (toward $\xi = 0$), starting from $\xi = 50$, and setting the boundary conditions there for $\bar{\rho}$ and v from the asymptotic conditions (7.57) with $A = 2.0001$, 2.005, 2.05, and 2.5. The *dotted lines* show the infall velocity profile $-v(\xi)$, the *solid lines* show the density profile $\bar{\rho}(\xi)$, and the *dash-dotted lines* show the internal mass $m(\xi)$ given by Equation (7.41). Note that for $A \to 2^+$, the infall speed develops a sharp drop in the neighborhood of $\xi = 1$, so that the material outside this point is essentially at rest, while the material inside it is dynamically infalling onto the central mass (the protostar). However, already for $A = 2.05$ the infall speed v does not exhibit a substantial drop at $\xi = 1$, and the whole structure is infalling at all times, albeit v does decrease smoothly outwards with a profile approaching the asymptotic solution $-v \propto \xi^{-1}$. The density profile, on the other hand, almost does not change as A is varied. It is worth noting here that the quantity $A - 2$ can be considered a measure of how much the initial condition deviates from a strict SIS.

On the other hand, the solution in the limiting case $A \to 2^+$ deserves to be considered separately. This solution was calculated numerically by S77, and is shown in the bottom panel of Figure 7.3, taken directly from S77's Table 2 for the interval $0 \leqslant \xi \leqslant 1$. It is seen to consist of two regimes: an inner regime with $0 < \xi < 1$, in which the gas is infalling, asymptotically approaching the free-fall regime for $\xi \to 0^+$, and an outer, *hydrostatic* regime given by the SIS for $\xi > 1$, with a transition at $\xi_* = 1$, which corresponds to a radius $r_* = c_s t$, with $t > 0$.

This is the famous *inside-out* solution, and S77 interpreted it as the collapse from an SIS as initial condition, which is then perturbed at $t = 0$ in its central parts.

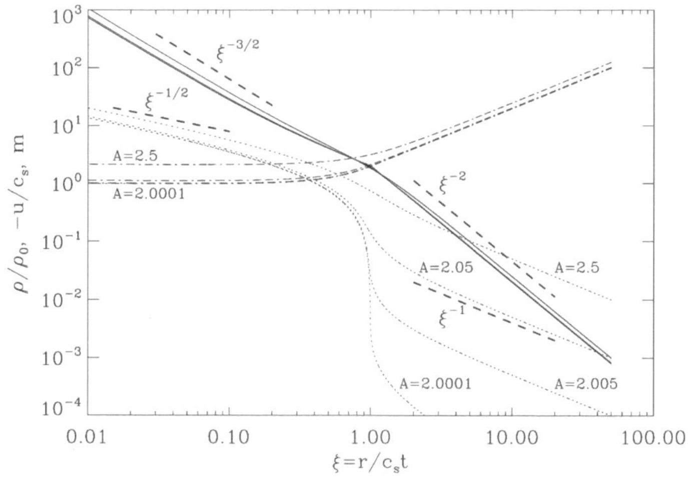

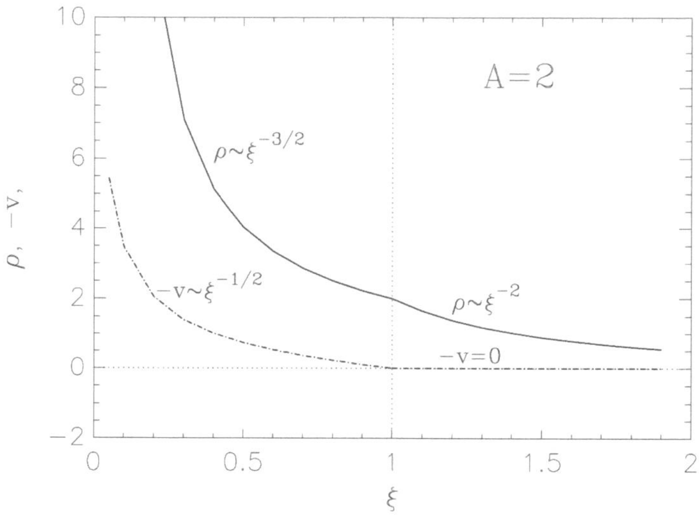

Figure 7.3. *Top panel:* Numerical solutions of the similarity Equations (7.42) and (7.43) satisfying the condition $v \to 0$ for $\xi \to \infty$ (i.e., for $t \to 0^+$), obtained by integrating inwards (toward $\xi = 0$), starting from $\xi = 50$, and setting $\bar{\rho}$ and v from the asymptotic conditions (7.57) for $A = 2.0001$, 2.005, 2.05 and 2.5 in Equation (7.57). The *dotted lines* show the infall velocity profile $-v(\xi)$, with A increasing from the bottom as indicated; the *solid lines* show the density profile $\bar{\rho}(\xi)$; the *dash-dotted lines* show the internal mass $m(\xi)$ given by Equation (7.41), and the *dashed lines* show the slopes of the asymptotic solutions given by relations (7.59) and (7.60). *Bottom panel:* The critical solution for $A = 2$, as given in Table 2 of S77. The *solid line shows* the density profile, while the *dashed line* shows the infall speed profile, which is strictly equal to zero for $\xi \geqslant 1$.

The inner regions then begin to collapse, causing their immediate outer neighbors to "lose support" and to begin collapsing themselves. Thus, a *wavefront of collapse* (or *rarefaction front*[5]), propagates outwards at the sound speed. Thus, at any time t, the

[5] This name arises because, inside this wave front, the density is lower than what would correspond to the hydrostatic SIS, since it is being drained by the infall onto the central star.

collapse front is located at the "sonic" radius $r_* = c_s t$ ($\xi = 1$). Outside this radius, the medium remains at rest, with the SIS density profile, because the wave of collapse has not reached it yet.

A number of features should be remarked concerning this solution:

1. Since the "initial condition" of the inside-out collapse is an SIS, which already contains a singularity at its center, the evolution corresponds exclusively to the *protostellar* stage. This implies that the whole *prestellar* stage is assumed to occur quasistatically.
2. At $t = 0$, the central collapsed object ("star") has $M = 0$, as can be seen from the second of Equations (7.35), in spite of the density diverging there. This is due to the r^{-2} density profile of the SIS, which increases more slowly than the volume decreases as $r \to 0$.
3. For $t > 0$, Equations (7.35) and (7.58) imply that the mass of the central star increases linearly in time as

$$M_c = \frac{c_s^3 m_0}{G} t, \tag{7.61}$$

so that the accretion rate onto the star is

$$\dot{M}_c = \frac{c_s^3 m_0}{G}. \tag{7.62}$$

This accretion rate is seen to depend only on the gas temperature (through the sound speed) and the mass of the central object m_0. As can be observed in the top panel of Figure 7.3, this central mass results also from the numerical solution of the similarity equations, through Equation (7.41). In the limit $A \to 2^+$, $m_0 \to 0.975$, but it is significantly larger for A larger than 2 (Table 7.1). We discuss the implications of this result for the accretion rate onto the central object in Section 7.3.4.

7.3.3.1 *Idealized Nature of the Critical Inside-Out Collapse Solution and Realistic Collapse Numerical Solutions*

The inside-out solution by Shu has been the paradigm for the gravitational collapse of molecular cloud cores for decades. However, this is not a regime that can be expected to arise in reality in nature. Rather, it is an idealized solution that can only serve as an initial reference point for understanding the actual regimes that are observed in numerical simulations.

Table 7.1. Relation between Constants A and m_0 (From Shu (1977)

$A =$	2.0^+	2.2	2.4	2.6	2.8	3.0	3.2	3.4	3.6	3.8	4.0
$m_0 =$	0.975	1.45	1.88	2.31	2.74	3.18	3.63	4.10	4.58	5.08	5.58

The main unrealistic features of the inside-out collapse are:

1. The assumed "initial condition" is an SIS, which, as we have discussed in Section 6.2.3, is the most unstable case of hydrostatic spherical equilibrium. This configuration is not expected to arise in nature because, being unstable, it cannot appear spontaneously through neither a dynamical nor a quasistatic evolution of a previously less concentrated sphere. Those previous states are already unstable, and will thus engage into dynamical collapse when perturbed, long before the central singularity appears. That is, *actual collapse does not start when the singularity appears, but rather much earlier, when the gas parcel becomes gravitationally unstable and starts to develop an infall flow*.
2. A quasistatic evolution was proposed to be possible in the presence of a strong magnetic field, such that the core is magnetically subcritical (cf. Section 9.2.1) and the contraction proceeds slowly, and only at the densest parts, as a consequence of *ambipolar diffusion* (e.g., Shu et al. 1987; Mouschovias 1991 and references therein). However, in this case, the symmetry departs from spherical, and also the magnetic support is lost long before the singularity arises, allowing the central parts of the core to collapse dynamically during the prestellar stage while the outer parts remain supported. Finally, for a centrally concentrated density profile under ideal MHD conditions, the mass-to-flux ratio is radius-dependent, typically increasing outwards (cf. Section 5.4.1), and thus an apparently subcritical core may actually just be the tip of the iceberg of a globally magnetically supercritical structure when its surrounding medium is considered. Thus, in general, again the initial hydrostatic condition cannot contain a singularity.
3. The inside-out solution itself is a *critical* solution, which passes through a critical point at $\xi = 1$, at which the infall speed vanishes. However, as seen in Section 7.3.3, any solution with $A \neq 2$ continues smoothly beyond $\xi = 1$, and the infall speed quickly becomes nearly continuous across it as soon as A differs even by a few percent from 2. Since dynamical collapse is expected to have started long before the singularity (protostar) forms, then by the time the protostar appears, the solution most appropriate to represent the ongoing collapse must correspond to A significantly larger than 2. In turn, this implies that the collapse is not expected to be confined to the region within $\xi = 1$ in real cases.

Realistic collapse is more likely to be similar to solutions starting from non-singular configurations, close to the critical BE sphere. Such solutions cannot be obtained from a similarity analysis, since the non-singular BE spheres are not self-similar, and thus have to be obtained by full numerical simulation. Foster & Chevalier (1993) showed that, in this case, the prestellar stage of the collapse proceeds from the outside-in—specifically, from a radius corresponding to $\xi \sim 1$—as shown in Figure 7.4. The solid line in the top left panel shows the infall speed, which those authors report to increase linearly with radius up to roughly one Jeans length ($\xi \sim 1$, as in the *settling* asymptotic solution of WS85). Beyond that point, it turns

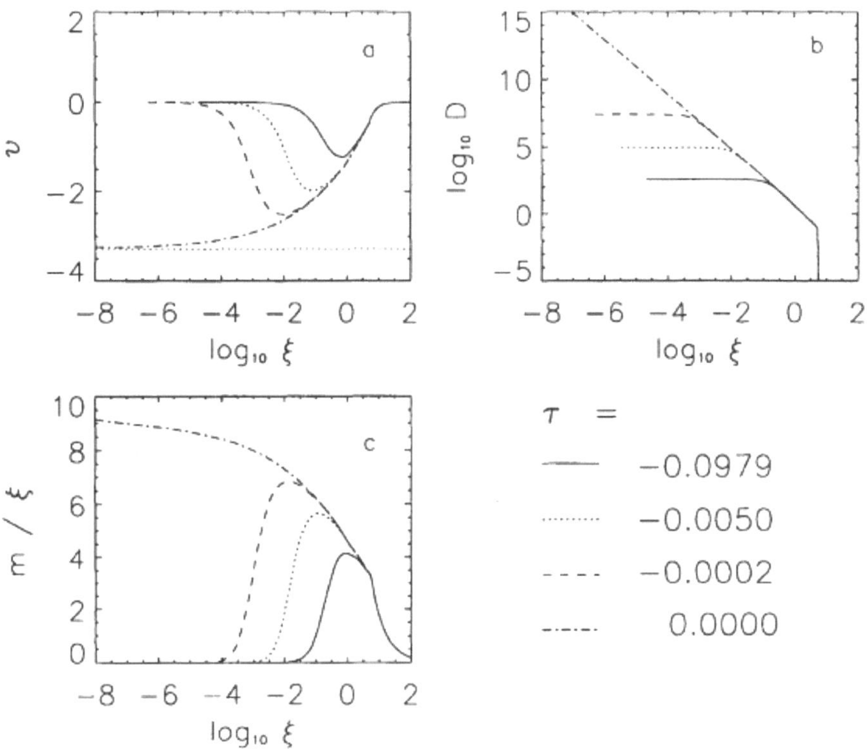

Figure 7.4. Numerical integration of the time-dependent hydrodynamic equations in spherical coordinates by Foster & Chevalier (1993) for prestellar collapse starting from a marginally unstable BE sphere immersed in a constant-pressure, diffuse medium, and terminating at the moment of protostar formation. In all panels, the cores's center is toward the left. The *top left* panel shows the infall speed, showing that the collapse *starts outside*, at roughly the Jeans length ($\xi \sim 1$), and propagates to the center. The *top right* panel shows the logarithm of the density, with the *dash-dotted* line having a logarithmic slope of -2, and the *lower left* panel shows the normalized mass within nondimensional radius ξ. The different line styles indicate different times, as indicated in the *bottom right* panel, with the final time being $t = 0$. Reproduced from Foster and Chevalier (1993).

around and begins to decrease.[6] The density, on the other hand, is seen to evolve according to the settling similarity solution, with a central flat part and an r^{-2} outer envelope. Simulations starting from generic, non-BE (Gaussian) hydrostatic conditions also develop similar density profiles (Naranjo-Romero et al. 2015; see also Section 7.4.1.1), implying that the r^{-2} density profile arises during the dynamical evolution, rather than from nearly quasistatic initial conditions, as suggested by S77. We discuss this further in Section 7.4.2.

[6] Note that Foster & Chevalier (1993) defined their ξ variable in terms of the constant *initial* central density ρ_c of the BE sphere as $\xi = (r/c_s)\sqrt{4\pi G\rho_c}$, rather than in terms of the *time-dependent* normalizing density given by Equation (7.36). Therefore, at later times in the evolution, the infall speed and the density in their graphs have their turnover points closer to the center at each time, rather than staying at the same value of the similarity variable as in the solutions by WS85.

7.3.4 Accretion Driven by the Central Object

An important result obtained by S77 concerned the accretion rate onto the central protostar. This follows from the asymptotic solution in the limit $\xi \to 0$ given by the third of Equations (7.58), $m \to m_0$. According to Equation (7.35), this implies that the central mass is given in dimensional form by

$$M(0, t) = \frac{m_0 c_s^3}{G} t, \tag{7.63}$$

so that the mass increases linearly with time. Differentiating with respect to time, this in turn implies that the *mass accretion rate* onto the central protostar is constant, given by

$$\dot{M} = \frac{m_0 c_s^3}{G}. \tag{7.64}$$

Through numerical integration, S77 found that, in the limit $A \to 2^+$, $m_0 \to 0.975$, as shown by the lowest dash-dotted line in the top panel of Figure 7.3 and Table 7.1. This implies that, *in the critical, inside-out case*, the accretion rate is of order c_s^3/G. However, this is the minimum possible accretion rate predicted by the spherical similarity solutions. For non-critical solutions with $A > 2$, m_0 is significantly larger, being already $\sim$5 times larger for $A = 4$, as shown in Table 7.1, reproduced from S77. Therefore, the accretion rate also increases rapidly with A.

It is important to note that generally it is considered that the accretion rate cannot be much larger than the minimum given by Equation (7.64) with $m_0 = 0.975$, because the typical luminosities of protostars are comparable to this value. Specifically, for a molecular cloud with $T = 10$ K and a mean molecular weight of $\mu = 2.36$, $c_s = 0.18$ km s^{-1}, and so Equation (7.64) with $m_0 = 0.975$ gives $\dot{M} \approx 1.6 M_\odot$ Myr^{-1}. In turn, the gravitational energy released by the accretion is of order

$$L_{\rm acc} \sim \frac{G M_* \dot{M}}{R_*}, \tag{7.65}$$

where M_* and R_* are, respectively, the protostellar mass and radius. Assuming $M_* \sim 1 M_\odot$ and $R_* \sim 10 R_\odot$, we find $L_{\rm acc} \sim 6 L_\odot$, where $M_\odot = 2.0 \times 10^{33}$ g, $R_\odot = 7.0 \times 10^{10}$ cm, and $L_\odot = 3.8 \times 10^{33}$ erg s^{-1} are, respectively, the solar mass, radius, and luminosity. This is of the order of the typical observed luminosities of Class 0 protostars, and several times larger than the luminosities of Class I protostars (e.g., Chapter 5 of Hartmann 2009).

The above result is often taken as an indication that the initial conditions for the collapse should be close to the SIS. However, since we have emphasized that this condition is unrealizable in nature, an alternative interpretation is necessary. One possible explanation is that the accreting material does not go directly to the protostar, but rather lands on a disk, whose radius is much larger than that of the protostar. Then the accretion from the disk onto the protostar need not be equal to the accretion rate onto the disk, if the disk is able to accumulate mass for a while, and then dump it abruptly onto the star via short accretion "bursts". Such bursts may correspond to the so-called FU Orionis objects. This is not the subject of the present book, but a detailed discussion can be found in (Chapter 9 of Hartmann 2009).

7.3.5 Radial Mass Flux across a Core

Being the densest regions within molecular clouds, the cores must have formed from a compression, as required by the continuity equation, Equation (2.74), which we rewrite here for convenience:

$$\frac{d\rho}{dt} = -\rho \nabla \cdot \boldsymbol{u}, \tag{7.66}$$

where, as usual, the operator $d/dt = \partial/\partial t + \boldsymbol{u} \cdot \nabla$ denotes the *total*, or *Lagrangian*, derivative, following a fluid parcel of fixed mass (cf. Section 2.2.1). This shows that, in a Lagrangian description, a negative divergence (i.e., a *convergence*, or a *compression*) is required in order to increase the density of a fixed-mass fluid parcel over time.

On the other hand, the continuity equation can also be written as

$$\frac{\partial \rho}{\partial t} = - \nabla \cdot (\rho \boldsymbol{u}). \tag{7.67}$$

Integrating this equation over an Eulerian volume V, fixed in space, we find

$$\begin{aligned} \int_V \frac{\partial \rho}{\partial t} dV' &= \frac{\partial}{\partial t} \int_V \rho dV' = \frac{d}{dt} \int_V \rho dV' = \frac{dM}{dt} \\ &= \int_V \nabla \cdot (\rho \boldsymbol{u}) dV' = \oint_S \rho \boldsymbol{u} \cdot \hat{n} dS, \end{aligned} \tag{7.68}$$

where M is the mass within volume V, S is the surface encasing V, and in the first line we have used the fact that the Eulerian volume does not depend on time. This shows that the "core" (the region within volume V) acquires mass via a *mass flux* $\rho \boldsymbol{u}$ passing through the surface enclosing this volume. Since, as discussed in Section 1.2.1, the boundaries of "clouds" or "cores" defined within a continuum are rather arbitrary, the volume V can have a range of sizes, as long as it contains a local density maximum, to be considered a "core". Therefore, *there must exist a net mass flux along a range of radii around any local growing density maximum*. This means that the core constitutes a sort of "funnel", through which mass flows from the cloud toward any protostars forming at the core's center. We refer to this as a *collapse flow*, which we discuss further in the following sections.

7.4 Collapse Flow: the Growth of Clouds and Cores

One fundamental property of the gravitational collapse of a non-uniform density fluctuation is that it occurs *non-homologously*. That is, while in the idealized case of the homologous collapse of a spherical, uniform-density region (Section 7.2.1) all of the material reaches the center at the same time, in realistic non-homologous collapse different parts of a collapsing cloud reach the center (and infinite density—i.e., a singularity) at different times (Section 7.3.2), a phenomenon which we will refer to as *asynchronous collapse*. Moreover, since the structures are not perfectly spherical, the objects collapse first along their shortest dimensions, leading to *anisotropic collapse* (Section 7.2.2). Therefore, as mentioned in the previous section, in general a *collapse flow* is established, with gaseous material continuously flowing from the peripheral, low-density regions to the more central, high-density ones. This flow moreover amplifies initial anisotropies of the collapsing regions.

In addition, a clear distinction arises between the *pre-* and *proto*stellar stages of the collapse, as discussed in Section 7.3.2 in the spherical case. The former stage corresponds to the growth of a density fluctuation, from being a relatively minor perturbation in the background medium, increasing its central density, to the moment when a singularity (in practice, a protostar) is formed. The second stage corresponds to the accretion of the gas onto the central object, which therefore grows in mass over time. We now discuss the mass growth during each stage.

7.4.1 The Spherical Case in a Uniform Background

For simplicity, in what follows we will consider spherical symmetry, and refer to the structure (cloud, clump, or core) simply as a "core", although the discussion will apply to any range of mass. Of course, in practice, none of those structures are spherical in general. Nevertheless, the spherical case provides a useful first approximation for understanding real structures, and, moreover, it may apply approximately to spherical averages of the physical quantities.

7.4.1.1 The Prestellar Stage: Outside-In and Inside-Out Parts of the Collapse

Within the spherical approximation, the most idealized case is that of very distant initial and boundary conditions, in which case the flow may approach a self-similar, asymptotic solution of the gas dynamics equations, as discussed in Section 7.3.2, and shown in Figure 7.2. That figure already indicates the growth of the central density of the collapsing region, recalling that the nondimensional density, shown in that figure, is defined as $\bar{\rho} \equiv 4\pi G t^2 \rho$. Thus, the constant value of the central nondimensional density corresponds to a diverging physical density as $t \rightarrow 0^-$ (recall the prestellar stage corresponds to $t < 0$ in the treatment of Section 7.3.2). The growth of the dimensional density with time is illustrated, for example, in the top right panel of Figure 7.4.

In order to quantify the increase in the mass of a core, however, we need to define a boundary and a distinct background medium for the core, and so the self-similar character needs to be abandoned. In early works, it was customary to assume that the core had a finite mass and extension, and was immersed in a warmer diffuse medium, which could provide a confining pressure for the core without adding mass to it which would destabilize it (Section 6.2.3). However, this is not an adequate assumption for young molecular cloud cores, which are generally part of a larger molecular cloud at roughly the same temperature as the core.[7]

A slightly more realistic boundary condition is to assume that the core is immersed in a static medium at the same temperature as the core, and with uniform density extending to infinity, so that initially the core is a minor density fluctuation on top of this uniform density. Since the background medium extends to infinity, when no perturbation is applied, it must remain in equilibrium, since the gravitational forces cancel out in all directions.[8]

Under these conditions, when a perturbation is added, it will begin to contract locally, leaving behind a low pressure region, and thus setting up an *accretion flow* onto it from the environment. Figure 7.5 shows the evolution of the number density n (solid green line) and infall speed in units of the sound speed (v/c_s, dashed blue line) in a numerical simulation of the collapse of an initially Gaussian spherical perturbation immersed in a uniform background, with periodic boundary conditions (Naranjo-Romero et al. 2015). Note that the initial Gaussian (rather than BE-like) profile prevents the possibility that a detailed pressure balance at all radii could have been pre-established in the initial condition.

In Figure 7.5 it is seen that, once the infall speed becomes supersonic, the outer sonic point (the outermost of the two points where the infall speed equals the sound speed), indicated by the vertical red lines, moves outwards. This is thus a similar phenomenon to the inside-out collapse proposed by Shu (1977), except that the outwards propagation starts from roughly one Jeans length from the center, rather than from the center itself, as initially proposed by Shu. On the other hand, within one initial Jeans length from the center, the collapse accelerates *inwards* (as indicated by the inwards motion of the inner sonic point, denoted by the vertical blue lines), amounting to an *outside-in* collapse.

The fact that the outer sonic point propagates outwards indicates the expansion of the accretion flow from the environment onto the original core. Furthermore, the radius at which the density has the value of the initial peak also moves outwards, together with the outer sonic point, indicating that the core's radius increases. Since, in addition, the central density of the core increases, the net result is that the core's mass, mean density, and size increase over time during the prestellar stage.

---

[7] It may apply, however, to the remains of cores immersed in H II regions around massive stars, since the warm ionized gas of the region does fulfill the requirement of having a much larger temperature and much lower density than those of the core. These cores may still be in the process of being photoevaporated by the ionizing radiation, while the less dense gas from the original cloud may have been already fully evaporated.

[8] Such an infinite medium can be modeled by replacing Poisson's Equation, (2.78), by

$$\nabla^2 \phi = 4\pi G(\rho - \rho_0), \tag{7.69}$$

where ρ_0 is the uniform density of the medium, so that only the density *fluctuations* generate a gravitational force. This is a common strategy when modeling flows with, for example, periodic boundary conditions, representing infinite media (or much larger than the core size) (e.g., Alecian & Léorat 1988). Another, much more realistic, possible way for a *local* collapse to occur inside a larger-scale nearly uniform medium at the same temperature, is that the latter is collapsing toward a *different* point, in a *conveyor-belt* type of flow (Longmore et al., 2014). This, however, requires abandoning the spherical symmetry, and we discuss it in Sections 7.5.5 and 11.4, and illustrate it in Figure 11.3.

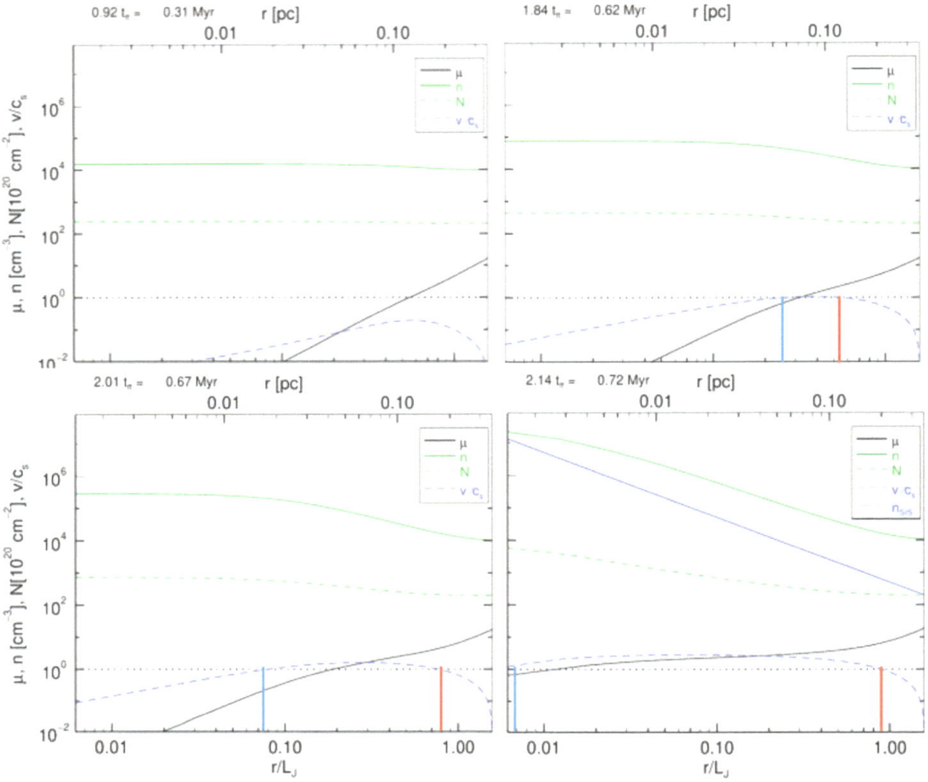

Figure 7.5. Evolution of the radial profiles of the density (n, solid blue line), column density (N, dashed green line), mass internal to radius, in units of the initial Jeans mass (μ, solid black lines), and infall speed, in units of the sound speed (v/c_s, dashed blue line), of an initially Gaussian spherical perturbation immersed in a uniform background, with periodic boundary conditions. The background density is $n = 10^4\ \mathrm{cm}^{-3}$, while the peak density is $n = 1.5 \times 10^4\ \mathrm{cm}^{-3}$. The temperature is 11.4 K, so that $c_s = 0.2\ \mathrm{km\ s}^{-1}$. The box size is $L = \sqrt{10}\ L_J \approx 0.71$ pc, where $L_J \approx 0.22$ pc is the Jeans length. The Gaussian has a full width at half maximum $\approx$0.06 pc. The time in units of both the background free-fall time and in Myr is indicated above the top left of each frame. The dotted horizontal line indicates the sound speed. It is seen that an outer sonic point (where the infall speed equals the sound speed—vertical red lines) moves outwards, indicating that the infall motions propagate outwards, similarly to the inside-out collapse, but starting from roughly one Jeans length from the center, rather than from the center itself, as is the case in that solution. The vertical blue lines denote the inner sonic point, which is seen to move inwards, indicating an outside-in collapse within the central Jeans length. Reprinted with permission from Naranjo-Romero et al. (2015). © 2015. The American Astronomical Society. All rights reserved.

7.4.1.2 The Protostellar Stage

As amply discussed in Section 7.3.2, the prestellar stage of collapse ends when the central density in the core diverges, corresponding to $t = 0$ in the treatment of that section. Later times correspond to the *protostellar stage*, in which the density profile becomes stationary, meaning that it does not change shape over time (see Equation (7.56)).

However, it is important to note that such stationarity can only occur for an r^{-2} density profile because, in this case, the mass accretion rate is independent of radius, and thus all the mass entering the core from the background medium is fully transferred to the protostar in the center, when the central mass is negligible. However, for different slopes of the density profile, or when the central mass is not negligible anymore, the two accretion rates differ, as will be discussed in the next section.

7.4.2 An Approximate Analytical Model of Core Evolution: Gas-driven Accretion and Gravitational Choking

An approximate model that allows calculation of the mass and size growth of cores, as well as the evolution of their density profile, was presented by Gómez et al. (2021) and Vázquez-Semadeni et al. (2024), and we follow their treatments here. The fundamental assumptions of this model are:

1. The core is embedded in a background, which can have either a uniform density, or a density profile with a different slope than that of the core itself.
2. The thermal pressure gradient is neglected. Therefore, at every radius, the gas is free-falling in response to the gravitational attraction of all the mass internal to that radius. During the prestellar stage, this comprises only the gaseous mass, while during the protostellar stage it comprises both the gaseous and the stellar mass.
3. The density profile at any time can be approximated as a single power law from the center of the core to where it merges with the background, of the form

$$\rho = \rho_0 \left(\frac{r}{r_0} \right)^{-p}, \tag{7.70}$$

where ρ_0 is the background density, and $r_0 \equiv L_J(\rho_0)$ is the Jeans length at that density. This means that the collapsing region is assumed to have the size of the background medium's Jeans length.

Assumption 3 above obviously breaks down within some inner radius, since in general the solutions discussed in Section 7.3.2 are characterized by *two* asymptotic profiles, not one. As discussed there, the inner part of the density profile during the prestellar stage is actually flat, with a finite central density, and the profile remains roughly uniform up to a radius of the order of the Jeans length for the central density and temperature (Equation (7.50)). Therefore, assumption 3 amounts to approximating the entire radial profile of the prestellar stage, including the central flat region, the outer r^{-2} envelope, and the region where the core's density profile merges into the background, by a single power law. In turn, for the protostellar stage of the collapse, this assumption amounts to taking an average slope between the two values of the inner and outer power-law regions.

In a sense, the single-slope model represents a core that is less idealized than the asymptotic solutions discussed in Section 7.3.2, since the very asymptotic nature of the latter implies that they are only applicable far from the boundaries and initial states. Moreover, the single-slope model, truncated at some radius r_0, does take into account the fact that the density profile's slope must become flatter both at its outer edge, where it merges into its background medium, and at the center, where becomes essentially zero in the central, flat-profile ($p = 0$) region. Therefore, the logarithmic slope of the density profile from the center to r_0 is, on average, $p < 2$. Of course, this treatment is also only an approximation, since the actual slope throughout this range is not constant, but it allows treatment of the initial and boundary condition effects, in particular the accretion onto the core and the resultant mass growth.

It is important to note that the total mass of a spherical core of radius r with the power-law density profile defined by Equation (7.70), is given by

$$M(r) = 4\pi\rho_0 r_0^p \int_0^r (r')^{2-p} dr', \tag{7.71}$$

and therefore remains finite for $p < 3$ in spite of the central divergence of the density, being equal to

$$M(r) = \frac{4\pi\rho_0 r_0^p}{3 - p} r^{3-p}. \tag{7.72}$$

7.4.2.1 Evolution of the Density Profile Slope

We can now derive an equation for the evolution of the density profile's logarithmic slope, p, during the collapse, assuming that it starts very small and it increases over time, as suggested by the numerical results described in Section 7.4.1. To this end, we consider the continuity equation, Equation (2.74), which we rewrite here as

$$\frac{\partial \rho}{\partial t} = -\nabla \cdot (\rho \mathbf{v}), \tag{7.73}$$

and assume that the radial inflow velocity at radius r is due to the gravitational attraction from infinity of the gaseous mass within r, so that, according to Equation (7.4) with $R \to \infty$, we have

$$v(r) = \sqrt{\frac{2GM(r)}{r}}. \tag{7.74}$$

Inserting Equation (7.72) in (7.74) we obtain

$$v(r) = -\left[\frac{8\pi G\rho_0 r_0^2}{3 - p}\left(\frac{r}{r0}\right)^{2-p}\right]^{1/2}. \tag{7.75}$$

In turn, inserting (7.75) in (7.73), assuming spherical geometry and only radial dependence, we obtain

$$\begin{aligned} -\nabla \cdot (\rho \mathbf{v}) &= -\frac{1}{r^2}\frac{\partial r^2 \rho v_r}{\partial r} \\ &= \left(3 - \frac{3p}{2}\right)\left[\frac{8\pi G\rho_0^3}{(3 - p)r_0^2}\right]^{1/2}\left(\frac{r}{r0}\right)^{-3p/2}. \end{aligned} \tag{7.76}$$

On the other hand, assuming that the density profile approximately evolves from one power law to another, so that p has a single value across the entire profile, and ρ is a function of p given by Equation (7.70), the left-hand side of Equation (7.73) is

$$\frac{\partial \rho}{\partial t} = -\rho_0 \left(\ln \frac{r}{r_0} \right) \left(\frac{r}{r_0} \right)^{-p} \frac{dp}{dt}. \tag{7.77}$$

Combining the latter two equations yields

$$\frac{dp}{dt} = \left(3 - \frac{3p}{2} \right) \left[\frac{8\pi G \rho_0}{3 - p} \right]^{1/2} \frac{(r/r_0)^{-p/2}}{-\ln(r/r_0)}. \tag{7.78}$$

Since we are considering the region where $r < r_0$, the sign of the right-hand side of Equation (7.78) is given exclusively by the factor $(3 - 3p/2)$. Therefore, if $p < 2$, p increases at all radii, while it decreases if $p > 2$, and reaches a steady state when $p = 2$. Thus, the value $p = 2$ is an *attractor* for the slope, meaning that the slope tends to converge to this value.

At this point it is important to note that Equation (7.78) shows that the slope p actually evolves at different rates at different radii because of the extra r-dependent factors in that equation, and thus the assumption that the profile remains a power law at all times is not strictly satisfied. Nevertheless, this equation provides an approximate means of understanding the evolution of the density profile, and its saturation at a slope of -2.

7.4.2.2 The Accretion Rate: Radial Dependence and Gravitational Choking

An important implication of the density profile slope is that it controls the accretion rate at radius r (the mass flux integrated over a spherical surface of radius r) due to the gravitational attraction of the gas mass internal to r. This rate is given by

$$\dot{M}(r) = 4\pi \rho(r) v(r) r^2, \tag{7.79}$$

where $v(r)$ is given by Equation (7.75) and $M(r)$ is given by Equation (7.72). Thus, the accretion rate becomes

$$\dot{M}(r) = \left(\frac{128\pi^3 G \rho_0^3 r_c^{3p}}{3 - p} \right)^{1/2} r^{\left(3 - \frac{3}{2}p\right)}. \tag{7.80}$$

This equation shows that, for $p = 2$, the accretion rate is independent of radius, and thus all the mass accreted by the core at its boundary is transferred to the stars in its center. However, for $p < 2$ the exponent of r is positive, and thus the mass accretion rate decreases with decreasing radius (i.e., the accretion decreases inwards). Therefore, not all of the mass entering the core through its outer boundary can be transferred to its central parts. Vázquez-Semadeni et al. (2024) referred to this mechanism as *gravitational choking*. It is important to note that this phenomenon does not involve any kind of support for the core; it is just that gravity cannot transfer mass at the required rate across the radius for a profile with $p < 2$. On the other hand, when $p = 2$, the accretion becomes independent of radius, as noted by

Li (2018), and thus all of the mass accreted by the core can be transferred to the central star. Also, according to Equation (7.78), the profile slope stops evolving, in agreement with the stationarity of the r^{-2} profile (Section 7.3.2.2).

7.4.2.3 Radial Growth of the Core after Slope Saturation

According to the results of the previous section, if for some reason the logarithmic slope of a core saturates at a value different from $p = 2$, the core's mass should continue to increase by gravitational choking. In fact, a recent compilation of the reported mean observed slopes in cores (Gómez et al. 2021) suggests that most cores have spherically-averaged slopes $p < 2$. The precise reason for this is unknown at the time of this writing, but Vázquez-Semadeni et al. (2024) suggested that the presence of filamentary accretion structures of nearly uniform density along their length may flatten the net spherically-averaged slope. This may explain the observed core growth both in the idealized, spherically symmetric simulation discussed in Section 7.4.1, which is also observed in more realistic simulations (Section 12.1.1).

7.4.3 The Unavoidability of a Central Unbound Region for $p < 2$

One very important result from the simple, single-slope model discussed in Section 7.4.2 is that *a collapsing core with a slope shallower than* -2 *necessarily contains an inner region that is locally Jeans-stable,* and thus contracts only because it is undergoing compression by the infall of the outer parts of the core. The presence of an inner Jeans-stable region can be seen by noting that the Jeans mass (Equation (6.16)) for an isothermal sphere with a power-law density profile of the form $\rho \propto r^{-p}$ scales with radius as $M_{\rm J}(r) \propto r^{p/2}$. On the other hand, the mass of the core itself scales with radius as $M(r) \propto r^{3-p}$. Therefore, the ratio of the core's mass to the Jeans mass scales as

$$\frac{M(r)}{M_{\rm J}(r)} \propto r^{3-\frac{3p}{2}}, \tag{7.81}$$

implying that, for $p < 2$, the ratio $M/M_{\rm J}$ decreases inwards. Therefore, even if the core as a whole is Jeans unstable, with a total mass $M_{\rm tot} > M_{\rm J}$, there exists some inner radius below which $M(r)$ becomes smaller than $M_{\rm J}(r)$, and the region internal to that radius is itself locally Jeans stable. This means that cores that appear Jeans-stable must be considered in the context of their surrounding clump. If the latter is unstable, then the central core is likely undergoing contraction by external compression, not by its own self-gravity.

7.5 Jeans Mass Reduction during Collapse: Hoyle's Gravitational Fragmentation

7.5.1 The Mechanism

Gravitational contraction in a nearly isothermal medium, such as molecular clouds, has the important implication that, as the mean density in the contracting region increases with time, the Jeans length and mass decrease as $\sim\rho^{-1/2}$, as dictated by

Equations (6.14) and (6.16) considered at constant temperature. The decrease of the Jeans mass implies that a certain Lagrangian (i.e., of constant mass) fluid parcel will contain an ever-larger number of Jeans masses as it contracts. Since every Jeans-mass parcel will have a tendency to begin contracting gravitationally on its own, this process, first discussed by Hoyle (1953), will tend to induce *gravitational fragmentation* of the cloud, clump, or core, meaning that it will tend to split into multiple pieces that begin contracting on their own. Each of these pieces will undergo a similar process, and therefore, this mechanism is a self-similar process of *collapses within collapses*. This process can only end when the fluid parcels become optically thick and begin trapping the heat generated by the collapse, therefore ceasing to be isothermal and increasing their effective polytropic exponent γ_e, until it becomes larger than 4/3, as discussed in Section 5.2.3.

The fragmentation process in this case can be seen as follows. By noting that, for a polytropic gas (cf. Section 5.2.3) with $P = P_0(\rho/\rho_0)^{\gamma_e}$, the sound speed is given by

$$c_s^2 = \left(\frac{\gamma_e P_0}{\rho_0^{\gamma_e}}\right)\rho^{\gamma_e - 1}. \tag{7.82}$$

It can then be readily checked that the Jeans length, Equation (6.14), is given by

$$L_J = \left[\frac{\pi\gamma_e P_0}{G\rho_0^{\gamma_e}}\left(\frac{3M}{4\pi}\right)^{\gamma_e - 2}\right]^{1/2} R^{3(2-\gamma_e)/2}. \tag{7.83}$$

Therefore, as the cloud contracts, the ratio of the cloud's radius to the Jeans length scales as

$$\frac{R}{L_J} \propto R^{1-3(2-\gamma_e)/2}.$$

We thus see that, for $\gamma_e < 4/3$, the exponent in the previous expression is negative, and *the ratio R/L_J increases as the radius decreases.* This means that an isothermal collapsing cloud will contain an ever increasing number of Jeans lengths (and of Jeans masses) as the collapse proceeds. Each Jeans-mass fragment may then begin to collapse itself, causing a hierarchical sequence of collapses within collapses. Note that the low-mass objects start to collapse later, because they have to wait until the global Jeans mass has decreased to the value of their own mass.

It important not to confuse the decrease of the Jeans mass as the cloud contracts discussed in this section from the inwards *increase* of the Jeans mass at any fixed time for a cloud or core with a spherically-averaged density profile with slope $p < 2$ discussed in Section 7.4.3. The latter refers to the *radial* variation of the Jeans mass at a *fixed* time, while the former refers to the *temporal* variation of the *average* Jeans mass within a cloud or core as it contracts over time.

7.5.2 Criticisms to Hoyle's Fragmentation and Resolution

In spite of being a very promising fragmentation mechanism, Hoyle's gravitational fragmentation was deemed unfeasible for some time after it was originally proposed because early numerical experiments showed that *linear* (i.e., low-amplitude) fluctuations within a spherical cloud do not have time to collapse themselves before the whole cloud completes its large-scale collapse (e.g., Tohline 1980). As discussed in Section 6.1.3, the reason for this is that the growth rate of collapse for the Jeans instability is largest for the largest-scale fluctuations (cf. Equation 6.12). However, it is important to note that the *linear* Jeans instability analysis, implicitly assumes that all density fluctuations have the same amplitude $\delta\rho$, independently of their wavenumber k, and that this amplitude satisfies $\delta\rho \ll \rho_0$, where ρ_0 is the background density. Therefore, all density fluctuations have essentially the same free-fall time, and thus their growth rate is determined exclusively by their wavenumber, according to Equation (6.12).

However, more recent three-dimensional simulations of self-consistent cloud formation by turbulent compressions or large-scale instabilities in the warm neutral medium (WNM, cf. Section 4.3.3) show that the clouds are capable of readily fragmenting. This may be due to at least two different properties of the clouds. The first is the fact that the density fluctuations in the clouds are in general *nonlinear* (i.e., of large-amplitude). These nonlinear fluctuations are the natural result of the presence of even moderately supersonic turbulence in the clouds, such as that produced by the cloud assembly mechanism (Section 10.1.3). Since molecular clouds are nearly isothermal, the typical relative amplitude of the density enhancements is $\delta\rho/\rho_0 \sim \mathcal{M}_s^2$, where $\mathcal{M}_s$ is the rms Mach number (cf. Section 3.9.2). Therefore, the typical free-fall time in the fluctuations, which is proportional to $\rho^{-1/2}$ (Equation (6.16)), is reduced by a factor $(\delta\rho/\rho_0)^{-1/2} \sim \mathcal{M}_s^{-1}$ in comparison to that of the background density, allowing them to collapse over significantly shorter timescales than those of the larger scales. In addition, the Jeans mass, which scales as $M_J \propto c_s^3 \rho^{-1/2}$ (Equation (6.16)), is reduced locally (at the density peaks) also by a typical factor $\mathcal{M}_s^{-1}$. Thus, for example, consider a supersonically turbulent isothermal flow with $\mathcal{M}_s \sim 3$, typical of the CNM clouds formed by transonic compressions in the WNM (Section 10.1.4). In this case, the typical turbulent fluctuations have densities ~ 9 times larger than the mean cloud density, and thus have *local* Jeans masses and free-fall times $\sim 1/3$ of the *mean* corresponding values for the entire cloud. Therefore, the cloud can readily fragment into clumps of mass $\sim$1/3 of the cloud's mass, and these clumps can collapse on a timescale $\sim 1/3$ of the cloud-scale free-fall time, even in near-spherical geometry.

The second property of clouds that may prevent the large-scale collapse from overwhelming the small-scale one (Section 7.5.2) is the *non-spherical shapes of the clouds*. In general, simulations of cloud formation in the WNM show that the clouds start out as thin CNM sheets, since they result from the condensation induced by the passage of a shock or one-dimensional compression wave (e.g., Heitsch et al. 2005, 2008; Vázquez-Semadeni et al. 2006, 2007; Heitsch & Hartmann 2008; Hennebelle et al. 2008; Inoue & Inutsuka 2009; Heiner et al. 2015; Fogerty et al. 2016). Indeed, the vast majority of turbulent compressions are one-dimensional (planar), since two- or three-dimensional ones would require the coincidence of three or more streams of gas at a given location at the same time, a very unlikely event, similar to three or more elementary particles colliding nearly simultaneously in a hot plasma.

For thin, sheet-like clouds, the growth rate of the instability, given by Equation (6.40), is maximum at twice the minimum unstable scale, and so fluctuations of this size can readily grow without being overwhelmed by the collapse of the whole cloud, as long as the cloud's shortest dimension is larger than twice the minimum unstable scale. In this case, the collapse generates intermediate filamentary structures (cf. Section 7.2.2). In Section 7.5.5 below, we discuss a further important feature of filamentation; namely, that it allows *conveyor-belt flow*, a form of non-concentric, hierarchical, multi-scale gravitational contraction in a finite medium.

7.5.3 Sequential Onset of Collapse of Subsequently Smaller-Scale Turbulent Fluctuations

An important feature of gravitational fragmentation that was not discussed by Hoyle (1953) is that *subsequently smaller fragments must begin to collapse at subsequently later times*, because they can only do it after the global Jeans mass has decreased to a value comparable to the fragments' masses. This can be seen as follows: consider a Lagrangian cloud (i.e., of fixed mass) that begins to contract when its mass equals the Jeans mass at its mean density and temperature. As it contracts and becomes denser, the mean Jeans mass within it becomes smaller, and thus lower-mass turbulent fluctuations within the cloud can now begin to contract themselves. But some amount of time is necessary for this to occur.

An idealized model for the sequential fragmentation of a collapsing spherical, isothermal region of radius $R(t)$, mean density $\rho(t)$, with constant mass (M) and turbulent Mach number ($\mathcal{M}_s$), can be constructed as follows (Vázquez-Semadeni et al. 2019). The treatment is simplified by using a simple fit to the evolution of the radius of a free-falling sphere obtained by Girichidis et al. (2014), and given by

$$R(\tau) = R_0(1 - \tau^2)^{a/3}, \tag{7.84}$$

where $\tau \equiv t/t_{\rm ff,0}$ is the time in units of the free-fall time for the initial density (ρ_0) of the object ($t_{\rm ff,0}$), R_0 is the initial radius, and $a \approx 1.8614$ is a parameter for which the fit remains within 0.5% of the actual solution during the whole collapse time.

Under the assumption of constant mass and spherical shape, the evolution of the cloud's mean density follows straightforwardly from Equation (7.84) as

$$\rho(\tau) = \rho_0(1 - \tau^2)^{-a}. \tag{7.85}$$

In turn, from the latter equation, the evolution of the Jeans mass $M_{\rm J}$, proportional to $\rho^{-1/2}(\tau)$ since the cloud is assumed isothermal, is given by

$$M_{\rm J}(\tau) = M_{\rm J}(\rho_0)(1 - \tau^2)^{a/2}. \tag{7.86}$$

The above equation gives the evolution of the cloud's *mean* Jeans mass. However, it is necessary to account for the fact that the medium is turbulent, and therefore populated by density fluctuations of rms amplitude $\rho_{\rm rms}$, given by

$$\rho_{\rm rms}(\tau) \sim \mathcal{M}_s^2 \rho(\tau). \tag{7.87}$$

This is important in avoiding the problem discussed in Section 7.5.2 that *linear*, small-scale density fluctuations collapse more slowly than the cloud itself due to the larger relative importance of the pressure gradient against gravity at smaller scales, according to the Jeans dispersion relation, Equation (6.12), which caused the dismissal of Hoyle's fragmentation as a viable fragmentation mechanism for near-Jeans clouds (Tohline 1980). Instead, in the *nonlinear* turbulent regime, the density fluctuations have a density $\rho_{\rm rms}$, significantly larger than the mean density even for moderately supersonic turbulence ($\mathcal{M}_{\rm s} \sim 3$), and therefore a significantly shorter free-fall time than that for the mean density. This allows the small-scale turbulent fluctuations to still collapse faster than the cloud as a whole.

In view of the above, we consider the evolution of the Jeans mass for the rms density fluctuation, given by

$$M_{\rm J,rms}(\tau) \approx \frac{M_{\rm J}(\rho_0)}{\mathcal{M}_{\rm s}}(1 - \tau^2)^{a/2}. \tag{7.88}$$

This equation can be then inverted to find the time $t_{M,\,\rm u}$ required for a mass $M = M_{\rm J,rms}$ to become unstable. In units of the initial free-fall time in the cloud, we find this time to be

$$\tau_{\rm M,u} \equiv \frac{t_{M,\,\rm u}}{t_{\rm ff,\,0}} \approx \left[1 - \mu^{2/a}\right]^{1/2}, \tag{7.89}$$

where $\mu \equiv \mathcal{M}_{\rm s} M / M_{\rm J}(\rho_0)$ is the fragment mass normalized to the initial fluctuation Jeans mass in the cloud. Figure 7.6 shows this time for destabilization (in units of the

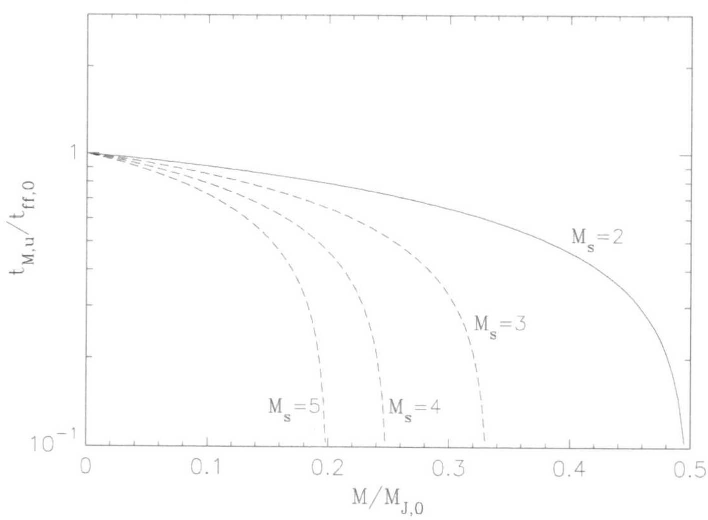

Figure 7.6. Time for destabilization (i.e., to the onset of local collapse after the onset of global collapse) of a fragment of density $\rho_{\rm rms}$ and mass M for various values of the turbulent Mach number, $\mathcal{M}_{\rm s}$, as given by Equation (7.89). Fragments of smaller masses start to collapse later, because they have to wait until the instantaneous Jeans mass for the whole cloud (Equation (7.86)) has decreased sufficiently to make them unstable. As the fragment's mass approaches zero, the time at which it becomes unstable approaches the free-fall time of the entire cloud. The time and mass are respectively normalized to the free-fall time and the Jeans mass at the mean initial density. Repoduced from Vázquez-Semadeni et al, (2024). CC BY 4.0 © 2024. Published by Oxford University Press on behalf of Royal Astronomical Society.

initial free-fall time) as a function of the fluctuation's mass M (in units of the initial Jeans mass) for various values of the Mach number. We consider only masses $\leqslant M_{J,0}/2$ assuming that fragmentation starts when there are at least two Jeans masses in the cloud. This figure shows that, for the same fluctuation density ρ_{rms}, fluctuations of larger masses become unstable at earlier times. As the fragment's mass approaches zero, the time at which it becomes unstable approaches the free-fall time of the entire cloud. Only fragments that satisfy the condition

$$\tau_{M,u} + \frac{t_{ff}(\rho_{rms})}{t_{ff,\,0}} < t_{ff,\,0} \tag{7.90}$$

manage to terminate their own collapse before the parent cloud does.

Of course, this model is extremely idealized, as it assumes spherical geometry, fixed fragment mass, and that all scales collapse on their free-fall time. Nevertheless, it provides a proof of concept for the delayed onset of collapse of the turbulent density fluctuations within a gravitationally contracting cloud, under spherical symmetry.

7.5.4 The Self-Similar Nature of Gravitational Fragmentation

A crucial feature of gravitational fragmentation is that, as long as the flow remains nearly isothermal (or, more generally, near-polytropic, with $\gamma_e < 4/3$), the process of fragmentation can continue indefinitely. This happens because, as discussed in Section 5.2.3, as the density in the inner parts of a collapsing Lagrangian (fixed mass) region increases, its Jeans mass decreases, and so the region contains an ever-larger number of Jeans masses. Thus, at early stages, the stable inner parts are relatively large, and so they remain coherent, as a unit. But once the Jeans mass becomes sufficiently smaller than their own mass, then these regions can undergo the same fragmentation process previously undergone by their parent structure. In fact, the fragmentation process only stops when the column density becomes large enough that the region becomes optically thick and begins to trap the radiation, thus becoming nearly adiabatic. At this time, γ_e becomes larger than 4/3, and the Jeans mass begins to *increase* with increasing density, eventually halting the collapse. This corresponds to the formation of the so-called *first hydrostatic (or Larson) core* (Larson 1969) during the formation of a protostar. Were it not for this transition to a nearly adiabatic regime, the final outcome of the collapse of an isothermal cloud would be an arbitrarily large number of vanishing-mass fragments.

7.5.5 Hierarchical Multi-Center Collapse and Conveyor-Belt Flow

The second feature of molecular clouds that restores the feasibility of Hoyle's gravitational mechanism mentioned in Section 7.5.2 is that they are non-spherical in general. This causes an amplification of anisotropies (Section 7.2.2), so that a cloud fragments into sheets and filaments, and the gravitationally-driven flow proceeds from the clouds to the sheets, from sheets to filaments, and from filaments to roundish collapse centers (cores) again. This flow sequence may start over after the cores accumulate many Jeans masses—possibly through the growth mechanism described in Section 7.4.2—and begin to collapse on their own. Numerical simulations (e.g., Gómez & Vázquez-Semadeni 2014) have revealed an important property of filaments formed by this mechanism: that the filaments tend to have roughly constant mean central density along their length, except at the cores. Therefore, the filaments provide a background medium in which the isolated collapse of cores can occur. If several cores of different masses form in a filament, the main flow along the filament will be directed toward the most massive core (the *hub*), while the collapses of the less massive cores will be *local*, riding on top of the large-scale, *global* flow toward the hub, in a *conveyor belt* fashion (Longmore et al. 2014). This mechanism is discussed further in Section 11.4, and illustrated in Figure 11.3.

It is important to note that the non-spherical, non-concentric nature of the conveyor-belt flow is crucial, as it provides a realistic mechanism for the appearance of a regime of *collapses-within-collapses*. This is not possible under the assumption of strict spherical geometry, since in this case the collapse of the uniform background medium would overtake that of any locally collapsing core immersed in it. Instead, a collapsing uniform filament has a longer contraction timescale than that of the core (Section 7.2.3), and furthermore the flow can be directed toward a different collapse center than that of the core. Thus, the two collapse flows can coexist, each on its own timescale and onto its own collapse center.

7.6 Highlights and Insight

In this chapter we have covered a long list of topics related to gravitational contraction in dense clouds, from standard concepts such as homologous collapse, the free-fall time, inside-out protostellar-stage collapse, and turbulent fragmentation, to less standard or somewhat-forgotten results from several decades ago, such as the collapse times for high-aspect ratio objects, the asymptotic inner and outer regimes for the pre- and protostellar stages, outside-in dynamical collapse initiated during the prestellar stage, and Hoyle's gravitational fragmentation and its applicability. The main, often-neglected conclusions from this chapter are:

- The free-fall time is a *lower limit* to the actual collapse time of a real marginally unstable object, since the net, unbalanced gravitational force driving the contraction is very small in the early stages of the collapse of a near-Jeans mass object. Real collapse always occurs on timescales of a few free-fall times.
- Similarly, homologous collapse is an idealization that cannot occur in nature, as any deviations from perfect sphericity and uniform density are amplified by free-fall collapse. Therefore, real collapse occurs anisotropically and asynchronously, generating a *gravitationally-driven flow*.
- The collapse of strongly anisotropic structures (sheets and filaments) takes a significantly longer time than the free-fall time for a spherical structure of the same density as that of the anisotropic structure, and a diameter equal to the structure's largest dimension. The collapse time of sheets is larger by a factor $\sim A^{1/2}$, where A is the aspect ratio, while that of filaments can be larger by a

factor $\sim A^{1/2}$ in the case of edge-on collapse, and by a factor $\sim A$ for homologous longitudinal contraction.

- Clouds formed by generic compressions in the WNM most likely start out as thin atomic sheets. Thus, molecular clouds are expected to be sheet-like in general. Spheroidal clouds are expected to be the exception rather than the norm.
- Non-homologous collapse generates a *flow*, so that material at initially different densities terminates its collapse (i.e., reaches diverging density) at different times. This is one important physical meaning of the *similarity variable* $\xi = r/c_s t$: material at a larger radial distance from the collapse center (and therefore typically at a lower density) can be considered to be at an earlier stage of its evolution toward reaching the collapse center, and takes a longer time to get there.
- The standard inside-out collapse solution for isothermal spherical collapse should be considered as an idealized situation, which could only occur if the pre-singularity (prestellar) evolution of the core ocurred quasistatically, and the dynamic collapse started from an SIS configuration, which is unattainable in reality.
- Realistic solutions must have started to collapse long before reaching the central singularity, and therefore must have a well-developed infall velocity profile by the time the singularity is formed. If they start from rest, their protostellar stage may be approximated by Shu's solution with $A > 2$, and thus the infall motions extend significantly further out than predicted by the inside-out, $A = 2$ solution (see, e.g. the dotted line corresponding to $A = 2.5$ in the top panel of Figure 7.3).
- As an isothermal region of fixed mass, M, collapses, the Jeans mass within it decreases, implying that the region contains an ever-larger number of Jeans masses, allowing it to undergo Hoyle gravitational fragmentation. For this to occur, however, it is necessary that the region is either significantly non-spherical (flattened or elongated) and/or contains nonlinear density fluctuations. Both conditions are feasible in molecular clouds because of the amplification of anisotropies caused by the collapse itself, and because of the presence of moderately supersonic turbulent fluctuations within the clouds.
- Hoyle fragmentation must occur sequentially in time, with the smallest-scale fluctuations (at a fixed density) initiating their collapse the latest.
- An important implication of the sequential destabilization of subsequently smaller scales during gravitational fragmentation is that, in this process, *collapse occurs from the outside-in*, since the largest scales are the first ones to become unstable.
- The collapse from the outside-in is contrary to the *inside-out collapse* proposed by S77. The reason for the difference is the assumption of a *singular* initial condition made by S77, which effectively bypasses the sequential destabilization process of the various scales within a core, so that all scales are already unstable at the initial condition.

References

Alecian, G., & Léorat, J. 1988, A&A, 196, 1
Bond, J. R., Kofman, L., & Pogosyan, D. 1996, Natur, 380, 603
Burkert, A., & Hartmann, L. 2004, ApJ, 616, 288
Fogerty, E., Frank, A., Heitsch, F., et al. 2016, MNRAS, 460, 2110
Foster, P. N., & Chevalier, R. A. 1993, ApJ, 416, 303
Girichidis, P., Konstandin, L., Whitworth, A. P., & Klessen, R. S. 2014, ApJ, 781, 91
Gómez, G. C., & Vázquez-Semadeni, E. 2014, ApJ, 791, 124
Gómez, G. C., Vázquez-Semadeni, E., & Palau, A. 2021, MNRAS, 502, 4963
Hartmann, L. 2009, Accretion Processes in Star Formation in 2nd ed. (Cambridge: Cambridge Univ. Press)
Heigl, S., Hoemann, E., & Burkert, A. 2022, MNRAS, 517, 5272
Heiner, J. S., Vázquez-Semadeni, E., & Ballesteros-Paredes, J. 2015, MNRAS, 452, 1353
Heitsch, F. 2013, ApJ, 769, 115
Heitsch, F., & Hartmann, L. 2008, ApJ, 689, 290
Heitsch, F., Burkert, A., Hartmann, L. W., Slyz, A. D., & Devriendt, J. E. G. 2005, ApL, 633, L113
Heitsch, F., Hartmann, L. W., Slyz, A. D., Devriendt, J. E. G., & Burkert, A. 2008, ApJ, 674, 316
Hennebelle, P., Banerjee, R., Vázquez-Semadeni, E., Klessen, R. S., & Audit, E. 2008, A&A, 486, L43
Hoyle, F. 1953, ApJ, 118, 513
Hunter, C. 1977, ApJ, 218, 834
Inoue, T., & Inutsuka, S.-i. 2009, ApJ, 704, 161
Larson, R. B. 1969, MNRAS, 145, 271
Li, G.-X. 2018, MNRAS, 477, 4951
Lin, C. C., Mestel, L., & Shu, F. H. 1965, ApJ, 142, 1431
Longmore, S. N., Kruijssen, J. M. D., Bastian, N., et al. 2014, Protostars and Planets VI, ed. H. Beuther (Tucson, AZ: Univ. Arizona Press) 291
Lyttleton, R. 1953, The Stability of Rotating Liquid Masses (Cambridge: Cambridge Univ. Press)
Mouschovias, T. C. 1991, The Physics of Star Formation and Early Stellar Evolution, ed. C. J. Lada, & N. D. Kylafis, NATO Advanced Study Institute (ASI) Series C (Dordrecht: Springer) 342, 449
Naranjo-Romero, R., Vázquez-Semadeni, E., & Loughnane, R. M. 2015, ApJ, 814, 48
Penston, M. V. 1969a, MNRAS, 145, 457
Penston, M. V. 1969b, MNRAS, 144, 425
Pon, A., Johnstone, D., & Heitsch, F. 2011, ApJ, 740, 88
Pon, A., Toalá, J. A., Johnstone, D., et al. 2012, ApJ, 756, 145
Shu, F. H. 1977, ApJ, 214, 488
Shu, F. H., Adams, F. C., & Lizano, S. 1987, ARA&A, 25, 23
Toalá, J. A., Vázquez-Semadeni, E., & Gómez, G. C. 2012, ApJ, 744, 190
Toalá, J. A., Vázquez-Semadeni, E., & Gómez, G. C. 2015, ApJ, 808, 105
Tohline, J. E. 1980, ApJ, 239, 417
Vázquez-Semadeni, E., Ryu, D., Passot, T., González, R. F., & Gazol, A. 2006, ApJ, 643, 245
Vázquez-Semadeni, E., Gómez, G. C., Jappsen, A. K., et al. 2007, ApJ, 657, 870

Vázquez-Semadeni, E., Palau, A., Ballesteros-Paredes, J., Gómez, G. C., & Zamora-Avilés, M. 2019, MNRAS, 490, 3061
Vázquez-Semadeni, E., Gómez, G. C., & González-Samaniego, A. 2024, MNRAS, 530, 3445
Whitworth, A., & Summers, D. 1985, MNRAS, 214, 1
Wyse, A. B., & Mayall, N. U. 1942, ApJ, 95, 24

Part II

Cloud Evolution and Star Formation

AAS | IOP Astronomy

Interstellar Flow and Star Formation

Enrique Vázquez-Semadeni

Chapter 8

Molecular Clouds and Their Star-forming Properties

8.1 Introduction

Most star formation (SF) in the present-day Universe occurs in molecular clouds (hereafter, MCs). Therefore, understanding SF requires an understanding of the origin, structure and evolution of MCs. Excellent reviews exist about them (e.g., Blitz 1993; Heyer & Dame 2015), and so in this chapter we present only a brief overview of the MC properties most relevant for the subsequent discussion.

8.2 What and Where They Are

8.2.1 Composition, Location and Structure

Molecular clouds and their substructures ("clumps", "filaments", "cores", and core "fragments") are the densest regions of the ISM, in which hydrogen exists predominantly in molecular (H_2) form (cf. Tables 1.1 and 1.2). Also, they are often surrounded by cold atomic gas envelopes which are, nevertheless, slightly warmer than the molecular gas ($T \sim$ 50–100 K; e.g., Wannier et al. 1983, 1991; Fukui et al. 2009; Pascucci et al. 2015; Zucker et al. 2021). As listed in Table 1.2, the densities of MCs start out at $\sim$ 100 cm^{-3}, although at these lower densities, typical of MC outskirts, the clouds can still contain significant amounts of cold HI and CO-dark molecular gas (e.g., Li & Goldsmith 2003; Xu et al. 2016). On the other hand, in their internal regions, their substructures can reach much higher densities.[1] Molecular cloud temperatures are $\sim$ 10–30 K, and their sizes are of the order of tens or even hundreds of parsecs, in the case of giant molecular clouds (GMCs).

Although most of the mass in MCs is in H_2 molecules, it is well known that this molecule is not observable because, at the typical temperatures of the clouds

[1] In fact, since stars form within the clouds, it can be said that there is a continuum of densities within MCs all the way up to stellar densities.

doi:10.1088/2514-3433/adcb28ch8

(~ 10 K), collisions among the molecules can only excite rotational transitions. However, once collisionally excited, an H_2 molecule cannot be radiatively de-excited by electric dipole transitions, because this molecule has zero electric dipole. Therefore, the main tracer to observe them is the molecule of carbon monoxide, CO, in spite of its abundance being much smaller than that of hydrogen ($[CO/H_2]. \approx 1.8 \times 10^{-4}$). Other molecules, with higher *effective* (Shirley 2015) excitation densities are used as tracers of the denser interiors of the MCs.

The location and apparent structure of the MCs is important for understanding their origin and evolution. The majority of the molecular gas in grand-design spiral galaxies is located at or near the spiral arms (see *left panel* of Figure 8.1), suggesting that the arms constitute the main molecular gas formation mechanism in this type of galaxies. Moreover, as clearly seen in this figure, in the parts of spiral arms closer to the Galaxy's center the molecular gas appears continuously distributed over several kiloparsecs, rather than existing in clearly isolated "clouds". On the other hand, clouds that are clearly isolated are discernable in the outer arms and in the interarm regions. In non-grand-design galaxies, the molecular gas still forms large, kpc-scale complexes, in spite of the absence of clearly defined large spiral arms.

Next in importance for molecular gas content appears to be the intersection of super-bubbles produced by collective supernova explosions (e.g., Fujii et al. 2021; Zucker et al. 2022; see *right panel* of Figure 8.1).

Structurally, MCs are highly complex structures, *far from any spherical symmetry*, which consist of a network of filaments with dense, massive condensations known as "hubs" at their intersections (see Myers 2009; Figure 8.2). In addition, the filaments themselves are complex, containing the majority of the lower mass condensations known as "cores" (Könyves et al. 2015) and sub-filaments, the latter discernable in position–velocity maps (Hacar et al. 2013). In recent years, it has been realized that the filaments and hubs appear to be parts of one larger-scale system referred to as a

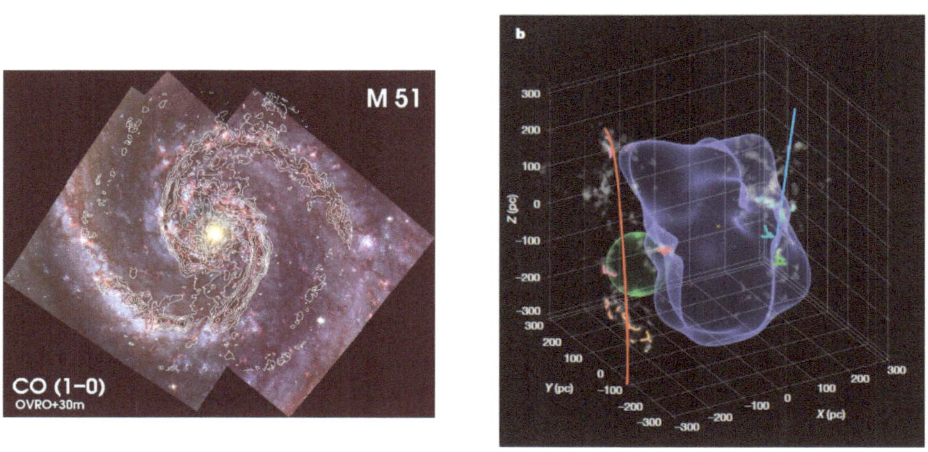

Figure 8.1. *Left:* Map of the Whirlpool galaxy, or M51, in visible light, with superposed contours of CO emission. The CO is seen to lie mostly at or near the spiral arms. CC BY 4.0. Credit: NRAO/AUI/NSF. *Right:* Schematic diagram showing the location of nearby MCs (in gray) in the Solar Neighborhood, showing their relation to the Local Bubble (purple) and the Per-Tau Superbubble (green). The Sun's location is marked by the "×" symbol. Reproduced from Zucker et al., (2022), with permission from Springer Nature.

Figure 8.2. Optical image of the Ophiucus molecular cloud complex, showing the embedded star cluster and the irregular and filamentary structure of the gas, seen as the dark lanes produced by the obscuration by the dust mixed with the gas. The white bar indicates a scale of 5 pc. Reprinted with permission from Myers (2009).

hub-filament system (HFS). Moreover, massive star formation appears to occur preferentially in these systems (e.g., Kumar et al. 2020). Finally, in recent years, some observational evidence has been surfacing that the molecular clouds may be flattened structures (Beaumont & Williams 2010; Storm et al. 2016; Veena et al. 2017; Kusune et al. 2019; Kabanovic et al. 2022; Rezaei & Kainulainen 2022).

8.2.2 Vagueness in Their Definition

As discussed in Section 1.2.1, it must be kept in mind that the definition of clouds is a highly ambiguous task, since they are part of a density continuum, with generally smooth, rather than sharp, physical and chemical transitions from the fully atomic to the fully molecular components. Observationally, MCs were initially defined by the detection of CO emission (e.g., Wilson et al. 1970; see also the discussion in Blitz 1993). However, the extent of the clouds in this case obviously depends on the sensitivity of the instrument and the integration time of the observation. Moreover, in the last couple of decades, it has been realized that a significant fraction of a MC's mass may not be bright in CO, and these regions of the clouds have been termed "CO-dark" (Grenier et al. 2005; Langer et al. 2010). This material can be traced in dust emission or absorption, or in other tracers such as the OH molecule and the C^+ ion. Therefore, the entire dense, cold density structures where MCs reside, including the CO-dark and the cold-HI regions, are likely to be significantly larger and more massive than the material directly observable in CO emission. This has the very important implication that the gravitational potential where the MCs are immersed is therefore determined by the whole cold and dense material, and not just by the regions observable in CO.

8.3 Physical Parameters

8.3.1 Masses, Sizes, Magnetizaton, and Kinematics

Molecular clouds are the coldest, densest, and most massive structures in the ISM. As seen from Table 1.2, Giant Molecular Clouds (GMCs) have masses $\sim 10^4$–10^6 $M_\odot$, densities $n \gtrsim 100$ cm^{-3}, and temperatures ~ 10 K. It is important to notice that, at these densities and temperatures, the adiabatic sound speed is $c_s \approx 0.2$ km s^{-1}, and thus the Jeans mass (cf. Equation (6.16)) is $\sim 45\ M_\odot$. Therefore, a GMC with a mass $M \sim 10^5\ M_\odot$ contains over 2000 Jeans masses. In Section 10.1.2 we discuss how they may attain such a strongly Jeans-unstable nature.

Molecular clouds are also highly dynamical structures. As listed in Table 1.2, the nonthermal velocity dispersions of giant molecular clouds (GMCs), derived from the molecular line widths, are $\sim$ 5–15 km s^{-1}, which correspond to Mach numbers $\mathcal{M}_s \gtrsim 20$. It is also important to note that Brunt et al. (2009) showed, using Principal Component Analysis (Heyer & Schloerb 1997), that the main contribution to the velocity dispersion in clouds is a "dipole-like" structure in the line-of-sight velocity, meaning that the line-centroid velocities vary from positive to negative velocities with respect to the cloud's centroid across the whole extension of the cloud; i.e., the clouds typically exhibit a large-scale velocity gradient (Figure 8.3). Such gradients have been known for several decades, and they have been generally interpreted as rotation (e.g., Goldsmith & Arquilla 1985; Goodman et al. 1993; Belloche 2013), although, they can also be interpreted as shear or inflow.

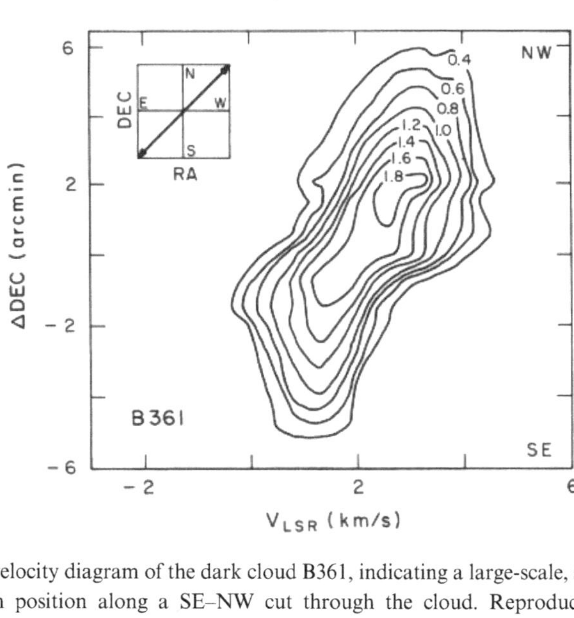

Figure 8.3. Position–velocity diagram of the dark cloud B361, indicating a large-scale, smooth variation of the velocity centroid with position along a SE–NW cut through the cloud. Reproduced from Arquilla and Goldsmith (1986).

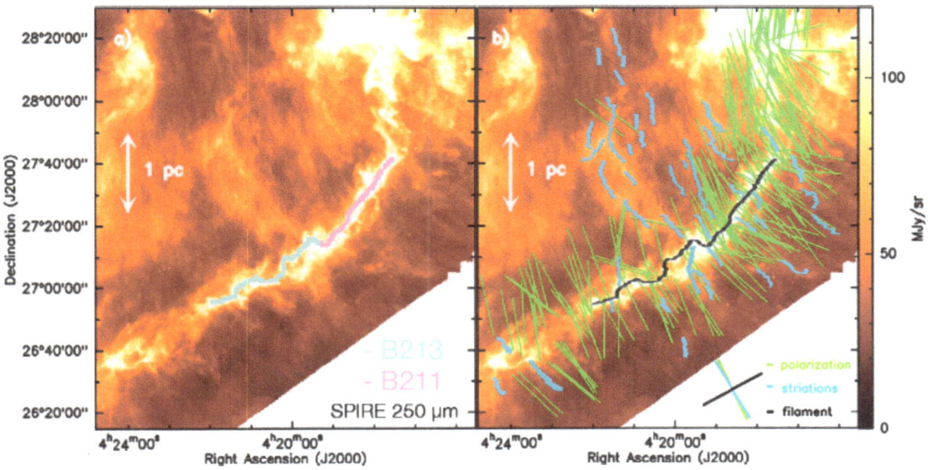

Figure 8.4. Herschel/SPIRE 250 μm image of the B211/B213/L1495 region in Taurus. The light blue and purple lines show the filaments identified in the region, and the green lines show the polarization orientation, here indicating the magnetic field direction. The field in the periphery of the main filament appears to be mostly perpendicular to it, and parallel to the low-density, secondary filaments ("striations"), which appear as "tributaries" to the main one. Palmeirim et al., (2013), reproduced with permission © ESO.

Furthermore, MCs are also in general permeated by magnetic fields, whose orientation in the neighborhood of molecular cloud filaments tends to be perpendicular to them (Figure 8.4). However, in the few currently existing observations sufficiently resolved to map the field in the interior of the filaments, the field appears to be preferentially aligned with them (Arzoumanian et al. 2021). The strength of the field is of the order of a few μG, and tends to increase with density, as we discuss below in Section 8.4.

8.4 Scalings among Variables

8.4.1 Larson's Relations, the Virial Parameter, the Larson Ratio, and the Keto–Heyer Diagram

A fundamental property of MCs is that their physical "parameters" (although they are actually physical variables) exhibit well defined scaling relations among themselves. The sizes (often denoted R or L, referring to radius or diameter, respectively), densities, velocity dispersions (as measured by the line widths of various molecular tracers such as CO) and magnetic field strengths of the clouds are not completely random in their magnitudes, but rather are observed to correlate with each other, as first noticed by Larson (1981). In a collection of clouds from the literature available at that time, he found the scalings (plotted in the top panels of Figure 8.5)

$$\langle n_{\rm H_2}\rangle \approx 3400\left(\frac{L}{\rm pc}\right)^{-1.1} {\rm cm}^{-3}, \tag{8.1}$$

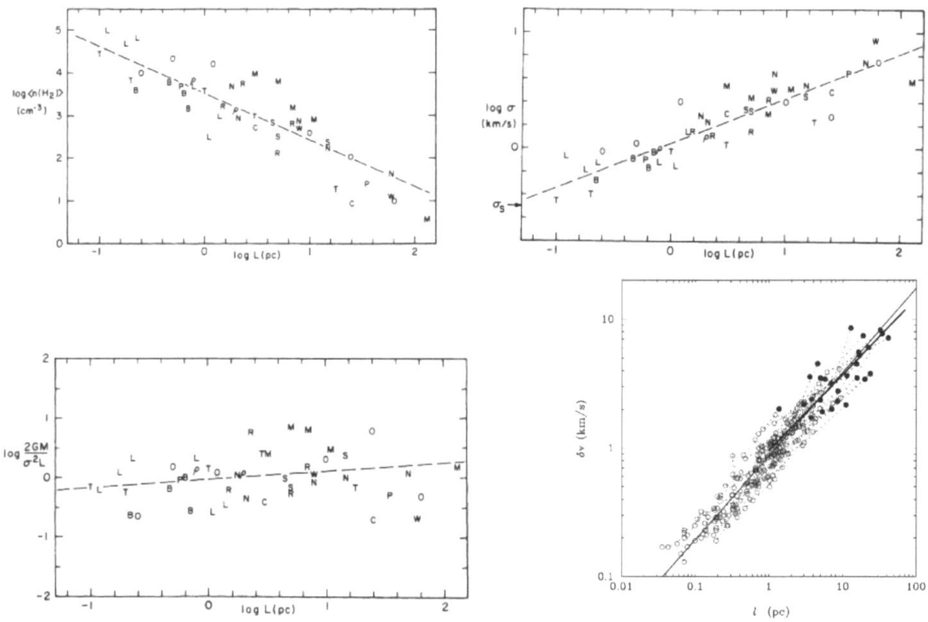

Figure 8.5. Original plots showing the scaling of the mean hydrogen density $\langle n_{\mathrm{H}_2} \rangle$ (*top left panel*); reproduced from Larson (1981) with permission from Oxford University Press on behalf the Royal Astronomical Society. Velocity dispersion $\sigma_{3\mathrm{D}}$ (or line width) (*top right panel*); reproduced from Larson (1981) with permission from Oxford University Press on behalf the Royal Astronomical Society. Gravitational-to-kinetic energy ratio (*bottom right panel*) with the approximate diameter L of a compilation of clouds; reprinted with permission from Heyer and Brunt (2004). © 2004. The American Astronomical Society. All rights reserved. Printed in U.S.A. (*Bottom right panel*: A more modern version of the velocity dispersion-size relation. Reproduced from Larson (1981) with permission from Oxford University Press on behalf the Royal Astronomical Society.

and

$$\sigma_{3\mathrm{D}} \approx 1.1\left(\frac{L}{\mathrm{pc}}\right)^{0.38} \mathrm{km\ s^{-1}} \tag{8.2}$$

where L is the cloud's approximate diameter, $\langle n_{\mathrm{H}_2} \rangle$ is the cloud's mean molecular hydrogen number density, and $\sigma_{3\mathrm{D}}$ is the three-dimensional velocity dispersion in the cloud. On the basis of more modern and homogeneous data sets, more recent studies have revised the velocity dispersion-size relation (also referred to as the *linewidth-size relation*) to (e.g., Solomon et al. 1987; Heyer & Brunt 2004; Heyer & Dame 2015)

$$\sigma \approx 1\ \mathrm{km\ s^{-1}}\left(\frac{L}{\mathrm{pc}}\right)^{0.5}, \tag{8.3}$$

which is shown in the *bottom right* panel of Figure 8.5. This relation has a slightly steeper logarithmic slope (larger exponent) than the one originally found by Larson, and is very similar to the scaling expected for strongly supersonic, compressible turbulence (Section 3.9.4.1, Equation (3.61)).

Larson also pointed out that, taken together, the density-size and the line width-size relations imply that the energy ratio $|E_g|/2E_k$ is given by

$$\frac{|E_g|}{2E_k} = \frac{2GM}{\sigma_{3D}^2 L} \approx 0.92\left(\frac{L}{\rm pc}\right)^{0.14}, \tag{8.4}$$

and it thus remains nearly constant and of order unity over the entire cloud ensemble, as seen in the *bottom left* panel of Figure 8.5.[2] This implies near equipartition between the kinetic and gravitational energies of the clouds. The kinetic energy, in principle, includes both thermal and nonthermal motions. However, in practice, since the observed line widths correspond to supersonic motions for size scales $L \gtrsim 1$ pc, the equipartition is in general considered to exist between the kinetic energy *in nonthermal motions* and the gravitational energy. These nonthermal motions are almost always interpreted as small-scale turbulence that can provide support against gravity, although this is not the only possibility and, as we will repeatedly mention in the remainder of this book, they can also correspond to large-scale systematic motions, such as infall.

As a consequence of the near equipartition, the virial parameter of molecular clouds ($\alpha_{\rm vir} = 2E_k/|E_g|$; cf. Equation (5.24)) is in general of order unity, and the Larson ratio $\mathcal{L} \equiv \sigma/R^{1/2}$ (cf. Equation (5.26)) is of the order of its virial value, $\mathcal{L} \approx (\pi G\Sigma/5)^{1/2}$, as initially shown by Keto & Myers (1986) and Heyer et al. (2009). However, more recent work has shown that $\alpha_{\rm vir}$ exhibits an additional dependence on the mass of the objects, decreasing as the object's mass increases (left panel of Figure 8.6). Similarly, the Larson ratio exhibits a dependence on the column density Σ, varying from significantly super-virial values at low Σ to nearly virial values at intermediate and high Σ. This is shown in the *right panel* of Figure 8.6, which we refer to as the *Keto–Heyer*, or KH, diagram. These trends extend to dense, massive clumps, so that denser and/or more massive clumps appear more *sub-virial* (i.e., with $\alpha_{\rm vir} < 1$; Figure 8.7).

It is important to note, however, that the $\alpha_{\rm vir}$-M and KH diagrams, taken together, imply that the low-Σ objects that have $\alpha_{\rm vir} \gg 1$ are the *least* massive objects in that column-density range. This can be seen in the *left panel* of Figure 8.8, which shows the $\alpha_{\rm vir}$-M scaling for samples of different classes of objects. All classes, each characterized by a certain typical column density, exhibit similar $\alpha_{\rm vir}$-M scalings, but displaced in mass. This implies that objects of similar masses belonging to different density regimes can have very different values of $\alpha_{\rm vir}$. That is, the $\alpha_{\rm vir}$-M scaling only holds for objects of similar density. This feature is also observed in numerical simulations, as illustrated in the *right panel* of Figure 8.8, which shows the distribution in the $\alpha_{\rm vir}$-M diagram of a sample of clouds and clumps extracted from

[2] Note that Larson wrote $E_g \approx 2GM^2/L$ for the gravitational energy, while in Section 5.2.5.1 we used the uniform-sphere value, $E_g = 3/5\, GM^2/R$, with $R = L/2$. Therefore, using Larson's definition, the condition $E_g/2E_k = 1$ implies that the virial parameter $\alpha_{\rm vir}$, as defined in Equation (5.24), takes the value $\alpha_{\rm vir} = 5/3$.

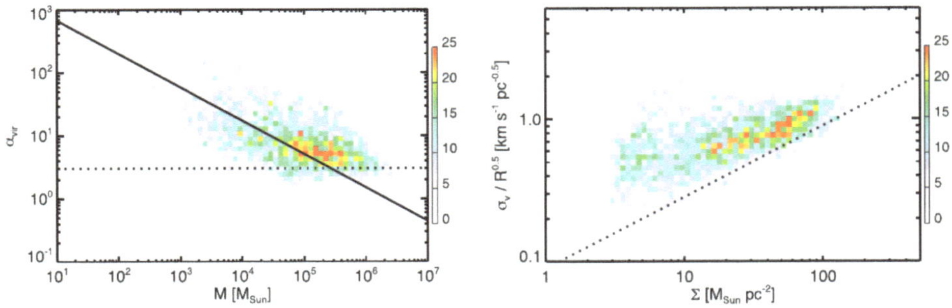

Figure 8.6. *Left:* Virial parameter $\alpha_{\rm vir}$ (cf. Equation (5.24)) *versus* mass for a sample of Milky Way molecular clouds by Miville-Deschênes et al. (2017). *Right:* The "Larson ratio", $\mathcal{L} \equiv \sigma/R^{1/2}$, *versus* column density Σ for the same sample. The dotted line corresponds to the threshold of "gravitationally bound" structures (which those authors define as objects with $\alpha_{\rm vir} = 3$) in both panels. Reprinted with permission from Miville-Deschenes et al. (2017).

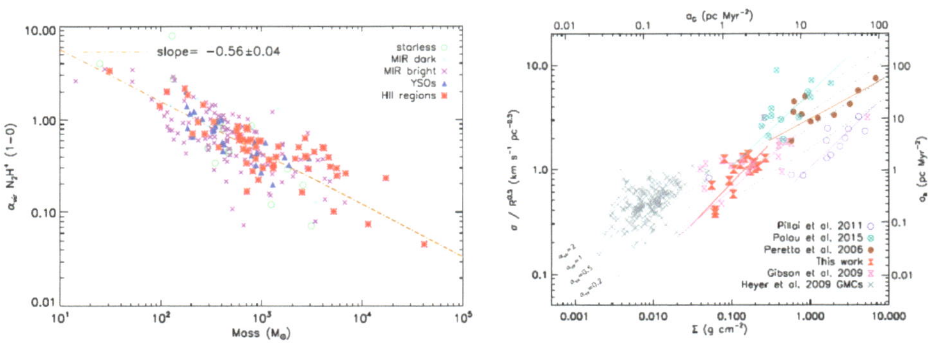

Figure 8.7. *Left:* The virial parameter $\alpha_{\rm vir}$ (cf. Equation (5.24)) as a function of mass for a sample of Milky Way massive, dense star-forming clumps from the Hi-GAL survey (Elia et al. 2017). Note the sub-virial ($\alpha_{\rm vir} < 1$) nature of clumps with masses $M \gtrsim 200 M_\odot$ (reproduced from Traficante et al., (2018b) with permission from Oxford University Press on behalf the Royal Astronomical Society. *Right:* The "Larson ratio", $\mathcal{L}$ as a function of column density Σ for a compilation of molecular clouds and massive clumps exhibiting sub-virial values of $\mathcal{L}$ at column densities $\Sigma \gtrsim 0.05$ g cm^{-2} $\sim 250\, M_\odot$ pc^{-2} (Reproduced from Traficante et al., (2018a) with permission from Oxford University Press on behalf the Royal Astronomical Society. Note that the clumps in this figure are systematically much denser and smaller than the clouds shown in Figure 8.6.

a numerical simulation of cloud formation and evolution (Camacho et al. 2016). The objects are identified as connected sets of voxels (three-dimensional pixels) above various density thresholds (nth), as indicated in the inset. The mean density of the objects is typically within a factor of 10 above the threshold, so objects identified at higher $n_{\rm th}$ are denser in general. It can be seen that the objects defined at each threshold are distributed along elongated regions in the diagram that have roughly the same slope, but are shifted to lower masses (or lower $\alpha_{\rm vir}$) as $n_{\rm th}$ increases.

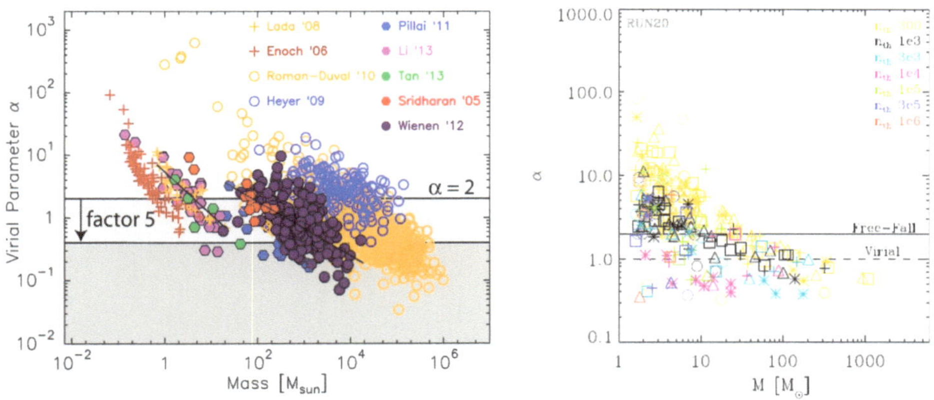

Figure 8.8. *Left panel:* The $\alpha_{\rm vir}$-M scaling for different classes of objects including entire molecular clouds (sizes $\gg$1 pc; open blue and yellow circles), massive clumps (sizes ~1 pc; filled symbols) and dense cores (sizes $\ll$1 pc; "plus" signs). Each class, characterized by a certain typical density, exhibits a similar $\alpha_{\rm vir}$-M scaling as the others, but displaced in mass. This shows that the tendency of $\alpha_{\rm vir}$ to decrease with increasing mass is repeated in each density interval. However, objects of similar mass but different densities can have very different values of $\alpha_{\rm vir}$. Reproduced from Kauffmann et al. (2013). *Right panel:* Distribution in the $\alpha_{\rm vir}$-M diagram of a sample of clouds and clumps defined as connected sets of voxels above a density threshold in a numerical simulation of the formation and subsequent gravitational contraction of a giant molecular cloud. The threshold $n_{\rm th}$ is indicated by the symbol colors. It is seen that objects defined at different thresholds occupy elongated regions with similar slopes, but shifted to lower masses at larger $n_{\rm th}$. Reproduced from Camacho et al. (2016).

In conclusion, in each density interval, the most massive objects (which are also the largest, since the density is roughly the same) are the ones with the lowest values of $\alpha_{\rm vir}$, and therefore, the most strongly bound.[3]

Finally, an important implication of the above discussion is that the mass or column density alone are not enough to determine the gravitational binding of a cloud or clump. Instead, *pairs* of quantities like (mass, size), (mass, density) or (density, size) are necessary to completely specify the gravitational binding.

8.4.2 Scaling of the Magnetic Field with Density

Molecular clouds are in general permeated by magnetic fields, whose intensity tends to increase with the density of the region, albeit with strengths that in most cases are insufficient to provide support against self-gravity (i.e., corresponding to super-critical mass to flux ratios). This can be seen in Figure 8.9, in which column densities $N_H \gtrsim 10^{21}\ {\rm cm}^{-2} \approx 10\ M_\odot {\rm pc}^{-2}$ correspond to molecular clouds and their substructures (clumps, filaments and cores). On the other hand, number column densities

[3] This is an important observation, since it is often erroneously claimed that the largest clouds are gravitationally unbound, because they typically have the lowest column densities, and low-density clouds have the largest scatter in the Keto–Heyer diagram. However, it is only the *small*, low-density clouds that are unbound. The largest GMCs, though also of low mean density, tend to have $\alpha_{\rm vir} \sim 1$, and are thus strongly gravitationally bound, even including the kinetic energy in nonthermal motions, often described as "turbulence" (although it will be argued in the subsequent chapters that the nonthermal motions may contain important—and perhaps dominant—contributions from infall motions.)

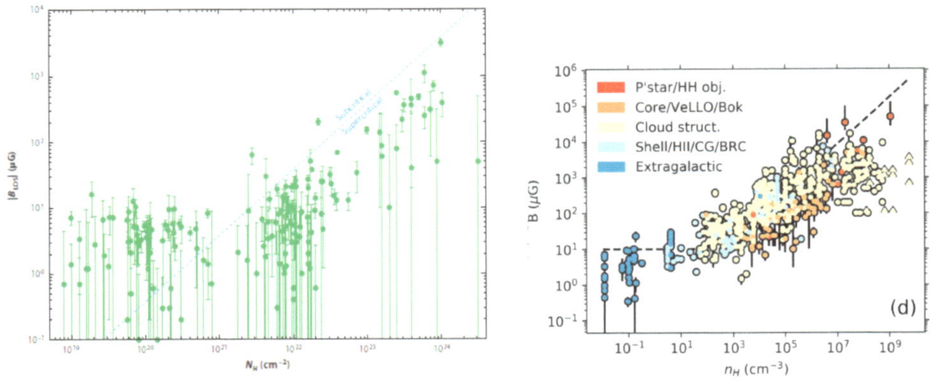

Figure 8.9. *Left panel:* Compilation of magnetic field strengths in (atomic and molecular) clouds and cores of various total column densities $N_H = N_{HI+2N_{H_2}}$. The dashed blue line corresponds to the critical mass-to-flux ratio $M/\varphi = 3.8 \times 10^{-21} N_H/B$ (see Equation (5.34)). Reproduced with permission from Crutcher, (2012). © Annual Reviews. *Right panel:* An extended compilation by Pattle et al. (2023), plotting field strength B versus *volume* density n. The colors indicate different types of objects: protostars and Herbig-Hora objects (red), pre- and protostellar cores, core-embedded very-low-luminosity infrared stars (Vellus) and Bek globules (orange), molecular clouds and their substructures (yellow), various structures under stellar feedback (cyan), and extragalactic structures (blue). Reproduced with permission from Pattle et al (2023). © Copyright 1988 - 2025 - Astronomical Society of the Pacific.

$N_H \lesssim$ a few $\times\ 10^{21}\ \mathrm{cm}^{-2}$ typically correspond to HI clouds, and for these, the magnetic field seems to vary more slowly with density, or even remain nearly constant, and exhibits a large scatter.

The general trend seen in both panels of Figure 8.9 is consistent with the cloud-buildup scenario of Hartmann et al. (2001), in which moderately supersonic compressions in the WNM nonlinearly trigger a phase transition that creates a CNM cloud, whose mass and column density grow as it accretes material from the WNM (see Section 4.5). This increase causes the cloud to eventually become molecular, Jeans-unstable, and magnetically supercritical when it reaches a column density $N \sim 10^{21}\mathrm{cm}^{-2}$ (for Solar Neighborhood typical values of the mean density, mean magnetic field strength, and metallicity). While the gas remains magnetically subcritical, it mostly slides along the magnetic field lines, while, after becoming supercritical, it begins to contract gravitationally and to compress the magnetic field lines, increasing the magnetic field strength. This scenario will be further discussed in Section 10.2.1.

8.5 Molecular Cloud Star Formation Activity

8.5.1 The Star Formation Rate and the Star Formation Efficiency

In order to quantify the rate at which a cloud (or any region from a galaxy to a dense core) forms stars, it is necessary to define various estimators. The *star formation rate* (SFR) is defined as

$$\mathrm{SFR}(t) \equiv \dot{M}_*(t) \equiv \frac{dM_*}{dt}, \tag{8.5}$$

where $M_*(t)$ is mass of stars formed up to time t in a certain cloud or region, taking $t = 0$ at the moment of the formation of the first star in the region. The *instantaneous* (i.e., time-dependent) *star formation efficiency* (SFE) of a molecular cloud is defined, in star formation studies in the Galaxy,[4] as

$$\mathrm{SFE}(t) \equiv \frac{M_*(t)}{M_{\mathrm{mol}}(t) + M_*(t)}, \tag{8.6}$$

where $M_{\mathrm{mol}}(t)$ is the gas mass of the molecular cloud at time t.

An important related quantity is the *star formation efficiency per free-fall time*, ϵ_{ff} (Krumholz & McKee 2005), which was defined as a proxy to the fraction of a cloud's mass that is converted into stars over its *mean* free-fall time, and is thus given by

$$\epsilon_{\mathrm{ff}} \equiv \frac{\dot{M}_*}{M_{\mathrm{mol}}} \langle \tau_{\mathrm{ff}} \rangle, \tag{8.7}$$

where

$$\langle \tau_{\mathrm{ff}} \rangle \equiv \left(\frac{3\pi}{32 G \langle \rho \rangle} \right)^{1/2}, \tag{8.8}$$

is the free-fall time corresponding to mean density of the cloud, $\langle \rho \rangle$. Importantly, the total stellar mass in a cloud at time t,

$$M_*(t) = \int_0^t \dot{M}_* dt, \tag{8.9}$$

is frequently approximated by

$$M_* \approx \langle \dot{M}_* \rangle \tau_{\mathrm{SF}}, \tag{8.10}$$

where $\langle \dot{M}_* \rangle$ is the average SFR during a cloud's star-forming stage, and τ_{SF} is the duration of this stage.

Note, however, that ϵ_{ff} is not really an "efficiency", since it does not necessarily measure the *fraction* of the gas mass that is turned into stars in a free-fall time, and in fact it can take values larger than unity if the instantaneous SFR were to exceed that *free-fall SFR*, $M_{\mathrm{mol}}/\tau_{\mathrm{ff}}$. Instead, strictly speaking, it is simply the ratio of the actual SFR to the free-fall SFR (Zamora-Avilés et al. 2024).

8.5.2 The Schmidt–Kennicutt SF Law

One fundamental result concerning the SFR in galaxies is that it correlates with the gaseous mass. This correlation is known as the *Schmidt–Kennicutt* (SK) law (Schmidt 1959; Kennicutt 1998). Observationally, the SFR is often derived from the intensity of tracers arising from HII regions, such as the Hα line, while the gas

[4] Note that in extragalactic studies, the star formation efficiency is defined differently, as the inverse of the gas depletion time, i.e., $\mathrm{SFE}_{exgal} \equiv 1/\tau_{\mathrm{dep}}$ (see, e.g., Section 1.2 of Kennicutt & Evans (2012).)

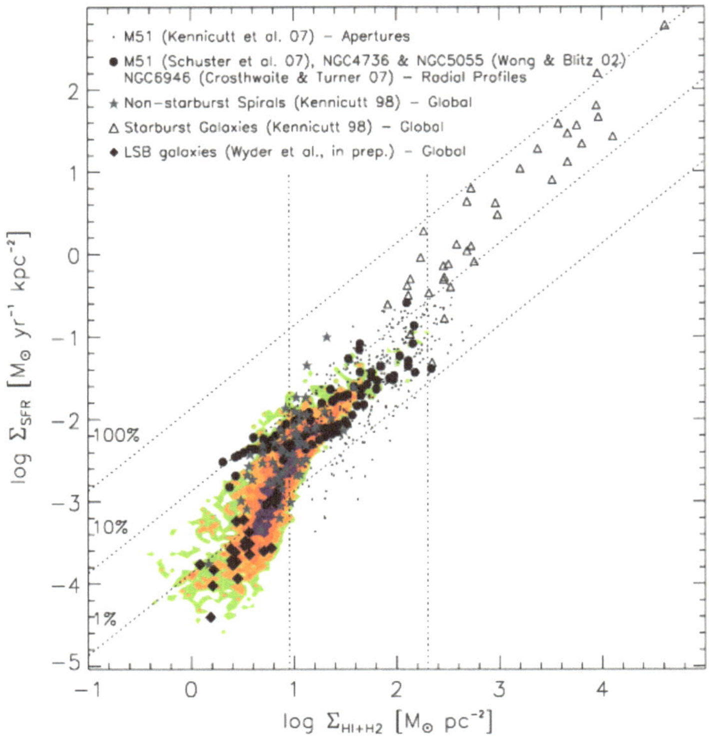

Figure 8.10. The *Schmidt–Kennicutt* (SK) law between the SFR per unit area and the gas surface density in clouds and subregions of a number of nearby external galaxies. The diagonal dotted lines indicate lines of constant SFE (in extragalactic form, i.e., SFE $=1/\tau_{\rm dep} = \dot{M}_*/M_{\rm g}$; see Section 8.5.1) corresponding to the SFR necessary to consume the indicated fraction of the gas in 10^8 yr. Reproduced from Bigiel et al. (2008).

mass can be derived from the intensity of molecular gas tracers such as the ^{12}CO rotational lines. The surface density of SFR ($\Sigma_{\rm SFR}$, typically with units of $M_\odot$ yr^{-1} kpc^{-2}) is found to correlate with the gas surface density $\Sigma_{\rm g}$. Early determinations of the SK law were performed over entire disks of external galaxies, but in more recent times this correlation has also been searched for in kpc-sized subregions of those disks, and in local Galactic molecular clouds (see, e.g., the review by Kennicutt & Evans 2012), as shown in Figure 8.10. A related quantity is the *depletion time*, $\tau_{\rm dep} \equiv M_{\rm g}/\dot{M}_*$, where $M_{\rm g}$ is the total (HI + H_2) gas mass, which is often determined as $\tau_{\rm dep} = \Sigma_{\rm g}/\Sigma_{\rm SFR}$.

8.5.3 The Stellar Initial Mass Function

Finally, the stellar initial mass function (IMF, denoted Φ, first introduced by Salpeter 1955), is defined as the differential mass distribution of the stars in a star-forming region at the time when they reach the zero-age main sequence (ZAMS). It is usually considered per logarithmic mass interval, and therefore written as

$$\Phi(\log M) \equiv \frac{dN(\log M)}{d\log M}, \tag{8.11}$$

where M is the mass of a star and $N(\log M)$ is the number of stars (in a ZAMS population) with masses in the interval $(\log M, \log M + d\log M)$. Of course, not all stars in the region form nor reach the main sequence at exactly the same time, because the high-mass stars evolve much more rapidly than the low-mass ones, and therefore there is a slight intrinsic ambiguity in the definition of the IMF.

The determination, origin, and possible universality of the IMF remain matters of active research, and we do not discuss these issues in this book, and instead we refer the reader to various excellent reviews concerning it (Bastian et al. 2010; Offner et al. 2014; Hennebelle & Grudić 2024). Here, we just assume a typical functional form, and mention that it is often modeled as either a lognormal function or as a piecewise-power law (Figure 8.11). Arguably, its most relevant properties are the mass at which it peaks ($M_{\rm peak} \sim 0.2 M_\odot$) and the logarithmic slope Γ *of its high-mass tail*, so that the IMF approaches the power law

$$\Phi(\log M) \approx \alpha \left(\frac{M}{M_0}\right)^{-\Gamma} \qquad \text{(for } M \gg M_{\rm peak}\text{)}, \tag{8.12}$$

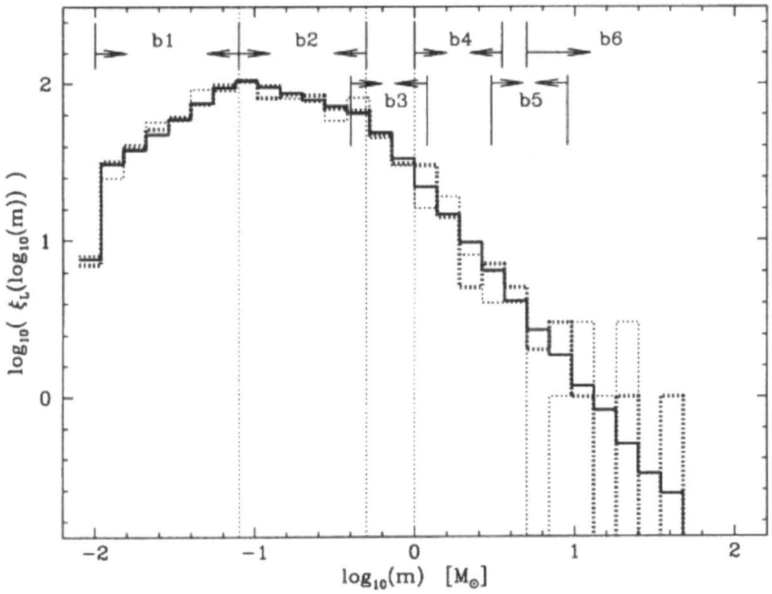

Figure 8.11. A piecewise-power-law representation of the IMF for 10^6 stars, according to Kroupa (2001). The mass ranges over which power-law functions are fitted are indicated by the six regions delimited by arrows, while the thin vertical dotted lines indicate the masses at which the logarithmic slope changes. The thin and bold dotted-line histograms show two random renditions of this IMF with 10^3 stars. Reproduced from Kroupa (2001) with permission from Oxford University Press on behalf the Royal Astronomical Society.

where M_0 is a reference mass and α is a normalization coefficient. Note that, for such a power-law distribution, the number of stars per linear (rather than logarithmic) mass interval is given by

$$\frac{dN(M)}{dM} \approx \frac{\alpha}{M_0}\left(\frac{M}{M_0}\right)^{-(\Gamma+1)} \qquad (\text{for } M \gg M_{\text{peak}}). \tag{8.13}$$

Some convenient quantities derived from the IMF that will be useful in Chapter 12 are:

- *The total number of stars*, given by:

$$N_{\text{tot}} = \int_{-\infty}^{\infty} \Phi(\log M)\; d\log M; \tag{8.14}$$

- *The total mass in stars*, given by:

$$M_{\text{tot}} = \int_{-\infty}^{\infty} M\; \Phi(\log M)\; d\log M; \tag{8.15}$$

- *The masses of the least and most massive stars in a population*, which are given by the conditions

$$\int_{-\infty}^{\log M_{\min}} \Phi(\log M) d\log M = 1, \tag{8.16}$$

and

$$\int_{\log M_{\max}}^{\infty} \Phi(\log M) d\log M = 1. \tag{8.17}$$

I.e., these are the minimum and maximum masses for which at least one star can be found.

References

Arquilla, R., & Goldsmith, P. F. 1986, ApJ, 303, 356

Arzoumanian, D., Furuya, R. S., Hasegawa, T., Tahani, M., et al. 2021, A&A, 647, A78

Bastian, N., Covey, K. R., & Meyer, M. R. 2010, ARA&A, 48, 339

Beaumont, C. N., & Williams, J. P. 2010, ApJ, 709, 791

Belloche, A. 2013, Angular Momentum Transport During Star Formation and Evolution, ed. P. Hennebelle, & C. Charbonnel, (Les Ulis: EDP Sciences) 62, 25

Bigiel, F., Leroy, A., Walter, F., et al. 2008, AJ, 136, 2846

Blitz, L. 1993, Protostars and Planets III, ed. E. H. Levy, & J. I. Lunine (Tucson, AZ: Univ. Arizona Press) 125

Brunt, C. M., Heyer, M. H., & Mac Low, M. M. 2009, A&A, 504, 883

Camacho, V., Vázquez-Semadeni, E., Ballesteros-Paredes, J., et al. 2016, ApJ, 833, 113

Crutcher, R. M. 2012, ARA&A, 50, 29
Elia, D., Molinari, S., Schisano, E., et al. 2017, MNRAS, 471, 100
Fujii, K., Mizuno, N., Dawson, J. R., et al. 2021, MNRAS, 505, 459
Fukui, Y., Kawamura, A., Wong, T., et al. 2009, ApJ, 705, 144
Goldsmith, P. F., & Arquilla, R. 1985, Protostars and Planets II, ed. D. C. Black, & M. S. Matthews (Tucson, AZ: Univ. Arizona Press) 137
Goodman, A. A., Benson, P. J., Fuller, G. A., & Myers, P. C. 1993, ApJ, 406, 528
Grenier, I. A., Casandjian, J.-M., & Terrier, R. 2005, Sci, 307, 1292
Hacar, A., Tafalla, M., Kauffmann, J., & Kovács, A. 2013, A&A, 554, A55
Hartmann, L., Ballesteros-Paredes, J., & Bergin, E. A. 2001, ApJ, 562, 852
Hennebelle, P., & Grudić, M. Y. 2024, ARA&A, 62, 63
Heyer, M., & Dame, T. M. 2015, ARA&A, 53, 583
Heyer, M., Krawczyk, C., Duval, J., & Jackson, J. M. 2009, ApJ, 699, 1092
Heyer, M. H., & Brunt, C. M. 2004, ApL, 615, L45
Heyer, M. H., & Schloerb, P. F. 1997, ApJ, 475, 173
Kabanovic, S., Schneider, N., Ossenkopf-Okada, V., et al. 2022, A&A, 659, A36
Kauffmann, J., Pillai, T., & Goldsmith, P. F. 2013, ApJ, 779, 185
Kennicutt, R. C., & Evans, N. J. 2012, ARA&A, 50, 531
Kennicutt, R. C. 1998, ARA&A, 36, 189
Keto, E. R., & Myers, P. C. 1986, ApJ, 304, 466
Könyves, V., André, P., Men'shchikov, A., et al. 2015, A&A, 584, A91
Kroupa, P. 2001, MNRAS, 322, 231
Krumholz, M. R., & McKee, C. F. 2005, ApJ, 630, 250
Kumar, M. S. N., Palmeirim, P., Arzoumanian, D., & Inutsuka, S. I. 2020, A&A, 642, A87
Kusune, T., Nakamura, F., Sugitani, K., et al. 2019, PASJ, 71, S5
Langer, W. D., et al. 2010, A&A, 521, L17
Larson, R. B. 1981, MNRAS, 194, 809
Li, D., & Goldsmith, P. F. 2003, ApJ, 585, 823
Miville-Deschênes, M.-A., Murray, N., & Lee, E. J. 2017, ApJ, 834, 57
Myers, P. C. 2009, ApJ, 700, 1609
Offner, S. S. R., Clark, P. C., Hennebelle, P., et al. 2014, Protostars and Planets VI, ed. H. Beuther, et al. (Tucson, AZ: Univ. Arizona Press) 53
Palmeirim, P., André, P., Kirk, J., et al. 2013, A&A, 550, A38
Pascucci, I., Edwards, S., Heyer, M., et al. 2015, ApJ, 814, 14
Pattle, K., Fissel, L., Tahani, M., Liu, T., & Ntormousi, E. 2023, in Protostars and Planets VII, 534, ed. S. Inutsuka, et al. (Tucson, AZ: Univ. Arizona Press) 193
Rezaei, Kh, S, & Kainulainen, J. 2022, ApL, 930, L22
Salpeter, E. E. 1955, ApJ, 121, 161
Schmidt, M. 1959, ApJ, 129, 243
Shirley, Y. L. 2015, PASP, 127, 299
Solomon, P. M., Rivolo, A. R., Barrett, J., & Yahil, A. 1987, ApJ, 319, 730
Storm, S., Mundy, L. G., Lee, K. I., et al. 2016, ApJ, 830, 127
Traficante, A., Fuller, G. A., Smith, R. J., et al. 2018a, MNRAS, 473, 4975
Traficante, A., Duarte-Cabral, A., Elia, D., et al. 2018b, MNRAS, 477, 2220
Veena, V. S., Vig, S., Tej, A., Kantharia, N. G., & Ghosh, S. K. 2017, MNRAS, 465, 4219
Wannier, P. G., Lichten, S. M., & Morris, M. 1983, ApJ, 268, 727

Wannier, P. G., Andersson, B. G., Morris, M., & Lichten, S. M. 1991, ApJS, 75, 987
Wilson, R. W., Jefferts, K. B., & Penzias, A. A. 1970, ApL, 161, L43
Xu, D., Li, D., Yue, N., & Goldsmith, P. F. 2016, ApJ, 819, 22
Zamora-Avilés, M., Camacho, V., Ballesteros-Paredes, J., et al. 2024, arXiv:2409.11588
Zucker, C., Goodman, A., Alves, J., et al. 2021, ApJ, 919, 35
Zucker, C., Goodman, A. A., Alves, J., et al. 2022, Natur, 601, 334

AAS | IOP Astronomy

Interstellar Flow and Star Formation

Enrique Vázquez-Semadeni

Chapter 9

Historical Perspective of Molecular Cloud and Star Formation Models

9.1 The Discovery of Molecular Clouds and Their Supersonic Internal Motions

Molecules were discovered in the ISM in the 1930s, when Dunham (1937) reported an unidentified interstellar absorption line in the optical spectra of early-type stars, which shortly afterwards was identified by Swings & Rosenfeld (1937) as pertaining to the CH molecule. However, the detection of molecular emission lines in radio frequencies did not occur until 1963, when Weinreb et al. (1963) observed lines pertaining to OH. A detailed historical perspective on the discovery of interstellar molecules is given by Herzberg (1988).

The very important CO molecule, our main tracer of molecular gas in the ISM, was discovered by Wilson et al. (1970), who reported the observation of the ^{12}CO ($J = 1 - 0$) rotational emission line at 2.6 mm. Already in this observation it was evident that the full width at half maximum was of the order of a few km s^{-1} (Figure 9.1), while the sonic linewidth at the temperature of molecular gas ($\sim$ 10 K) is $\sim$ 0.2 km s^{-1}. This implies a strongly supersonic linewidth.

9.1.1 1974: The Rapid Rise and Fall of the Early Collapse Interpretation: The Star Formation Conundrum

The earliest interpretation of such supersonic linewidths was that they corresponded to gravitational collapse in the clouds. In observations of ^{12}CO, ^{13}CO and CS of the Orion A molecular cloud, Liszt et al. (1974) noted

1. The existence of a large linewidth ($\sim$ 4–6 km s^{-1});
2. A systematic velocity gradient across the cloud;
3. The coincidence of the intensity peaks of the optically thick ^{12}CO and the optically thin ^{13}CO, and
4. The absence of self-absorption in the ^{12}CO emission line profiles.

doi:10.1088/2514-3433/adcb28ch9

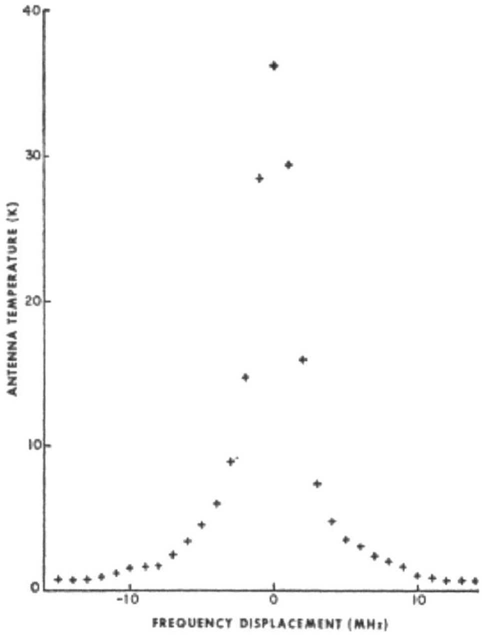

Figure 9.1. First detection of the ^{12}CO ($J = 1 - 0$) emission line at 2.6 mm. The reported full width at half maximum was 6.2km s^{-1}. Reproduced from Wilson et al. (1970).

From this evidence, they concluded that the outer layers of the cloud should have a different line-of-sight (LOS) velocity than its inner regions, and ruled out the possibility that the large linewidth could be due to *small-scale*, supersonic turbulence. The large-scale gradient would allow even the optically thick ^{12}CO line to see deep into the cloud and avoid the formation of a self-absorption dip in this line, since, being at a different velocity, the outer layers could not obscure the inner parts. However, these authors noted that the velocity difference could imply either collapse or expansion of the cloud, and their interpretation became generically referred to as the "radial motion" (RM) model. The results by Liszt et al. (1974) motivated Goldreich & Kwan (1974) to interpret the radial motions as collapse, and, through a calculation of the optical thickness and emission rates from various lines, to balance the resulting cooling with the energy injected by the collapse, thus explaining the observed cloud temperatures.

In addition, the gravitational fragmentation proposed by Hoyle (1953, see discussion in Section 7.5) should be applicable to molecular clouds, given their approximate isothermality. Thus, as the clouds contract globally, the mean Jeans mass within them decreases due to the increase in the cloud's mean density, and each Jeans-mass subregion is in principle able to begin contracting on its own, thus producing fragmentation of the original cloud.

However, in the same year, the collapse idea was deemed unfeasible by Zuckerman & Palmer (1974), who argued that it would have the following consequences:

1. If all the molecular gas mass in the Galaxy underwent collapse in the free-fall time corresponding to its mean density, the total Galactic star formation rate (SFR) would be much larger than the currently observed rate. Indeed, with present-day estimates, the total molecular gas mass in the Galaxy is $M_{\rm mol} \sim 1.5 \times 10^9 M_\odot$ (cf. Table 1.1). Assuming a typical density of ~ 100 cm^{-3}, the free-fall time, Equation (7.8), is ~ 3.35 Myr. Therefore, the SFR corresponding to free-fall is $M_{\rm mol}/\tau_{\rm ff} \sim 450 M_\odot$ yr^{-1}. Instead, current estimates of the Galaxy's SFR are ~ 2 $M_\odot$ yr^{-1} (e.g., Chomiuk & Povich 2011), more than two orders of magnitude lower. In what follows, we will refer to this problem as *the SFR conundrum.*
2. If the cloud envelopes are generally falling onto the central regions, and these are populated by H II regions, the absorption lines produced in the envelopes, viewed against the H II regions, should be systematically redshifted with respect to the recombination emission lines produced in the H II regions themselves. However, this systematic redshift is not observed in general.
3. If the clouds are collapsing, one should expect the velocity dispersion (line width) to be largest when one looks through the cloud center, and to decrease as the LOS passes farther away from the center. This effect again is not observed.

In view of these problems, Zuckerman & Evans (1974, hereafter ZE74) proposed that the large-scale motions—i.e., at the scale of the whole clouds—have small velocities compared to those of the small-scale, "turbulent" motions of small individual gas parcels with masses ~ 1000 $M_\odot$, and that the latter are responsible for the observed ^{12}CO linewidths. In this scenario, then, there is no concern for an excessive SFR nor for the absence of the systematic velocity shifts between the inner and outer parts of the clouds, since the clouds are not contracting with any significant speed. Also, in this scenario, the cloud remains optically thick, and so one cannot see through it. They supported this implication by noting that, when an H II region can be confidently established to lie deep within a cloud, often a ^{12}CO peak is *not* observed to coincide with the H II region. Therefore, ZE74 concluded that the observations of Liszt et al. (1974) could be understood in terms of supersonically turbulent clouds, *in which the turbulent motions had to be restricted to scales much smaller than the cloud size.*

Note, however, that the nature of the "turbulence" proposed by ZE74 is opposite to that of Kolmogorov turbulence. In the latter, the largest velocities occur at the largest scales (see Section 3.12), while ZE74's proposed turbulence has the largest velocities at scales small compared to the size of the clouds. Therefore, this hypothetical turbulence could be considered as an "additional temperature" that provides pressure in addition to the thermal one, rather than as realistic turbulence.

Another reason for the early dismissal of the cloud-collapse scenario was that, as also described in Section 7.5.2, objections were raised against the gravitational fragmentation mechanism (Hoyle 1953) around 1980, as early numerical simulations and analytical studies of the collapse of *near-Jeans* (i.e., marginally unstable) spherical clouds with *linear* (small-amplitude) perturbations showed that these structures did not fragment, as the collapse of the largest scales (the sphere as a whole) proceeded faster than the collapse of the perturbations (e.g., Tohline 1980, 1982). This can be easily understood by the Jeans dispersion relation (Equation (6.12)), which shows that the fastest growing speed ω of unstable perturbations corresponds to those with the smallest wavenumber k; i.e., to perturbations with the longest wavelength $\lambda = 2\pi/k$. When an isolated sphere of radius R is considered, then the largest available dimension is the diameter of the sphere, and so the growth rate of this mode overwhelms that of all other fluctuations of size smaller than R. That is, the sphere terminates its full-scale collapse earlier than any smaller-scale linear fluctuations it may contain.

As a result of the considerations above, the possibility of freely collapsing clouds was dismissed, and the notion that clouds must be supported by some agent(s) became the starting point for modeling molecular clouds by the mid 1970s. Furthermore, the detection of Zeeman splitting of OH lines (Crutcher & Kazès 1983) in the early 1980s allowed the consideration of magnetic support (Mestel & Spitzer 1956) for molecular clouds, in addition to that from turbulence.

9.2 Cloud-Support Models

Despite the general agreement that giant molecular clouds (GMCs) could not be in a state of gravitational collapse, it was well known that in general they are strongly Jeans-unstable. As already mentioned in Section 8.3.1, this can be seen by noting that the Jeans mass (Equation (6.16)) for clouds of density $n \sim 100\,\mathrm{cm}^{-3}$ and temperature $T \sim 10$ K, is $\sim 50 M_\odot$, while typical GMC masses are $M \sim 10^4$–$10^6 M_\odot$ (cf. Table 1.2). Therefore, GMCs contain thousands (or more) of Jeans masses, and are thus extremely Jeans-unstable. In consequence, cloud models that do not collapse require some form of support, and we refer to them as *cloud support* models. As mentioned above, the two main mechanisms of support that have been considered in the literature are magnetic fields and turbulence. We now briefly discuss each one.

9.2.1 Magnetic Support Model

Although the initial proposal by ZE74 was that molecular clouds are permeated by strongly supersonic, small-scale turbulent motions, the ubiquitous presence of such strong motions was questioned by Goldreich & Kwan (1974) because turbulence is known to dissipate rapidly (in roughly one turbulent crossing time; cf. Section 3.5.1). This was made worse by the requirement that the turbulent motions should be confined to scales much smaller than the cloud or clump scale, implying that the crossing time for these scales is itself short (see Section 2.3.1 of Shu et al. 1987, hereafter SAL87).

SAL87 further argued that, since magnetic fields are not observed to be strongly tangled over the scale of molecular clouds, then the turbulence should be more wave-like than eddy-like. This would have the added advantage of explaining the observation that the velocity dispersion in clouds appears to correspond to approximate virial equilibrium (Larson 1981). Their argument was that, if clouds have masses comparable to the critical mass for magnetic support, given by Equation (5.30), then the Alfvén speed $v_A \equiv \sqrt{B^2/4\pi\rho}$ is of the order of the virial velocity $v_{vir} \equiv \sqrt{GM/R}$, where ρ, M, and R are, respectively, the density, mass, and radius of the cloud. Actually, this is not surprising since, if the cloud is in magnetic virial equilibrium and the wave speed is taken as the velocity dispersion, this implies that the cloud is in kinetic virial equilibrium as well.

In addition, it was suggested that the "turbulence" would consist mostly of Alfvén waves since, being transverse, these waves do not produce shocks, and were therefore thought to be much less dissipative than the other (fast and slow) wave modes, which, being longitudinal, *can* generate shocks. So, it was conjectured that the fast and slow modes would rapidly dissipate, leaving only the Alfvén waves as the dominant contribution to the "turbulence".

9.2.1.1 Bimodal Star Formation: The High-Mass, Magnetically Supercritical Mode

In the magnetic support model it was proposed that star formation (SF) is a bimodal process (see Sections 2.1 and 3.1 of SAL87, and references therein): high-mass star-forming regions were thought to correspond to magnetically supercritical regions, which collapse dynamically. Instead, low-mass regions correspond to magnetically subcritical clouds, which are supported by the magnetic field against their self-gravity, and were thought to undergo gravitational contraction only in their central, densest parts, and on a long timescale, due to the mechanism of *ambipolar diffusion* (see below). Since low-mass star-forming regions are much more numerous than high-mass ones, it was assumed that most clouds were magnetically subcritical (see, e.g., Mouschovias 1991).

In the *high-mass mode*, the clouds, being magnetically supercritical, undergo global collapse, albeit at a somewhat slower rate than in the non-magnetic case, because the magnetic field, even if unable to prevent the collapse, is capable of retarding it. That is, the cloud behaves as if having a smaller ("reduced") mass due to the combined thermal and magnetic support (cf. Section 5.2.8).

The contraction along the magnetic field direction, however, is not opposed by the field, and so the cloud contracts faster along this direction. Nevertheless, since the contraction along this direction is essentially one-dimensional and the cloud is nearly isothermal, thermal pressure is able to eventually halt this contraction[1] when the cloud's thickness is of the order of the Jeans length[2] of the densest central gas. The density of this central region is in turn determined by the contraction perpendicular to the field at each moment in time. That is, in the direction parallel to the field, the contraction can be thought of as occurring through a succession of

[1] As discussed in Section 10.1.2, thermal pressure can always halt gravitational contraction occurring in one dimension in isothermal gas.

[2] Note that, when thermal pressure is able to prevent the collapse, the Jeans length represents a *stable* equilibrium.

hydrostatic equilibria determined at each moment in time by the ongoing perpendicular collapse.

Therefore, in this scenario, the collapse of the cloud consists of a genuine, two-dimensional gravitational contraction perpendicular to the magnetic field, and a reduction of the hydrostatic-equilibrium thickness, due to the reduction of the cloud's Jeans length as its density increases. This mode of contraction produces a flattened cloud of increasing *aspect ratio* (ratio of largest to smallest dimension; in this case, of radius to thickness). This can be understood as follows: let us first assume that the cloud's thickness remains constant. In this case, the density variation due to the radial contraction is $\rho \propto R^{-2}$. Therefore, $L_J \propto \rho^{-1/2} \propto R$, and so the ratio of Jeans length to cloud's radius remains constant too. But the decrease in the Jeans length implies that the cloud's thickness also decreases, and so the density increases *faster* than R^{-2} as R decreases. In turn, this implies that the Jeans length decreases *faster* than R, meaning that the aspect ratio increases. Therefore, the number of Jeans lengths within the cloud's radius increases. Moreover, the mass-to-flux ratio of any subregion with smaller radius but the same thickness as the cloud is the same as that of the whole cloud, and thus these subregions are supercritical too. Therefore, the cloud is prone to *fragmentation*.

In summary, the supercritical mode of collapse within the magnetic model was very similar to the original collapse model of Goldreich & Kwan (1974), together with Hoyle (1953) fragmentation, except with a reduced collapse rate due to the combined thermal and magnetic support, and applying only to a small subset of the clouds (the magnetically supercritical ones).

9.2.1.2 Bimodal Star Formation: The Low-Mass, Subcritical Mode

On the other hand, within the magnetic support scenario, the *low-mass mode* of star formation was thought to occur in magnetically subcritical clouds, which are supported by the magnetic field in the direction perpendicular to it, and, again, are in hydrostatic equilibrium between self-gravity and the thermal pressure gradient in the direction parallel to the field. As mentioned above, this is possible because the contraction along the magnetic field is one-dimensional, and, for isothermal flow, such contraction causes the Jeans length to *increase* with increasing density. Therefore, even if the cloud starts out larger than the Jeans length, the contraction eventually causes the cloud's thickness to become equal to the Jeans length, and the contraction is halted. So, the support of subcritical clouds is actually provided by the magnetic field in the two directions perpendicular to it, and by thermal pressure in the direction parallel to the field.

Nevertheless, a very slow contraction, especially in the densest parts of the cloud should still occur, due to the process of *ion-neutral slip*, also known as *ambipolar diffusion*. This process consists of the gradual slip between the neutral and ionized gas particles, since only the latter are dynamically coupled to the magnetic field. Thus, under the pull of a cloud's self-gravity, the neutrals "percolate" through the ions, and the cloud effectively contracts, albeit on a long timescale, which is determined by the collision rate among ions and neutrals (see, e.g., Section 2.3.6 of SAL87). This timescale was originally estimated to be much longer than the local free-fall time, and thus this mechanism would allow a much lower SFR than that implied by the free-fall timescale. Moreover, since it was speculated that the majority of molecular clouds were magnetically subcritical, their long corresponding contraction timescales thus provided a convenient solution to the Star Formation Conundrum. However, as observations began to show that the mass-to-flux ratio of dense cores was close to unity, the timescale was revised to be close to the free-fall time (e.g., Ciolek & Basu 2001), and the usefulness of magnetic support to resolve the Star Formation Conundrum decreased.

The magnetic support model became the paradigm of star formation until the turn of the 21st century, when new observational evidence and numerical results began to question the validity of its assumptions, and therefore, of its applicability to real molecular clouds, as described in the next section.

9.2.2 The Turbulent Support (TS) Model

9.2.2.1 Problems with the Magnetic Model

Two of the main assumptions of the magnetic support model were that *(a)* the supersonic turbulence in molecular clouds consists mostly of MHD Alfvén waves, since these were though to be less dissipative than the slow and fast MHD waves, which would produce shocks and thus dissipate strongly, and *(b)* most molecular clouds are magnetically subcritical. However, around the turn of the 21st century, several numerical simulations investigating the decay of MHD turbulence showed that it dissipates just as rapidly as hydrodynamic turbulence, and that supersonic and subsonic turbulence also dissipate at the same rate, namely, as t^{-1} (Mac Low et al. 1998; Stone et al. 1998; Padoan & Nordlund 1999; see Section 3.5.1, Equation (3.34)). That is, turning to MHD turbulence does not help in reducing its dissipation rate, invalidating assumption *(a)*.

In addition, an increasing number of Zeeman determinations of the magnetic field strength in regions of various mean densities began to show clearly that most regions are magnetically supercritical—i.e., not magnetically supported. Crutcher (2012) summarized the observations available at the time, showing that the overwhelming majority of objects with column densities typical of molecular gas, $N_H \gtrsim 10^{21}\ \mathrm{cm}^{-2}$ (Figure 8.9), are supercritical, and thus invalidating assumption *(b)*. More recent data still suggest this trend (Pattle et al. 2023).

Moreover, it became evident that the majority of low-mass stars are formed in regions that also form massive stars (Lada & Lada 2003), and so the prediction of having two separate modes of star formation for low- and high-mass stars, corresponding to sub- and supercritical clouds, became unnecessary, or at least, the magnetically supercritical case became the norm, rather than being the exception.

9.2.2.2 The Energy Sources for Turbulence

As a consequence of the problems mentioned above with the magnetic support model, a return to the proposal of ZE74 ensued, so that the linewidths in molecular clouds are assumed to correspond to fully turbulent motions. These were assumed to prevent the collapse of the clouds as a whole, maintaining them in near virial equilibrium. The latter condition arose from the observation that molecular clouds tend to exhibit a near-virial relation between their internal kinetic and gravitational energies (e.g., Larson 1981; Heyer et al. 2009; Figure 8.5), so that

$$\sigma^2 \approx \frac{GM}{R}, \tag{9.1}$$

(see Sections 5.2.5 and 8.4.1.) However, in view of the realization that MHD turbulence could not resolve the issue of rapid dissipation, it became evident that a continuous driving agent was necessary for the turbulence. The various sources of energy and its relative importance has been discussed extensively by Mac Low & Klessen (2004), and here we just mention a few important additional considerations.

One important source of driving is the energy injection from stellar sources, including, from low to high power, bipolar outflows, expanding HII regions and winds from massive stars, and either isolated or clustered supernova (SN) explosions (e.g., Gentry et al. 2017). It is important to note that outflows, HII regions and winds constitute *internal* sources for the turbulence in molecular clouds, while SN explosions will generally be *external* to the clouds. Specifically, even in the case of Type II supernovae [SNe], which are the result of single massive stars exploding after a few to several million years, the explosions are expected to occur in the neighborhood (but not inside) of molecular clouds, because the photoionizing radiation from the progenitor stars can generally disperse the clouds before the explosion occurs (e.g., Chevance et al. 2020). In the case of Type I SNe, which are the result of accretion onto a white dwarf in binary systems, and therefore have long lifetimes, the explosions essentially occur at random locations in the ISM, completely unrelated to their birth sites. Being external and producing (compressive) shock waves, which in turn generate vorticity, the SN explosions can either *form* or disrupt molecular clouds, and their effect on the latter is highly sensitive to the relative placement of the explosions with respect to the clouds, and possibly not very efficient (e.g., Iffrig & Hennebelle 2015, 2017).

Other important sources of turbulence are the action of several combined instabilities during the assembly of the clouds by external compressions (see, e.g., Heitsch et al. 2006), and the turbulence amplification during gravitational contraction (cf. Section 3.11). These mechanisms are clearly internal to the clouds.

9.2.2.3 Turbulent Support and the Dual Role of Turbulence

Under the premise that there are plenty of mechanisms capable of maintaining the turbulence in molecular clouds, the scenario of *turbulent support* (e.g., Vazquez-Semadeni et al. 2000; Mac Low & Klessen 2004; McKee & Ostriker 2007; Hennebelle & Falgarone 2012, hereafter, TS) proposes that *strongly supersonic* ($\mathcal{M}_s \gtrsim 10$) turbulence is pervasive in molecular clouds, which is expected to have a *dual role* on them: First, it is expected to provide an effective "pressure" that can prevent the global collapse of the clouds. Moreover, being strongly supersonic, this turbulence can explain the *linewidth-size* (or, equivalently, the *velocity dispersion-size*) scaling relation found by Larson (1981) (cf. Section 8.4.1 and *left* panel of

Figure 8.5), which, as mentioned in that section, is currently observationally determined to be (e.g., Heyer & Brunt 2004)

$$\sigma \approx 1 \text{ km s}^{-1} \left(\frac{L}{\text{pc}} \right)^{0.5}, \tag{9.2}$$

and is therefore consistent with the scaling $\sigma_\ell \propto \ell^{1/2}$ expected for Burgers turbulence (Equation (3.61)). Therefore, under the turbulent support scenario, Larson's line-width-size scaling is the signature of the strongly compressible turbulence pervading molecular clouds.

In addition, this turbulence is hypothesized to be the result of virialization at all scales, meaning that the near-virial equilibrium condition,

$$\frac{2E_k}{|E_g|} = \frac{\sigma^2 L}{2GM} \approx 1, \tag{9.3}$$

where the first equality has been taken from Equation (8.4), is assumed to be approximately satisfied in molecular clouds and their substructures. If this condition is satisfied, then the other relation reported by Larson (1981), the *density-size scaling*, follows. Indeed, writing $L = 2R$, $M = 4/3\pi\rho R^3$, $\rho = \mu m_\text{H} n$, where n is the number density, m_H is the hydrogen mass and $\mu = 2.36$ is the mean particle mass for molecular gas, and then introducing Equation (9.2) in Equation (9.3) and solving for the number density, we obtain

$$n \approx 3.8 \times 10^3 \text{ cm}^{-3} \left(\frac{L}{\text{pc}} \right)^{-1}, \tag{9.4}$$

which is indeed very close to the density-size scaling, given by Equation (8.1), and shown in the *right* panel of Figure 8.5. In turn, this relation implies that molecular clouds should all have roughly the same column density, of order

$$N \approx nL \approx 1.2 \times 10^{22} \text{ cm}^{-2} \quad \Rightarrow \quad \Sigma \approx 220 \; M_\odot \text{pc}^{-2}. \tag{9.5}$$

The interpretation of these relations is therefore that giant molecular clouds have column densities of the order given by Equation (9.5), and are near virial equilibrium, with the kinetic energy consisting of strongly supersonic turbulence, characterized by the linewidth-size relation given by Equation (9.2), which is consistent with the expectation for Burgers' turbulence, and is driven by one or more of the mechanisms mentioned above. Thus, the first role of turbulence in the TS scenario is to support the clouds, maintaining them near virial equilibrium.

The second role of turbulence in this scenario is to produce strong *local* density fluctuations within the molecular clouds, resulting from the field of shocks generated by the random supersonic turbulent motions. Since the gas in the molecular clouds is roughly isothermal, the typical density fluctuation is expected to have a density enhancement of a factor $\sim \mathcal{M}_\text{s}^2$ over the mean density (see Section 3.9.2). Typical rms Mach numbers are of order $\mathcal{M}_\text{s} \sim 10$, so the density fluctuations are expected to be of order $\delta\rho/\rho \sim 100$. Therefore, if the GMCs have densities $n_\text{GMC} \sim 100 \text{ cm}^{-3}$, the turbulent density fluctuations are expected to have peak densities $n_{\rm core} \sim 10^4\ {\rm cm}^{-3}$, thus being consistent with the properties of dense molecular cloud clumps (see Table 1.2).

Moreover, the approximate isothermality of molecular gas implies that the Jeans mass (Equation (6.16)) scales roughly as $\rho^{-1/2}$, and therefore, the local Jeans mass in the dense post-shock gas is expected to depend on the Mach number as $M_{\rm J} \sim 1/\mathcal{M}_{\rm s}$. This means that the Jeans mass within the density fluctuations is expected to occasionally become smaller than the fluctuation's own mass, allowing the latter to undergo local collapse. Thus, while the turbulence is thought to prevent the global collapse of entire GMCs, it is considered to induce *local* collapses. In this way, the SFR conundrum of Zuckerman & Palmer (1974) is directly addressed, because only a small fraction of the total mass in GMCs can undergo collapse and form stars.

It is worth noting that a certain similarity exists between the magnetic support and TS models, since in both cases the molecular clouds are supported globally by the corresponding agent (magnetic fields and turbulence, respectively), and in both cases a small fraction of the clouds' mass escapes the support—due to ambipolar diffusion in the case of magnetic support, and to the local compressions in the turbulent case.

9.3 Global Hierarchical Collapse Model

9.3.1 Problems with the Turbulent Support Model

Although the TS model appears to effectively address many features of molecular clouds and their star formation activity, similarly to what happened to the magnetic support model, during the second decade of the 21st century a number of problems began to be apparent with TS, as we list below:

- Ever since early after its discovery, the Larson density-size scaling has been questioned as possibly being the result of various possible observational selection effects (e.g., Kegel 1989; Scalo 1990). At present, it is indeed considered that this scaling arises from the steep decline of the column density PDF at large values, which means that most of the clouds' mass is at or near the density threshold used to define them (Ballesteros-Paredes et al. 2012; Beaumont et al. 2012).
- If all clouds adhered to a unique linewidth-size relation, then this relation could be understood as the natural velocity dispersion-size scaling, $\sigma \propto R^{1/2}$, produced by strongly supersonic, Burgers-like turbulence. However, the Burgers-like scaling is actually *inconsistent* with virial equilibrium as soon as clouds (and smaller clumps) of different column density are considered, as demonstrated by Heyer et al. (2009) and Ballesteros-Paredes et al. (2011), and illustrated in the *right* panel of Figure 8.6. Indeed, in this case, the scaling observed is $\sigma \propto (\Sigma R)^{1/2}$, while the Burgers scaling does not involve any dependence on the column density. In other words, Larson's linewidth-size relation is only consistent with virial equilibrium when objects of roughly the same column density are considered. Early molecular cloud samples indeed

were biased toward constant column density. However, when this condition is relaxed, because observational biases are removed, Larson's linewidth-size relation must be generalized to Heyer's relation.

- The fact that the nonthermal motions in the clouds, generally interpreted as turbulence, appear to be in general close to the virial value given by Equation (8.4) imposes tight constraints to the plausible mechanisms that drive the turbulence. If the driving comes from stellar sources, such as protostellar outflows, massive-star winds and ultraviolet (UV) photoionizing radiation, and supernova explosions, then there is no obvious reason why the clouds and clumps should almost always appear virialized over a wide range of column densities. In fact, the observational evidence suggests that the turbulence level is in fact independent of whether the clouds contain or do not contain massive stars capable of driving the turbulence (e.g., Maddalena & Thaddeus 1985; Williams & Blitz 1998; Kawamura et al. 2009), and that the onset of early massive-star feedback (winds and photoionizing radiation) rapidly disperses the clouds (e.g., Chevance et al. 2022), rather than maintaining them in near virial equilibrium.

  On the other hand, if the turbulence is driven by various instabilities during the cloud assembly and collapse itself (Heitsch et al. 2006; Klessen & Hennebelle 2010), then numerical simulations of these processes suggest that the turbulence thus generated is insufficient to halt or significantly slow down the collapse (e.g., Vázquez-Semadeni et al., 2007; Heitsch & Hartmann 2008; Guerrero-Gamboa & Vázquez-Semadeni 2020), because of the strongly dissipative nature of turbulence (cf. Section 3.5.1). Therefore, either way, it does not appear feasible that the available driving mechanisms for the turbulence are capable of maintaining the clouds in near virial equilibrium.
- The apparent near-virialization of the clouds is easier to understand if the nonthermal motions are a manifestation of infall motions, because in this case their kinetic energy must satisfy $E_k \sim |E_g|$ at all scales, because the motions are continuously driven by gravity. Since this condition is very similar to that of virial equilibrium, $2E_k = |E_g|$, the apparent virialization becomes a necessary condition for clouds undergoing infall.
- The TS model, being based on the assumption of near-equilibrium, is by construction a *stationary* model. However, there is clear evidence that molecular clouds evolve both in their physical properties and their star formation activity (e.g., Kawamura et al. 2009; Chevance et al. 2020; Demachi et al. 2024).

9.3.2 Global Hierarchical Collapse Postulates

As a consequence of the above problems with the TS scenario, a new scenario of *global hierarchical collapse* (GHC, Vázquez-Semadeni et al. 2009, 2019) of cloud formation and evolution of molecular clouds has been recently proposed, that incorporates the effects of *moderately* supersonic turbulence generated self-consistently during the assembly of the clouds, subsequent gravitational collapse, and massive-star feedback that destroys the clouds early during their collapse, thus preventing most of their mass to be turned into stars. This scenario postulates that:

1. Cold atomic clouds and the GMCs within them are initially produced by compressions in the WNM that induce a phase transition to the CNM. The compressions generate moderately supersonic turbulence ($\mathcal{M}_s \sim 3$) in the compressed gas via various instabilities (Vishniac 1994; Koyama & Inutsuka 2002; Heitsch et al. 2006), generating nonlinear density fluctuations of amplitude $\delta\rho/\rho \sim 10$ (rather than the $\delta\rho/\rho \sim 100$ assumed in the TS model).

2. Upon the transition to the CNM, the temperature drops by a factor $\sim$100 and the density increases by roughly the same factor. As a consequence, the Jeans mass, $M_J \propto T^{3/2}\rho^{-1/2}$ (cf. Section 6.1.2), drops by a factor $\sim 10^4$ (Gómez & Vázquez-Semadeni 2014), causing the clouds to rapidly have masses much larger than their Jeans mass, and to begin to undergo gravitational contraction.

3. The large contracting cloud, however, contains nonlinear density fluctuations. As the global, cloud-scale contraction proceeds and the mean density increases, the mean Jeans mass decreases, causing density fluctuations of ever smaller masses to sequentially become unstable and to undergo local collapse themselves at subsequently later times. Moreover, each local collapse center increases its density, and thus further decreases its local Jeans mass, allowing for subsequent fragmentation, which produces new collapsing regions within the previous ones. This process generates a hierarchy of *collapses within collapses*, and amounts to Hoyle fragmentation (Hoyle 1953), though seeded by nonlinear, rather than linear, density fluctuations (cf. Section 7.5).

4. Since the clouds are undergoing large-scale contraction and are also internally turbulent, the molecular linewidths, the velocity dispersion, and the virial parameter $\alpha_{\rm vir}$, contain contributions from both the large-scale infall motions and from the internal turbulent motions.

5. Because the Jeans mass decreases in each collapse center, gravity continues to become stronger than thermal support, and eventually the large regions become nearly pressureless. In consequence, the collapse amplifies anisotropies, generating sheets and filaments (see Lin et al. 1965, and Section 7.2.2 here), and developing a *conveyor-belt-like accretion flow* along them toward the multiple collapse centers.

6. The cores and hubs at the collapse centers grow in mass, producing ever more massive stars, until the feedback from the most massive stars becomes capable of dispersing the clouds. Due to the non-homologous character of the collapse, which causes denser gas to reach the stars sooner, stars with sufficient mass to disperse the clouds can form long before much of the

clouds' mass manages to be turned into stars, thus avoiding the SFR conundrum (cf. Zuckerman & Palmer 1974; Section 9.1.1). The dispersed gas may be returned to the hot and warm diffuse phases of the ISM, or remain dense, in the form of small cloud "shreds".

7. The hierarchical structure and kinematics of the clouds are imprinted on the structure and kinematics of young stellar groups, associations, and clusters produced by the clouds.

In the GHC model, therefore, the interpretation of Larson's relations for molecular clouds is then that the kinetic energy E_k is dominated by the gravitational contraction motions, and thus satisfies the condition of near *equipartition* expected for free-fall, $E_k \sim E_g$, where E_g is the gravitational energy. In this case, the Keto–Heyer (KS) scaling $\sigma \propto (R\Sigma)^{1/2}$ is satisfied. In addition, objects defined by a column density threshold criterion tend to have mean column densities close to the threshold value. This causes the column density of cloud samples to be roughly constant, in which case the KS scaling reduces to Larson's linewidth-size relation.

Also, in the GHC scenario, the role of turbulence within molecular clouds is only to maintain a population of density fluctuations, some of which eventually become unstable as the cloud evolves and its global Jeans mass is reduced, but not directly as a consequence of turbulent compression at the early stages of evolution. Moreover, the turbulence produces no strong *stabilizing* effect for the clouds. Instead, when stellar feedback becomes important, the clouds are disrupted altogether, rather than maintained in near equilibrium as in the TS scenario.

In the following chapters we discuss the presently available observational and numerical knowledge on the formation and evolution of GMCs, and, when appropriate, we present the viewpoints of both the TS and the GHC scenarios, on the basis of the theoretical concepts described in Part I.

References

Ballesteros-Paredes, J., Hartmann, L. W., Vázquez-Semadeni, E., Heitsch, F., & Zamora-Avilés, M. A. 2011, MNRAS, 411, 65
Ballesteros-Paredes, J., D'Alessio, P., & Hartmann, L. 2012, MNRAS, 427, 2562
Beaumont, C. N., Goodman, A. A., Alves, J. F., et al. 2012, MNRAS, 423, 2579
Chevance, M., Kruijssen, J. M. D., Hygate, A. P. S., et al. 2020, MNRAS, 493, 2872
Chevance, M., Kruijssen, J. M. D., Krumholz, M. R., et al. 2022, MNRAS, 509, 272
Chomiuk, L., & Povich, M. S. 2011, AJ, 142, 197
Ciolek, G. E., & Basu, S. 2001, ApJ, 547, 272
Crutcher, R. M. 2012, ARA&A, 50, 29
Crutcher, R. M., & Kazès, I. 1983, A&A, 125, L23
Demachi, F., Fukui, Y., Yamada, R. I., et al. 2024, PASJ, 76, 1059
Dunham, T. J. 1937, PASP, 49, 26
Gentry, E. S., Krumholz, M. R., Dekel, A., & Madau, P. 2017, MNRAS, 465, 2471
Goldreich, P., & Kwan, J. 1974, ApJ, 189, 441
Gómez, G. C., & Vázquez-Semadeni, E. 2014, ApJ, 791, 124

Guerrero-Gamboa, R., & Vázquez-Semadeni, E. 2020, ApJ, 903, 136
Heitsch, F., & Hartmann, L. 2008, ApJ, 689, 290
Heitsch, F., Slyz, A. D., Devriendt, J. E. G., Hartmann, L. W., & Burkert, A. 2006, ApJ, 648, 1052
Hennebelle, P., & Falgarone, E. 2012, A&AR, 20, 55
Herzberg, G. 1988, JRASC, 82, 115
Heyer, M., Krawczyk, C., Duval, J., & Jackson, J. M. 2009, ApJ, 699, 1092
Heyer, M. H., & Brunt, C. M. 2004, ApL, 615, L45
Hoyle, F. 1953, ApJ, 118, 513
Iffrig, O., & Hennebelle, P. 2015, A&A, 576, A95
Iffrig, O., & Hennebelle, P. 2017, A&A, 604, A70
Kawamura, A., Mizuno, Y., Minamidani, T., et al. 2009, ApJS, 184, 1
Kegel, W. H. 1989, A&A, 225, 517
Klessen, R. S., & Hennebelle, P. 2010, A&A, 520, A17
Koyama, H., & Inutsuka, S.-i. 2002, ApL, 564, L97
Lada, C. J., & Lada, E. A. 2003, ARA&A, 41, 57
Larson, R. B. 1981, MNRAS, 194, 809
Lin, C. C., Mestel, L., & Shu, F. H. 1965, ApJ, 142, 1431
Liszt, H. S., Wilson, R. W., Penzias, A. A., et al. 1974, ApJ, 190, 557
Mac Low, M.-M., & Klessen, R. S. 2004, RvMP, 76, 125
Mac Low, M.-M., Klessen, R. S., Burkert, A., & Smith, M. D. 1998, PhRvL, 80, 2754
Maddalena, R. J., & Thaddeus, P. 1985, ApJ, 294, 231
McKee, C. F., & Ostriker, E. C. 2007, ARA&A, 45, 565
Mestel, L., & Spitzer, L. J. 1956, MNRAS, 116, 503
Mouschovias, T. C. 1991, The Physics of Star Formation and Early Stellar Evolution, 342, ed. C. J. Lada, & N. D. Kylafis (Dordrecht: Springer) 449
Padoan, P., & Nordlund, Å. 1999, ApJ, 526, 279
Pattle, K., Fissel, L., Tahani, M., Liu, T., & Ntormousi, E. 2023, Protostars and Planets VII, 534, ed. S. Inutsuka, Y. Aikawa, T. Muto, K. Tomida, & M. Tamura (Tucson, AZ: Univ. Arizona Press) 193
Scalo, J. 1990, Physical Processes in Fragmentation and Star Formation, 162, ed. R. Capuzzo-Dolcetta, C. Chiosi, & A. di Fazio (Dordrecht: Kluwer Academic) 151
Shu, F. H., Adams, F. C., & Lizano, S. 1987, ARA&A, 25, 23
Stone, J. M., Ostriker, E. C., & Gammie, C. F. 1998, ApL, 508, L99
Swings, P., & Rosenfeld, L. 1937, ApJ, 86, 483
Tohline, J. E. 1980, ApJ, 239, 417
Tohline, J. E. 1982, FCPh, 8, 1
Vazquez-Semadeni, E., Ostriker, E. C., Passot, T., Gammie, C. F., & Stone, J. M. 2000, Protostars and Planets IV, 3, ed. V. Mannings, A. P. Boss, & S. S. Russell (Tuscon, AZ: Univ. Arizona Press)
Vázquez-Semadeni, E., Gómez, G. C., Jappsen, A. K., et al. 2007, ApJ, 657, 870
Vázquez-Semadeni, E., Gómez, G. C., Jappsen, A. K., Ballesteros-Paredes, J., & Klessen, R. S. 2009, ApJ, 707, 1023
Vázquez-Semadeni, E., Palau, A., Ballesteros-Paredes, J., Gómez, G. C., & Zamora-Avilés, M. 2019, MNRAS, 490, 3061
Vishniac, E. T. 1994, ApJ, 428, 186

Weinreb, S., Barrett, A. H., Meeks, M. L., & Henry, J. C. 1963, Natur, 200, 829
Williams, J. P., & Blitz, L. 1998, ApJ, 494, 657
Wilson, R. W., Jefferts, K. B., & Penzias, A. A. 1970, ApL, 161, L43
Zuckerman, B., & Evans, N. J. I. 1974, ApL, 192, L149
Zuckerman, B., & Palmer, P. 1974, ARA&A, 12, 279

AAS | IOP Astronomy

Enrique Vázquez-Semadeni

Chapter 10

Molecular Cloud Formation and Growth

Molecular clouds are the birthplace of stars, and therefore, an understanding of their formation and evolution is a crucial step in the understanding of the star formation process. In this chapter, we discuss the mechanisms that lead to their formation and early evolution.

10.1 Clouds as Dynamic Density Fluctuations: The Physics of Compressions

10.1.1 The Need for Flow Convergence

"Clouds" in general (either atomic or molecular) are regions of higher-than-average density in the Galactic ISM. This implies that the surrounding flow must have *converged* onto them to assemble them, as indicated by the continuity (mass conservation) Equation (2.74), which we rewrite here as

$$\frac{d\rho}{dt} = -\rho \, \nabla \cdot \boldsymbol{u}, \tag{10.1}$$

where d/dt, as given by Equation (2.81), is the total (or Lagrangian) time derivative, following the flow. This equation describes the fact that, in order for a given fluid parcel to increase its density, the velocity divergence must be negative there; a negative divergence indicates a *convergence* of the flow onto that parcel. Such convergence can be accomplished by either large-scale instabilities involving gravity and thermal condensation, combined with inertial compressions that trigger the instabilities, such as those induced by passing supernova or HII region shock fronts, collisions of such fronts, stellar winds, or simply generic compressions in the diffuse gas.

An important point to note is that a converging flow viewed from a *Lagrangian* standpoint (traveling with the flow) implies that the parcel is undergoing *compression*

doi:10.1088/2514-3433/adcb28ch10

(a reduction of its volume), while viewed from an *Eulerian* perspective (a fixed region in space), the converging flow can be regarded as an *accretion* flow crossing the region's boundaries.

Also, note that the requirement of a converging flow to form a cloud implies that, whenever we consider "a cloud in equilibrium with its environment" for the purpose of modeling, we are neglecting the fundamental mechanism by which the cloud itself first came into existence. Furthermore, there is no guarantee that a cloud will ever come to equilibrium with its environment before it is destroyed. That is, clouds may be transient, out-of-equilibrium objects (cf. Section 1.2.1). Even more, they may not be "objects" at all (i.e., entities consisting of the same material during most of their lifetime), but rather they may be "regions" in the flow, through which the gas passes, adopting a specific set of physical conditions. Some analogies may be the "rapids" section of a river, where the water moves faster due to a narrowing of the riverbed, or a waterfall, where the water accelerates as it falls down a cliff. In both cases, the rapids section or the waterfall remain at their location for long times, but the water only transits through them during a short time. In what follows, we will refer to these regions as *flow features*.

10.1.2 Cooling and Dimensionality Restrictions on the Triggering of Gravitational Instability

10.1.2.1 The Role of Cooling and the Strongly Super-Jeans Nature of Molecular Clouds

When a region of the ISM increases its density, it generates a deeper gravitational potential, as indicated by Poisson's equation, $\nabla^2\phi = 4\pi G\rho$ (Equation (2.78)). In fact, as described in Section 8.3, MCs contain hundreds or even thousands of Jeans masses, implying that they are strongly Jeans-unstable with respect to thermal support. Thus, an important question about the evolution of MCs is how they attain such a strongly super-Jeans nature.

As shown in Section 6.1.2, the Jeans mass depends both on density and temperature (through the sound speed) as

$$M_{\mathrm{J}} \equiv \frac{\pi^{5/2}}{6}\frac{c_{\mathrm{s}}^3}{G^{3/2}\rho_0^{1/2}};$$

that is, the Jeans mass decreases for increasing density and/or decreasing temperature. Therefore, in order to form a Jeans-unstable cloud, it is necessary that the density and temperature of the cloud vary in such a way that its Jeans mass M_{J} becomes smaller than its actual mass M.

Of course, a simple way for a cloud to become Jeans-unstable and to acquire a large number of Jeans masses is to allow it to cool down at approximately constant density. However, this can only occur for very large regions, which normally have very low densities and high temperatures ($\sim 10^6$ K), and thus can undergo the isochoric mode of thermal instability.[1] Thus, rather than being dense "clouds", these regions are large, hot "voids", such as supernova-carved bubbles.

Another possibility is that the cloud is formed by compression by some mechanism, such as its own self-gravity, the gravity of an externally-generated gravitational potential (e.g., from the stellar spiral arm or from the cloud's surrounding gas), or a mechanical compression, such as a passing supernova or expanding-HII region shock front or a generic turbulent compression. In this case, it is necessary that the cloud's temperature does not increase so much upon the compression as to offset the effect of the increasing density.

The necessary variation of the temperature with density in order to allow a decrease of the Jeans mass (or, equivalently, the Jeans length) with respect to the cloud size, is most easily seen in the case of a polytropic gas, in which $P \propto \rho^{\gamma_e}$, where γ_e is the *effective polytropic exponent*, which represents the net effect of the heating and cooling processes. In this case, the cooling is assumed to happen infinitely rapidly, so that the parcel instantaneously reaches the equilibrium temperature between heating and cooling corresponding to its density (cf. Section 4.2). Indeed, as described in that section, inserting the polytropic law into the ideal-gas equation of state, one can eliminate the temperature, and write the pressure as a function of the density only, thus eliminating the need to solve the thermal energy equation, Equation (2.76), to compute the pressure.

For a polytropic gas, it was shown in Section 5.2.3 that $\gamma_e = 4/3$ is the critical value of the polytropic exponent separating regimes of stability and instability upon three-dimensional compressions of fluid parcels with a fixed mass. Indeed, for $\gamma_e > 4/3$, the ratio of the thermal to gravitational energy of a fluid parcel *increases* as the parcel reduces its size. This implies that the parcel's Jeans mass *increases* as the parcel is compressed. Thus, if the compression was of gravitational origin because the parcel's mass was initially larger than its Jeans mass, the compression itself will eventually cause the Jeans mass to exceed the parcel's mass, and the contraction will terminate. Therefore, the parcel is gravitationally *stable* upon compressions. The opposite occurs if $\gamma_e < 4/3$. In this case, the compression induces a *reduction* of the Jeans mass, causing the cloud's mass to become increasingly larger than its Jeans mass, and so the cloud will continue to collapse with ever less opposition from thermal pressure, making the collapse effectively pressureless for size scales much larger than the Jeans length.

10.1.2.2 The Role of the Dimensionality of the Compression

It should be noted, however, that the critical value of γ_e for compression-induced collapse discussed above implicitly assumes that the compression occurs in 3D, since it assumes spherical geometry and a reduction of the radius. However, it must be

[1] As discussed in Section 4.3.1, the cooling will cause a drop of pressure in the cloud, and so, if initially it was in pressure equilibrium with its surroundings, after the cooling it will be at a lower pressure than its surroundings, which will then compress the cloud at roughly the sound speed. So, the cloud can only maintain a roughly constant density over timescales shorter than the sound crossing time across it, hence the requirement of a very large size of the cloud. This process then tends to occur in very large regions of gas at high temperature.

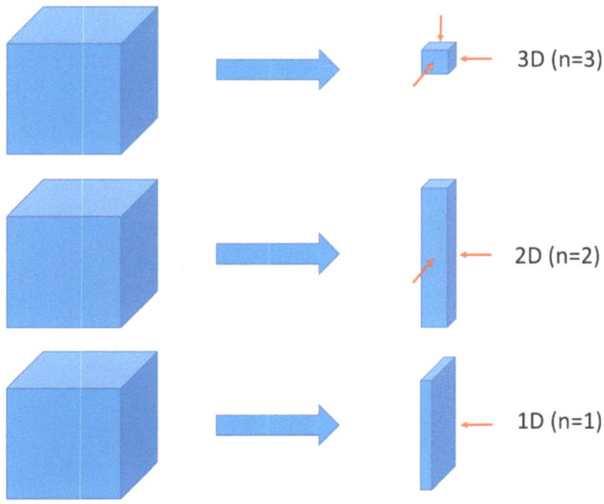

Figure 10.1. Schematic illustration of compressions in 3D, 2D, and 1D. Passing shocks and converging flows will generally compress in one dimension only.

recognized that the compression can occur in fewer than three dimensions. This is in fact the case when the compression is produced by a piston-like agent, such as a passing shock, or the convergence of oppositely-directed streams of gas, both of which compress the gas in one dimension only (see Figure 10.1).

As it turns out, the required value of the polytropic exponent for inducing gravitational instability through a compression depends on the number of dimensions, n, over which the compression occurs. This can be seen in the polytropic case as follows (Vazquez-Semadeni et al. 1996). We start by writing the polytropic condition (the *effective equation of state*) in generic form as

$$\frac{P}{P_0} = \left(\frac{\rho}{\rho_0}\right)^{\gamma_e}, \tag{10.2}$$

where P_0 and ρ_0 are, respectively, the pressure and the density of the initial state. Under this condition, the sound speed in the medium is

$$c_s^2 \equiv \frac{dP}{d\rho} = \frac{\gamma_e P_0}{\rho_0}\left(\frac{\rho}{\rho_0}\right)^{\gamma_e - 1} \equiv c_{s,0}^2\left(\frac{\rho}{\rho_0}\right)^{\gamma_e - 1}, \tag{10.3}$$

where $c_{s,0} \equiv \sqrt{\gamma_e P_0/\rho_0}$ is the sound speed in the initial state (see, e.g., Equation (6.6)). Introducing this in the expression for the Jeans length, Equation (6.14), we obtain

$$L_J = \left[\frac{\pi c_{s,0}^2}{G\rho_0}\left(\frac{\rho}{\rho_0}\right)^{\gamma_e - 2}\right]^{1/2} = L_{J,0}\left(\frac{\rho}{\rho_0}\right)^{\frac{\gamma_e - 2}{2}}, \tag{10.4}$$

where $L_{J,0}$ is the Jeans length in the initial state.

We can consider explicitly the dimensionality of the compression by writing the volume of the gas parcel as $V = L^n L_0^{3-n}$, where L_0 is the dimension that remains constant, while L represents the dimension(s) that is (are) compressed (simultaneously). Assuming a fixed mass, this then implies that $\rho \propto L^{-n} L_0^{n-3}$, while $\rho_0 \propto L_0^{-3}$. Thus,

$$\frac{\rho}{\rho_0} = \left(\frac{L}{L_0}\right)^{-n},$$

and therefore, from (10.4),

$$\frac{L_{\rm J}}{L_{\rm J,0}} = \left(\frac{L}{L_0}\right)^{-\frac{n}{2}(2-\gamma_{\rm e})}. \tag{10.5}$$

The volume becomes marginally Jeans stable when the length L equals the Jeans length. We thus set the ratio $L_{\rm J}/L$ to unity, to get

$$\frac{L_{\rm J}}{L} = 1 = L_{\rm J,0} \frac{L^{-\frac{n}{2}(2-\gamma_{\rm e})-1}}{L_0^{-\frac{n}{2}(2-\gamma_{\rm e})}}.$$

In order for the right-hand side to be a constant (equal to unity), it is necessary that the exponent of L vanishes, and so we obtain

$$\gamma_{\rm e} = 2\left(1 - \frac{1}{n}\right). \tag{10.6}$$

Therefore, the critical values of the polytropic exponent for compressions in n dimensions are

$$\gamma_{\rm e} = \begin{cases} \dfrac{4}{3} & n = 3; \\ 1 & n = 2; \\ 0 & n = 1. \end{cases} \tag{10.7}$$

Equation (10.7) indicates that simultaneous compressions in different numbers of dimensions n require different values of the polytropic exponent in order to be able to induce gravitational instability by the compression. For $n = 3$, we recover the well-known value of $\gamma_{\rm e} = 4/3$ we derived in Section 5.2.3. However, for compressions in less than three dimensions, a "softer" effective equation of state (i.e., one with smaller values of $\gamma_{\rm e}$, so that the flow is less resistive to compressions) is necessary for a compression to induce gravitational instability. Specifically, two-dimensional compressions ($n = 2$) require a softer-than-isothermal regime ($\gamma_{\rm e} < 1$) in order to induce gravitational instability, and one-dimensional compressions require a softer-than-isobaric one ($\gamma_{\rm e} < 0$; i.e., thermally unstable under the isobaric mode).

It is important to note that one-dimensional compressions are necessarily the most common form of compressions that can be driven by stellar energy injection, such as supernova explosions, expanding HII regions, bipolar flows, etc. This is because these processes produce shock *fronts*, which locally compress in just one dimension, except when two or three fronts collide. On the other hand, *body forces*, such as gravity, which are applied over the entire volume of a fluid parcel, can in principle exert three-dimensional compressions. However, even in this case, gravitational contraction occurs first along the shortest dimension of a non-spherical object (Section 7.2.2), and therefore, even gravitational contraction ends up acting primarily in one dimension at any given time.

Finally, note that the above calculations assume a fixed gas mass M, and the conditions may differ in the presence of accretion, which increases the gas parcel's mass. A first step toward considering this case was taken by Hunter & Fleck (1982), who showed that the Jeans mass in 3D is reduced in the presence of an external compressive velocity field.

10.1.3 Turbulence Generation during Cloud Assembly: The Nonlinear Thin-Shell Instability (NTSI)

In Section 4.5.1 we described the mechanism of nonlinear triggering of the thermal instability (TI) by transonic compressions in the warm neutral medium (WNM) in one dimension, and mentioned that this mechanism produces a pair of shocks, and a dense layer roughly one cooling length behind them, which is bounded by two *phase-transition fronts* (Figure 4.8). In more than one dimension, however, the dense layer becomes unstable to the combined action of various instabilities. These are the well-known Kelvin–Helmholtz and Rayleigh–Taylor instabilities (see, e.g., Shu (1992), for a description) the thermal instability (Sections 4.3–4.5; see also Koyama & Inutsuka, 2002) and the so-called *nonlinear thin-shell instability* (NTSI; Vishniac, 1994). Since the latter is not so well known and understood as the former two, we discuss it somewhat further here, in a qualitative manner (see also the discussion by Heitsch et al., 2006).

10.1.3.1 NTSI: The Isothermal Case

Let us first consider the collision of streams in the isothermal case. In this case, there are no phase transition fronts, and the dense layer is directly bounded by the shock fronts. The NTSI is a *bending-mode* instability, which arises when *shape* perturbations are applied on the surface of the shock-bounded layer, causing it to bend as illustrated in the *left* panel of Figure 10.2, in which d is the thickness of the layer and Δ is the displacement of its central plane from the equilibrium position. The latter corresponds to $\Delta = 0$, and is referred to as the *middle plane*. The bending causes the incoming flows to contain a component parallel to the surface, thus transporting material and x-momentum from the peaks to the troughs of the bent surface. Two effects arise from this transport. On the one hand, the transport of mass to the troughs tends to decrease the bending amplitude, as the troughs are less distant from

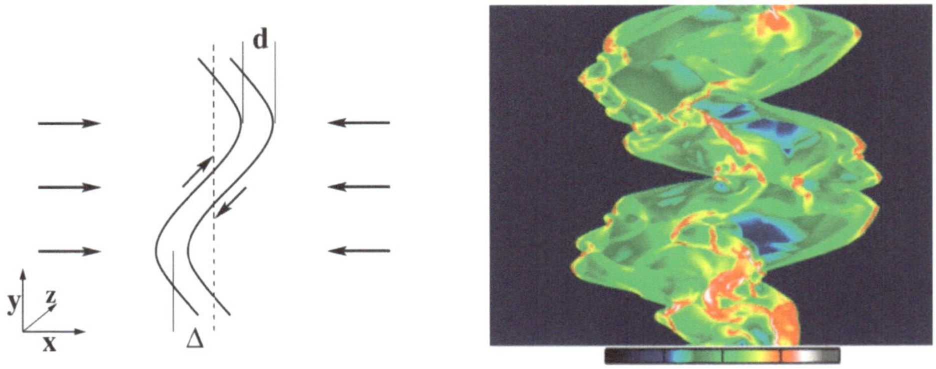

Figure 10.2. *Left:* Schematic representation of the development of the nonlinear thin-shell instability (NTSI) in a compressed layer between two supersonic convergent streams. A sinusoidal *bending* perturbation of the layer induces the generation of flow tangential to the layer's surface, generating vorticity and density enhancements at the peaks of the sinusoidal perturbation. The *dashed line* indicates the "middle plane", defined by $\Delta = 0$. Reprinted with permission from Heitsch et al. (2006). © 2006. The American Astronomical Society. All rights reserved. Printed in U.S.A. *Right:* Density field in a numerical simulation of supersonically colliding streams at an upstream Mach number $\mathcal{M}_s = 22$, illustrating the generation of turbulence in the dense layer. The color bar indicates the range $0 \leqslant \log \rho \leqslant 4$. Folini and Walder., A&A, 459, 1, 2006, reproduced with permission © ESO.

the middle plane (defined by $\Delta = 0$) than the peaks (see the figure). On the other hand, the transport of x-momentum to the peaks tends to increase the amplitude, as it tends to increase their velocity away from the middle plane.

Vishniac (1994) showed that the necessary condition for instability to occur (growth of the bending amplitude) in the isothermal case is $\Delta > d$; i.e., the midplane displacement Δ needs to be larger than the thickness, d. Moreover, he pointed out that this condition is easier to achieve for larger density contrasts between the layer and the external inflows, since this amounts to making the layer thinner, thus making it easier to satisfy the condition $d < \Delta$. This is why the instability is *nonlinear*, since it requires that the perturbation amplitude be large compared to the layer thickness. In turn, since the flow is isothermal, the required large density jump implies that the Mach number must also be large, and therefore the instability is known to require stream collisions with large Mach numbers in the isothermal case.

Finally, it is noteworthy that this flow also generates shear across the layer, since the gas is moving in opposite directions on both sides of the layer. If strong enough, this shear induces Kelvin–Helmholtz instability in the layer, causing it to become turbulent, as illustrated in the *right* panel of Figure 10.2.

10.1.3.2 NTSI: The Thermally Unstable Case

The specific case studied by Vishniac (1994) considered a strongly supersonic flow under the isothermal assumption which, as discussed in Section 4.2, carries the implicit assumption of infinitely fast cooling. However, in the presence of more realistic cooling, with a finite cooling time $\tau_c > 0$ (cf. Equation (4.1)), the gas passing through a shock takes a finite time to return to the thermal-equilibrium values of the

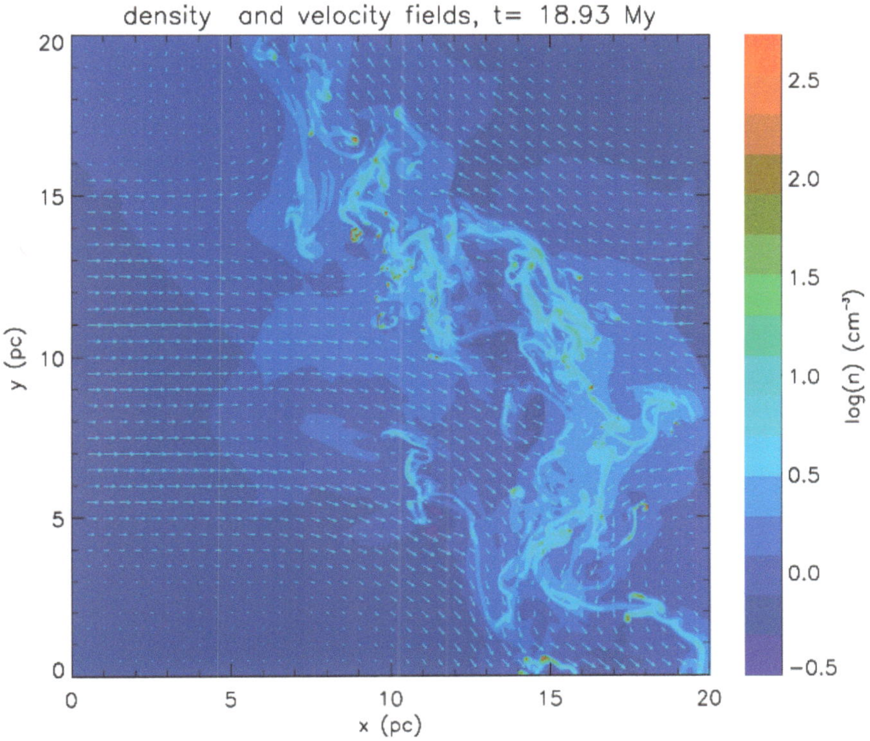

Figure 10.3. Density cut in a two-dimensional, thermally unstable simulation of two converging gas streams entering the box at ± 17 km s^{-1} (Mach number $\mathcal{M}_s \sim 1.5$ in the diffuse gas, of density $\sim$ a few cm^{-3}) at the left and right boundaries of the numerical box, respectively. The shock fronts are seen as the boundary of the first density level higher than the background. The condensation fronts are the boundaries of the thin structures with density $\log[n/\mathrm{cm}^{-3}]. \approx 1$. Reproduced from Audit and Hennebelle (2005), with permission from Springer Nature.

pressure and temperature. During this time, the flow traverses a certain distance, known as the *cooling length*, given by

$$l_c = u_2 \tau_c, \tag{10.8}$$

where u_2 is the post-shock velocity (see Figure 4.8). Therefore, in the case with finite cooling times and sub-isothermal behavior (i.e., with $\gamma_e < 1$, including, in particular, the thermally unstable case under the isobaric mode for which $\gamma_e < 0$; see Section 4.2), the development of the NTSI exhibits some important differences with respect to the isothermal case:

1. Rather than being coincident, the shock and the condensation fronts are separated by roughly one cooling length, which may be significant in the diffuse gas, where τ_c is large. The flow in the vicinity (both fore and aft) of the condensation front may even be entirely subsonic, in spite of the presence of the strong density jump across the condensation front (Vázquez-Semadeni et al. 2006), and the shock may be significantly distant upstream of the condensation front.

2. In the thermally unstable case, because the flow spontaneously *condenses* (i.e., cools and contracts) to a density roughly 100× higher than that of the pre-shock gas, the dense layer is very thin, and therefore the condition $\Delta > d$ is satisfied even for moderately supersonic upstream speeds. In other words, the thermal instability allows the flow to be NTSI-unstable even for transonic compressions. This generates turbulence in the clouds even for moderately supersonic compressions (see Figure 10.3).

10.1.4 Thermal and Ram Pressure Balance

Numerical simulations of the formation of dense layers by supersonic compressions in strongly cooling media often find that the rms Mach number of the turbulence *within* the cold dense gas is only moderately supersonic.[2] This can be understood in terms of an argument similar to that put forward by Banerjee et al. (2009): For a *thermally-bistable* flow (i.e., mediated by the isobaric mode of thermal instability), we expect the thermal pressures of the cold and warm media to be roughly the same, and so

$$\rho_c c_{s,c}^2 \approx \rho_w c_{s,w}^2, \tag{10.9}$$

where the subindices "c" and "w", respectively, represent the cold and warm phases.

In addition, in a quasi-stationary situation such as that depicted in Figure 4.8, the total (thermal+ram) pressures of the cold and warm phases must also be similar (otherwise the dense layer would be expanding or contracting). Therefore, we also have that

$$\rho_c\left(c_{s,c}^2 + \sigma_c^2\right) \approx \rho_w\left(c_{s,w}^2 + v_w^2\right), \tag{10.10}$$

where v_w is the inflow speed in the warm gas, and σ_c denotes the turbulent velocity dispersion in the cold gas, so that $\rho\sigma_c$ is the turbulent ram pressure in this phase. But then, from Equations (10.9) and (10.10) we obtain

$$\rho_c \sigma_c^2 \approx \rho_w v_w^2. \tag{10.11}$$

It then follows from relations (10.9) and (10.11) that the turbulent Mach number in the cold phase, $M_{s,w} \equiv \sigma_w / c_{s,c}$, and the inflow Mach number, $M_{s,w} \equiv v_w / c_s$, must be similar:

$$M_{s,c} \approx M_{s,w}. \tag{10.12}$$

Since the velocity dispersion in the WNM, ~15 km s^{-1}, is only slightly larger than the sound speed there $c_{s,w} \sim 10$ km s^{-1}, this result implies that the motions induced *inside* the CNM are also only moderately supersonic, in agreement with the observed values in simulations.

[2] Note that when the dense layer fragments into smaller clumps, the clump-to-clump velocity dispersion may be strongly supersonic with respect to the Mach number in the cold gas (e.g., Koyama & Inutsuka 2002; Audit & Hennebelle 2005). However, the turbulence *inside* the dense clumps remains only *moderately* supersonic (see, e.g., Figure 9 of Audit & Hennebelle 2005).

10.2 The Formation and Early Evolution of Molecular Clouds: The Converging-Flow Model

In the previous section we have seen that the formation of clouds requires compressions (i.e., converging flows) in their parent diffuse medium. The compressions may be *inertial* (i.e., due to random turbulent motions in the medium), or driven by larger-scale instabilities, or a large-scale gravitational potential, such as that, for example, of a stellar spiral arm. In this section we now describe the sequence of events that occur in order for a cold, dense, gravitationally bound, molecular cloud to be formed, as described by Hartmann et al. (2001) (see also Franco & Cox 1986), in what can be called the *converging-flow model* of molecular cloud formation.

10.2.1 Coevolution of Magnetic Criticality, Molecular Fraction and Gravitational Binding

As mentioned in Section 10.1.2, compressions are most likely to occur preferentially in one dimension, and so the following discussion assumes 1D compressions, such as that shown in Figure 4.8. Furthermore, we assume that, if a magnetic field is present, it is parallel to the compression direction. This assumption is justified by numerical simulations that show that compressions in which the field is oblique to the compression direction evolve by reorienting either the magnetic of the velocity field depending on the magnetic field strength, in such a way that the compression ends up being aligned with the magnetic field (e.g., Hennebelle & Pérault 2000; Fogerty et al. 2016).

Under these conditions, Hartmann et al. (2001) showed that compressions in the warm neutral medium (WNM), with physical and chemical conditions typical of the Solar Neighborhood in the Galaxy, produce three different effects roughly simultaneously as the column density of the dense accumulated gas increases:

10.2.1.1 Atomic to Molecular Transition

Chemical evolution calculations applicable to star-forming regions (e.g. van Dishoeck & Blake, 1998) show that the minimum necessary hydrogen column density for effective shielding of H_2 and CO molecules from photodissociating UV radiation is

$$N_{\rm H,min} \approx 1\text{–}2 \times 10^{21}\,{\rm cm}^{-2}. \tag{10.13}$$

Reaching this column density by accumulation of material starting from a mean density of $n \sim 1\ {\rm cm}^{-3}$, representative of the WNM in the Galactic spiral arms, at a velocity $v \sim 10\ {\rm km\ s}^{-1}$ (roughly the velocity dispersion in that gas) requires a time

$$\tau_{\rm Nshield} \sim 30\ {\rm Myr}\left(\frac{N}{10^{21}\ {\rm cm}^{-2}}\right)\left(\frac{n}{1\ {\rm cm}^{-3}}\right)^{-1}\left(\frac{v}{10\ {\rm km\ s}^{-1}}\right)^{-1}. \tag{10.14}$$

We can compare this timescale to the characteristic timescale for molecule formation. To do this, we use the result that the H_2 formation rate per unit volume on dust grain surfaces is (Jura 1975; see also Klessen & Glover 2016 and Section 5.1.1 here)

$$R_{H_2} \sim 3 \times 10^{-17} \left(\frac{n}{\text{cm}^{-3}}\right)\left(\frac{n_H}{\text{cm}^{-3}}\right) \text{cm}^{-3}\,\text{s}^{-1}, \tag{10.15}$$

where n and n_H are, respectively, the total and the atomic hydrogen number densities. Therefore, the characteristic molecular hydrogen formation timescale is

$$\tau_{\text{mol}} \equiv \frac{n_H}{R_{H_2}} \sim 20 \left(\frac{n}{50 \text{ cm}^{-3}}\right)^{-1} \text{Myr}, \tag{10.16}$$

where we have normalized the density to a value typical of the CNM, $n \sim 50 \text{ cm}^{-3}$. We therefore see that the molecule formation timescale in this medium satisfies

$$\tau_{\text{mol}} \lesssim \tau_{\text{Nshield}}, \tag{10.17}$$

meaning that the gas has time to become molecular during the accumulation process.

The gas required to reach the molecule-formation column density is accumulated from a distance

$$L_{\text{acc}} = v\tau_{\text{Nshield}} \sim 300 \text{ pc}, \tag{10.18}$$

which is commonly referred to as the *accumulation length*. Note, however, that, since the density increases by a factor $\sim 100\times$ from the WNM to the CNM, by the time a cold atomic cloud reaches N_{mol}, the gas layer is only ~ 3.5 pc thick. Moreover, it can be only fractions of a parsec thick in its early growth stages. This is likely the reason why CNM clouds seem to be extremely thin, although quite extended (Heiles & Troland 2003): the fronts may extend over large areas, but the initial thickness of the cold cloud is very small.

10.2.1.2 Onset of Gravitational Instability

The gravitational stability of the cloud can be estimated by considering its virial balance, assuming spherical geometry. Assuming that the cloud is in near virial equilibrium between its self gravity and its internal and external pressures over a time period short compared to the characteristic time for mass growth due to accretion, we can apply Equation (5.20), which we rewrite here as:

$$P_{\text{ext,eq}} = P_{\text{int}} - \frac{3GM^2}{20\pi R^4} = P_{\text{int}} - \frac{3\pi G\Sigma^2}{20}, \tag{10.19}$$

where we have defined the mass column density as $\Sigma \equiv \mu m_H N = M/\pi R^2$, with N being the number column density, $\mu = 1.27$ being the mean particle mass in atomic units for neutral atomic gas, and m_H being the hydrogen mass. We also remind the reader that here $P_{\rm ext,eq}$ is the *required* (not the actual) value of the external pressure in order to confine the cloud. This required external pressure vanishes when

$$\frac{3\pi G\Sigma^2}{20} = P_{\rm int}. \tag{10.20}$$

However, because the cold and warm phases are mediated by the isobaric mode of the thermal instability, the internal pressure is of the order of the *actual* external pressure $P_{\rm ext}$ (cf. Equation (10.9)). We can therefore write

$$N = \frac{1}{\mu m_{\rm H}}\left(\frac{20P_{\rm ext}}{3\pi G}\right)^{1/2} = 9.9\times 10^{20}\left(\frac{P_{\rm ext}}{k_{\rm B}}\right)_3^{1/2} {\rm cm}^{-2}, \tag{10.21}$$

where $k_{\rm B}$ is Boltzmann's constant and $(P_{\rm ext}/k_{\rm B})_3$ is the value of $P_{\rm ext}/k_{\rm B}$ in units of 10^3 K cm^{-3}.

Equation (10.21) gives the value of the number column density N at which no external pressure is required to confine the cloud, beyond which no equilibrium is possible (Section 5.2.4). But, since the *actual* external pressure is nonzero, it compresses the cloud unimpededly, pushing it into gravitational contraction. This is precisely the case of the last stable sphere subject to self-gravity and both internal and external pressures, indicated by the line labeled $R_{(P_{\rm ext}=0)}$ in Figure 5.1. Noting that the typical value of the WNM pressure at the Solar Galactocentric radius is such that $(P_{\rm ext}/k_{\rm B})_3^{1/2} \sim 1.7$, and comparing with the value indicated in Equation (10.13), we see that, by the time the cloud is becoming molecular, it is also becoming gravitationally unstable.

Since the cold atomic cloud becomes gravitationally unstable, it is worth computing the free-fall time, Equation (7.8), for gas with typical CNM conditions. Considering an atomic cloud with a mean particle mass $\mu = 1.27$, the free-fall time is

$$\tau_{\rm ff} = 6.5\left(\frac{n}{50\ {\rm cm}^{-3}}\right)^{-1/2} {\rm Myr}. \tag{10.22}$$

This is a factor of several shorter than the time it takes to collect the cloud from the diffuse medium, $\tau_{\rm Nshield}$ (Equation (10.14)). Therefore, once the cloud becomes gravitationally unstable, the contraction process is quite rapid. Note, however, that this timescale is actually a lower limit, since, if the cloud is significantly flattened, its actual collapse timescale may be longer than the spherical free-fall time by factors of a few to several, as discussed in Section 7.2.3.

10.2.1.3 Onset of Magnetic Supercriticality

The third effect identified by Hartmann et al. (2001) to occur at column densities of the order of 10^{21} cm^{-2} (for solar neighborhood parameters) is that the clouds begin to become magnetically supercritical. This can be seen from Equation (5.34) for the critical value of the column density, which we rewrite here in terms of the column number density as

$$N_{cr} = 3.1 \times 10^{20}\left(\frac{B}{1\mu\mathrm{G}}\right)\mathrm{cm}^{-2}. \tag{10.23}$$

Comparing to Equations (10.13) and (10.21), we see that, by the time the cloud is becoming molecular and gravitationally unstable, it has also recently become magnetically supercritical. Therefore, by the time cold dense clouds become molecular, they are also losing thermal and magnetic support, and are capable of undergoing gravitational contraction.

10.2.2 Summary

The discussion in this chapter can be summarized as follows:

- Cloud formation from the WNM requires some form of compressive event, which can be induced by large-scale gravitational potential, such as that from the stellar spiral arms in the Galaxy, or from the self-gravity of the gas at larger scales, or else by at least moderately supersonic inertial compressions, such as the passage of a supernova or expanding HII region front, or simply random supersonic turbulent motions in the WNM.
- The inertial compressions are likely to occur preferentially in one dimension, because shock fronts are planar. Compressions driven by the *self* gravity of the clouds are also likely to occur in one dimension because pure gravitational contraction (unimpeded by any opposing processes) occurs first along the shortest dimension of the contracting object, amplifying anisotropies. Compressions driven by external gravitational potentials may occur along more than one direction simultaneously depending on the morphology of the attracting gravitational potential well.
- Under the conditions typical of the solar neighborhood, even moderately supersonic compressions in the WNM can trigger a phase transition to the CNM, accumulating dense material at the converging layer, so that a thin, dense, cold atomic cloud "appears" by condensation from the WNM.
- The dense compressed layer becomes turbulent due to the combined action of the Kelvin–Helmholtz, Rayleigh–Taylor, and nonlinear-thin-shell instabilities, as well as the continuation of the condensation process due to the thermal instability.
- Once the column density of the dense gas reaches a value $\sim 10^{21}\ \mathrm{cm}^{-2}$, three transitions occur roughly simultaneously:
  - The cloud begins to become molecular.
  - The cloud begins to become Jeans-unstable.
  - The cloud begins to become magnetically supercritical.
- The timescale for the cloud to become molecular is of order a few tens of megayears. Once the cloud becomes Jeans unstable, its collapse timescale is significantly shorter, of the order of several megayears.

It is important to note that the converging-flow model of Hartmann et al. (2001) is fully consistent with the observations of magnetic field strength in atomic and

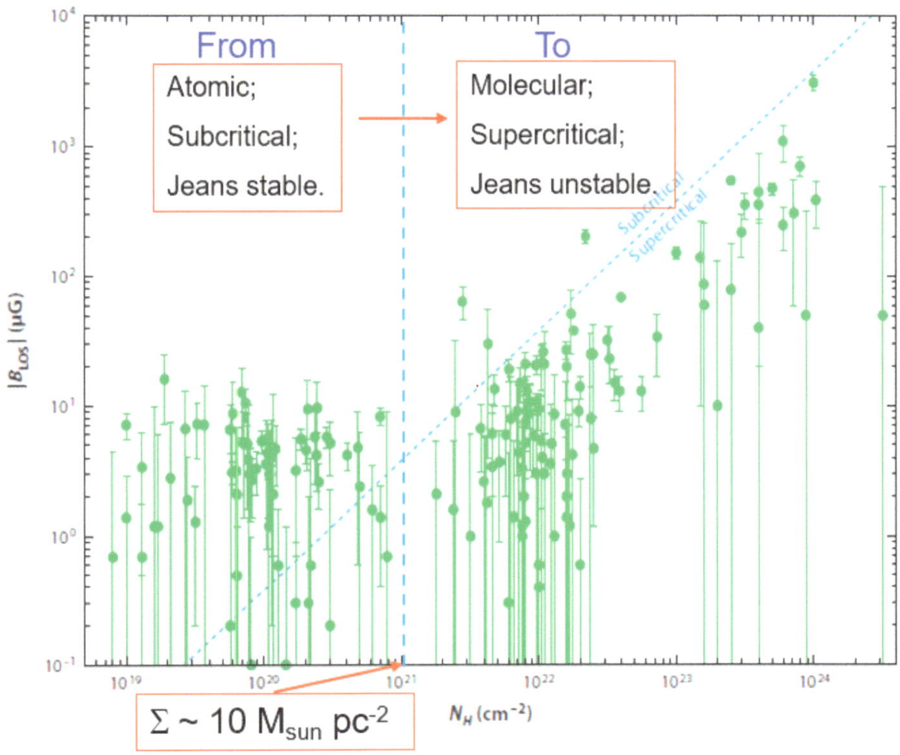

Figure 10.4. The same compilation of magnetic field strengths in (atomic and molecular) clouds and cores of various total column densities $N_H = N_{HI+2N_{H_2}}$ by Crutcher (2012), shown in the left panel of Figure 8.9, with the additional indication of the transitions occurring when low-density gas is collected by large-scale compressions to form high-density clouds, and indicating the transitional column density $N \sim 10^{21}\,\mathrm{cm}^{-2}$, corresponding to a mass surface density $\Sigma \approx 10M^{\odot}\ \mathrm{pc}^{-2}$, around which these transitions occur. The inclined dashed blue line corresponds to the critical mass-to-flux ratio $(N_H/B)_{\mathrm{cr}} \approx 3 \times 10^{20}\,\mathrm{cm}^{-2}\mu\mathrm{G}^{-1}$ (cf. Equation (10.23)). Reproduced with permission from Crutcher, (2012). © Annual Reviews.

molecular clouds and cores by Crutcher (2012), shown in Figure 8.9. Figure 10.4 shows the same plot, but augmented with the processes discussed in this section, indicating that, for solar neighborhood conditions, the column number density $N \sim 10^{21}\,\mathrm{cm}^{-2}$ ($\sim 10M^{\odot}\ \mathrm{pc}^{-2}$) marks the approximately simultaneous transition from atomic, sub-Jeans and magnetically subcritical to molecular, super-Jeans and magnetically supercritical gas. Reaching this critical column density is accomplished by compressive motions carrying the gas from distances a few hundred parsecs away to the location of the newly formed cloud. Note also that the critical nature of this column density (for Solar neighborhood parameters) was predicted by Hartmann et al. (2001) several years before its first observational confirmation by Crutcher et al. (2010), later extended by Crutcher (2012).

Finally, it is also important to note that supernova (SN) explosions are often considered as the drivers of turbulence in the ISM and, by inference, in molecular clouds, through a turbulent cascade. However, because the expansion of the

SN-driven fronts implies shocks that sweep the ambient gas, it must be kept in mind that they also constitute a mechanism for *forming* new clouds, not just stirring of pre-existing ones.

References

Audit, E., & Hennebelle, P. 2005, A&A, 433, 1

Banerjee, R., Vázquez-Semadeni, E., Hennebelle, P., & Klessen, R. S. 2009, MNRAS, 398, 1082

Crutcher, R. M. 2012, ARA&A, 50, 29

Crutcher, R. M., Wandelt, B., Heiles, C., Falgarone, E., & Troland, T. H. 2010, ApJ, 725, 466

Fogerty, E., Frank, A., Heitsch, F., et al. 2016, MNRAS, 460, 2110

Folini, D., & Walder, R. 2006, A&A, 459, 1

Franco, J., & Cox, D. P. 1986, PASP, 98, 1076

Hartmann, L., Ballesteros-Paredes, J., & Bergin, E. A. 2001, ApJ, 562, 852

Heiles, C., & Troland, T. H. 2003, ApJ, 586, 1067

Heitsch, F., Slyz, A. D., Devriendt, J. E. G., Hartmann, L. W., & Burkert, A. 2006, ApJ, 648, 1052

Hennebelle, P., & Pérault, M. 2000, A&A, 359, 1124

Hunter, J. H. J., & Fleck, R. C. J. 1982, ApJ, 256, 505

Jura, M. 1975, ApJ, 197, 575

Klessen, R. S., & Glover, S. C. O. 2016, Saas-Fee Advanced Course, 43, 85

Koyama, H., & Inutsuka, S.-i. 2002, ApL, 564, L97

Shu, F. H. 1992, Physics of Astrophysics Vol. II (Mill Valley, CA: Univ. Science Books)

van Dishoeck, E. F., & Blake, G. A. 1998, ARA&A, 36, 317

Vazquez-Semadeni, E., Passot, T., & Pouquet, A. 1996, ApJ, 473, 881

Vázquez-Semadeni, E., Ryu, D., Passot, T., González, R. F., & Gazol, A. 2006, ApJ, 643, 245

Vishniac, E. T. 1994, ApJ, 428, 186

AAS | IOP Astronomy

Interstellar Flow and Star Formation

Enrique Vázquez-Semadeni

Chapter 11

Onset of Gravitational Collapse, Fragmentation and Filamentation

In the previous chapter we have described the early evolution of molecular clouds (MCs), from the formation of their precursor cold atomic clouds, which are likely born as thermally and magnetically supported atomic clouds, with moderately supersonic turbulence, to their transition to molecular, thermally and magnetically *un*supported clouds. In this chapter we now describe the evolution of the clouds, from this stage, to the moment when massive star formation begins.

11.1 Onset of Self-Gravitational Instability and Collapse Flow

In Section 10.2.1.2 we showed that, beyond the transitional column density $N_{\rm H} \approx 10^{21}\,{\rm cm}^{-3}$, a cloud assembled by large-scale compressions in the ISM ceases to be in equilibrium between self-gravity and the internal and external pressures, and initiates gravitational contraction. In addition, noting that the Jeans mass scales as $M_{\rm J} \propto c_{\rm s}^3 \rho^{-1/2}$ (cf. Equation (6.16)), we see that it drops by a factor $\sim 10^4$ upon the transition from the WNM to the CNM, during which the temperature drops by a factor ~ 100 and the density increases by the same factor. Specifically, as seen in Table 1.1, the Jeans length and mass (Equations (6.14) and (6.16)) in the atomic gas drop from

$$L_{\rm J,WNM} \approx 1.6\,{\rm kpc}; \qquad M_{\rm J,WNM} \approx 3.3 \times 10^7 M_\odot \tag{11.1}$$

in the WNM, to

$$L_{\rm J,CNM} \approx 16\,{\rm pc}; \qquad M_{\rm J,CNM} \approx 3300 M_\odot \tag{11.2}$$

doi:10.1088/2514-3433/adcb28ch11

in the CNM. Therefore, cold atomic clouds with masses of a few thousand solar masses become Jeans unstable.[1]

Molecular clouds are even denser and colder than cold atomic clouds, and therefore, their Jeans lengths and masses are even smaller. Indeed, as also seen from Table 1.1, with densities $n \gtrsim 300\ \mathrm{cm}^{-3}$ and temperatures $T \sim 10$ K, molecular clouds have

$$L_{\mathrm{J,MC}} \lesssim 1.5\ \mathrm{pc}; \qquad M_{\mathrm{J,MC}} \lesssim 30 M_{\odot}. \tag{11.3}$$

To put things in perspective, this implies that a typical, moderate-mass, 10^4-$M_{\odot}$ giant molecular cloud (GMC) contains over ~300 Jeans masses.

11.2 The Turbulent Support and Global Hierarchical Collapse Scenarios

From this point onwards, wherever possible, we will present our discussion in terms of the two main views that exist at the time of this writing concerning the physical state and the fate of molecular clouds. On the one hand, the *turbulent support model* (TS) assumes that the clouds are formed by supersonic compressions in the ISM, in particular those due to supernova explosions, and that the clouds' self-gravity is globally balanced by *strongly* supersonic internal turbulence, which prevents global, cloud-scale collapse, and, in consequence, avoids an excessive star formation rate (SFR; see Section 9.1.1). Moreover, being strongly supersonic, this turbulence produces strong shocks, which induce the formation of large-amplitude local density fluctuations. These fluctuations constitute clumps, filaments, and cores. The cores, being the densest structures, can become locally dominated by their self gravity, and thus begin to contract gravitationally (e.g., Vázquez-Semadeni et al. 2000; 2003; Mac Low & Klessen 2004; Elmegreen & Scalo 2004; McKee & Ostriker 2007; Klessen & Glover 2016; Padoan et al. 2020).

On the other hand, the *global hierarchical collapse* (GHC) model assumes that massive molecular clouds are dominated by gravity throughout most of their volume, and may in fact be the result of gravitational contraction ever since their infancy as cold atomic clouds (e.g., Hartmann et al. 2001; Heitsch et al. 2006; Vázquez-Semadeni et al. 2007, 2009, 2019; Ibáñez-Mejía et al. 2016, 2017; Mac Low et al. 2017). The gravitational contraction is strongly chaotic and anisotropic, generating multiple centers of collapse in a hierarchical fashion: local collapse events occur on top of larger-scale collapse flows pointed toward different collapse centers. Moreover, as a result of the accelerated nature of the gravitational contraction, the SFR also accelerates during the early star-forming stages, until it becomes large enough for massive stars to appear. The massive stars disrupt the accretion flow, and reduce the SFR during the late stages of evolution, perhaps even shutting down the local star-formation activity, while possibly promoting it elsewhere. The excessive SFR expected

[1] Note that this is only a rough estimate, since cold atomic clouds are expected to be highly elongated or flattened (cf. Section 8.2.1), and so they may be larger than the Jeans length in some direction but smaller in another.

from collapsing clouds is avoided because most of the clouds' mass is dispersed by the feedback from the massive stars. In this view, the supersonic linewidths observed in molecular clouds are produced by a combination of an infall component and a *moderately* supersonic turbulent component, coexisting in different proportions at different evolutionary stages and different spatial and density scales (Chapter 13). See Chapter 9 for a historical perspective of the development of both models.

Since both models attempt to explain mostly the same sets of observed properties of molecular clouds, at the present time the existing observational evidence is rather ambiguous and does not allow to clearly discriminate among these models. Thus, in what follows, we discuss the view from both scenarios in regards to each topic when available.

11.3 The Transition from the Cold Atomic to the Molecular Regime

The most frequently discussed aspect of the formation of molecular clouds is the *chemical* transition from cold atomic to molecular gas. The details of this transition are controlled by microscopic (atomic, molecular, and dust-grain scales) processes, and are therefore out of the scope of this monograph, which focuses on the fluid dynamical aspects of cloud and star formation. We thus refer the reader to, for example, the extensive review by Klessen & Glover (2016), and here we just briefly mention that the formation of molecular hydrogen (H_2) requires self-shielding by the H atoms themselves from photodissociating radiation, while the formation of CO requires shielding by dust grains, and typically happens at greater depths within the clouds than H_2 formation (Bergin et al. 2004). Both of these mechanisms require that the cloud reaches a threshold column density $N_{\rm H,min} \sim 10^{21}\ {\rm cm}^{-2}$ (cf. Section 10.2.1).

On the other hand, the *physical* (dynamic and thermodynamic) transition from CNM to molecular gas, which involves a final step of cooling and compression since GMCs are typically a factor $\sim$ 5–10× denser and colder than CNM clouds (see Tables 1.1 and 1.2), has not been so frequently discussed in reviews and monographs. As a consequence of this final compression and cooling step, molecular clouds end up acquiring a large number of Jeans masses. Whether the onset of gravitational contraction occurs because the clouds have become molecular, or vice versa, is a question that to this date remains open, although some numerical and theoretical results suggest that gravitational contraction happens first and that molecule formation is a byproduct of it (e.g., Heitsch & Hartmann 2008; Mac Low & Glover 2012; Mac Low et al. 2017).

In any case, the fact that molecular clouds have masses orders of magnitude larger than their Jeans masses implies that they will have a strong tendency to contract gravitationally. Under the TS scenario, this tendency is suppressed by the strongly supersonic motions observed within them, which are interpreted as turbulence that produces a large enough pressure to balance the self-gravity of the clouds, or even disperse them, with only the small scales ($\lesssim$1 pc) being capable of collapsing. Under GHC, the gravitational contraction proceeds at the entire cloud scale until a sufficient number of massive stars has formed to then destroy or disperse the rest of the cloud, thus preventing most of its mass from being converted to stars, as discussed in Chapter 9.

11.4 Fragmentation, Filament Formation, and "Conveyor Belt" Flow

As described in Chapter 8, molecular clouds are far from being homogeneous structures. Instead, they are strongly amorphous and inhomogeneous, containing filamentary substructures, which in turn contain roundish, denser regions. The latter are referred to as *hubs* when they are the point of convergence of several filaments, or *cores*, when they are smaller and contained within just one filament. These regions are collectively referred to as hub-filament systems (HFSs; Myers 2009), and have total sizes of a few tens of parsecs (top left panel of Figure 11.1).

The proposed origin of HFSs is different in the TS and the GHC models. In the former, HFSs are *speculated*[2] (top right panel of Figure 11.1) to be formed by turbulent, *inertial* compressions in the molecular clouds as a consequence of the strongly supersonic turbulence present in the clouds, in turn driven by external, relatively nearby supernova explosions (Padoan et al. 2020). Instead, in the GHC scenario, numerical simulations of clouds undergoing this process (Gómez & Vázquez-Semadeni 2014) have shown that these structures can arise by *anisotropic gravitational contraction* (Section 7.2.2; bottom left panel of Figure 11.1).

Indeed, as discussed in Section 7.2.2, gravitational contraction is extremely anisotropic because in general the collapse proceeds fastest along the shortest dimension of a non-spherical object, amplifying its anisotropy. Moreover, the collapse is non-homologous, meaning that lower-density material contracts on a longer timescale than higher-density gas, since the latter has a shorter free-fall time (Equation (7.8)). Therefore, a *collapse flow* is established such that a fluid parcel increases its density as it *travels* from low- to high-density regions, with parcels at initially lower densities taking longer times to reach a diverging density (a singularity; cf. Section 7.3) than parcels at initially higher densities. Moreover, it should be noticed that the collapse flow may last a long time, as long as there is a supply of low-density gas from the surroundings.

The picture under GHC is that the collapse flow may be likened to the flow of a river (cf., bottom right panel of Figure 11.1), in which water at higher altitudes (corresponding to the low-density gas in a molecular cloud) flows toward lower altitudes (corresponding to the higher-density material in a filament), until it reaches a lake or a sea (corresponding to a dense hub or core). Evidence for this two-step accretion, from cloud to filament and from filament to hub, was first reported by Kirk et al. (2013).

Note that, in this analogy, the river is not a material "object"; i.e., it is not the water itself, but rather it is the *riverbed* (the locus, a "channel", or "funnel") through which the flow passes on its way to the sea. The riverbed remains stationary, or changes on long, geological timescales, but the water molecules pass by on a much

---

[2] We say "speculated" because the simulations by Padoan et al. (2020) *did* include self-gravity, and the suggestion that the flow that forms the HFSs is purely inertial was based on indirect evidence—that systems of this size scale tend to be super-virial—, but no tests were made on whether purely turbulent flows without self-gravity are capable of forming HFSs.

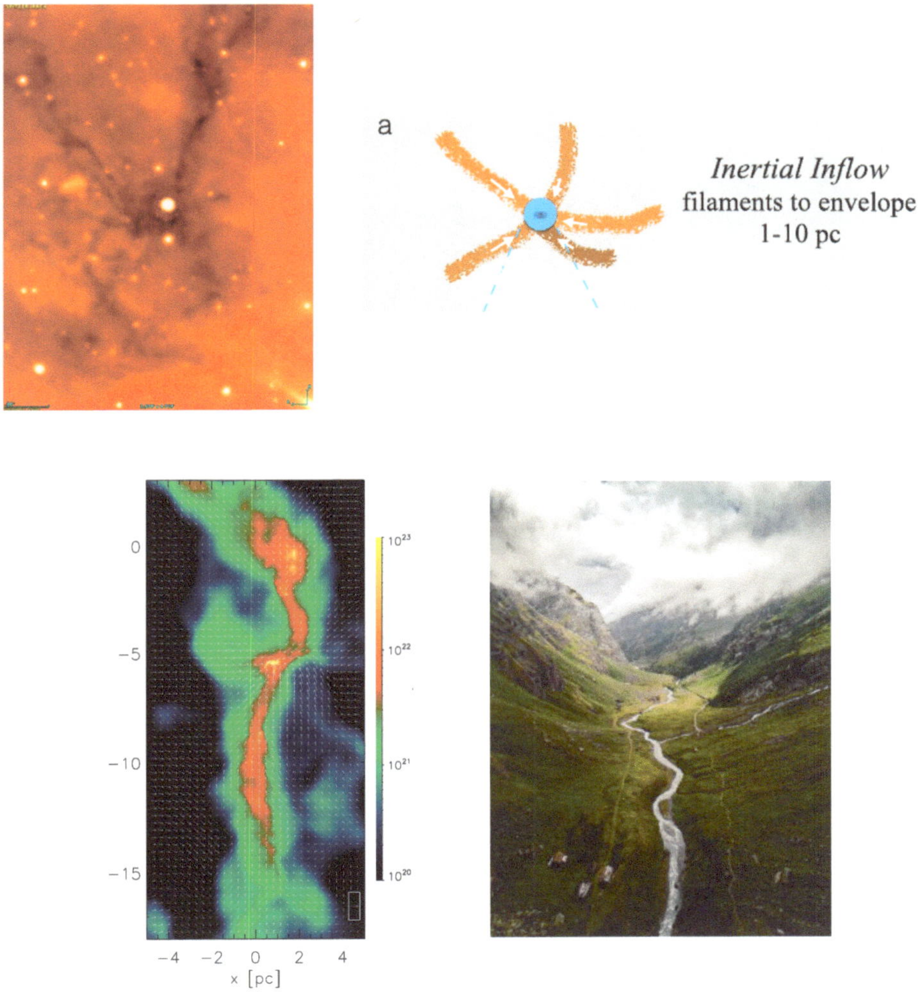

Figure 11.1. *Top left:* A typical hub-filament system (HFS) as viewed by the *Spitzer* telescope at 8 μm, where the *hub* is the central, roundish structure, and the filaments extend out to several parsecs. The roundish, darker spots along the filaments are the cores. (Kumar et al., A&A, 642, A87, 2020, reproduced with permission © ESO. *Top right:* Schematic diagram of the hypothetical inertial flow (due to turbulence) that would be responsible for forming a hub-filament system (Reproduced from Padoan et al., (2020). © 2020. The American Astronomical Society. All rights reserved.). *Bottom left:* Column density (color scale) and velocity field (arrows) of a hub-filament system arising spontaneously in a numerical simulation of a cloud undergoing global, anisotropic and hierarchical gravitational contraction (available as an animation in the electronic edition of this book https://doi.org/10.1088/2514-3433/adcb28). The arrow at the lower-right inset indicates a velocity of 2 km s^{-1} (Reproduced from Gómez & Vázquez-Semadeni (2014). © 2014. The American Astronomical Society. All rights reserved.). *Bottom right:* A riverbed illustrating how it collects water downsliding from the mountain tops (top of the potential well), and then conducting it along the riverbed (intermediate potential) down to a lake or an ocean (bottom of the potential well). CC BY 4.0.

shorter timescale. Moreover, note that the water carves its own path, making it deeper and thus prone to capture even more water. A similar process occurs with a filament: as it attracts more material, it becomes denser, thus deepening the gravitational potential well (Figure 11.2). Finally, note that the river can persist

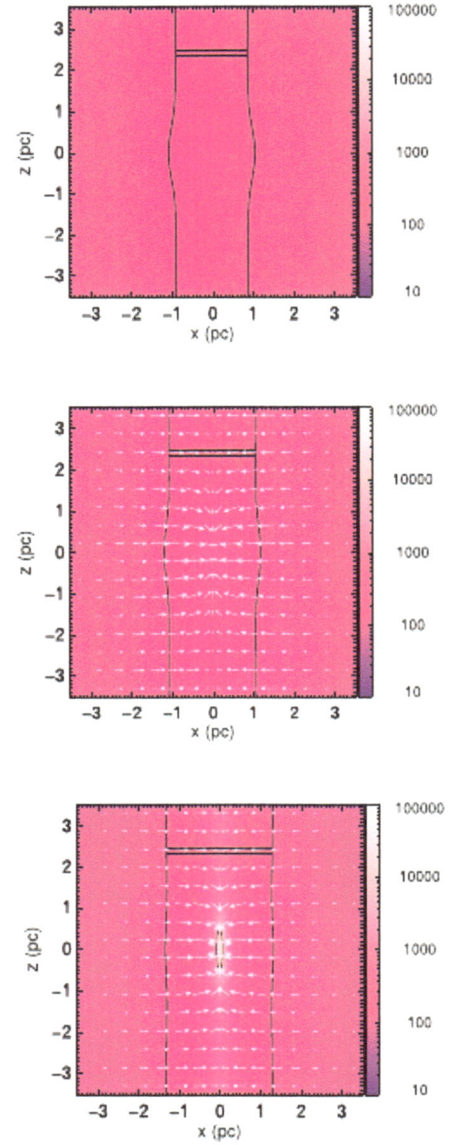

Figure 11.2. Numerical simulation of the gravity-driven evolution of a filamentary perturbation with a central spherical enhancement, representing a hub-filament system. The images show a slice along the axis of the filament at times at $t = 0, 0.53$ and $1.02\tau_{\rm ff}$, where $\tau_{\rm ff}$ is the free-fall time for the mean density in the simulation. The vertical solid lines indicate the isodensity contours where the density is 30% larger than the background. In the bottom panel, the central contour indicates the locus where the density is 30% larger than that at the filament's axis. The arrows indicate the accretion flow, which is seen to *not* become radial, but rather to mainly proceed from the cloud to the filament and from the filament to the hub, causing both the filament and the hub to increase their respective densities over time. Reproduced from Naranjo-Romero et al., (2022) with permission from Oxford University Press on behalf the Royal Astronomical Society.

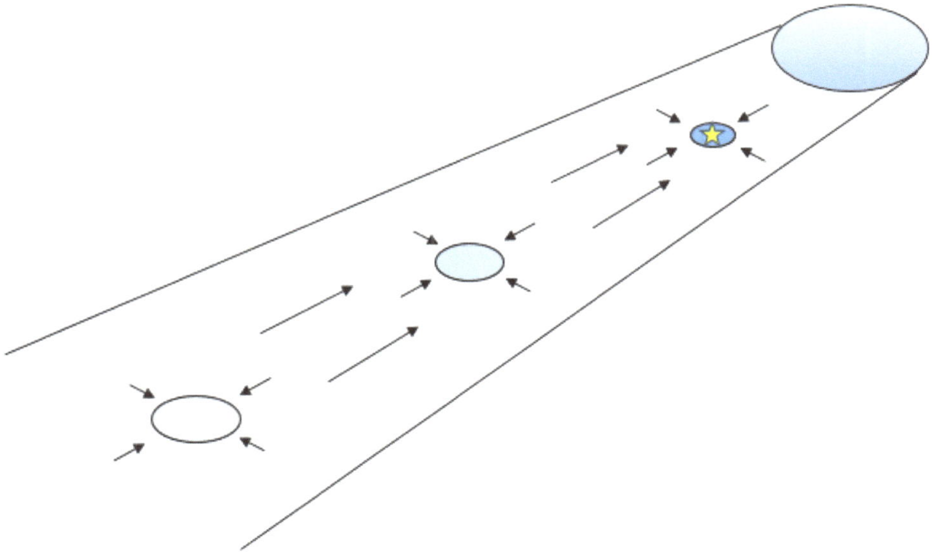

Figure 11.3. A schematic illustration of a *conveyor-belt* type of flow. The large-scale collapse flow is directed toward a main collapse center, while, simultaneously, secondary smaller-scale collapses occur within the large-scale flow, amounting to a hierarchical, collapse-within-collapse flow. This allows for a local collapse to occur within a globally gravitationally unstable medium without the latter falling onto the same collapse center as the small-scale one.

for as long as the water supply in the high mountains remains. Equivalently, a filament can persist for as long as it can accrete from its surrounding cloud, until the time when it is destroyed by the feedback from the stars it itself forms.

In such a river-like flow, in which the filamentary flow is directed toward a main hub, secondary collapse may occur as the gas flows along the filament, producing smaller cores that ride on top of the filamentary flow. This type of flow is often referred to as *conveyor-belt* flow (Longmore et al. 2014), and is schematically illustrated in Figure 11.3. This is a simple form of a *collapse-within-collapse* flow, which is characterized by the collapse centers being non-concentric, so that the small-scale, local collapse flow evolves while it itself flows toward the large-scale, global collapse center.

11.5 Sequential Collapse of Subsequently Smaller Scales under GHC

Under the TS scenario, density fluctuations of scales larger than ~ 0.1 pc (the dense molecular cloud core scale) are assembled by supersonic turbulent compressions, and typically only structures of scales smaller than that are "captured" by self-gravity, and begin to collapse locally. Therefore, gravitational collapse in the TS scenario is restricted to scales $\lesssim 1$ pc, while scales larger than that are assumed to be either supported by turbulent motions against collapse, or just transient turbulent density fluctuations.

Instead, under GHC, gravitational contraction is allowed to occur on all scales up to the entire-cloud scale, and therefore there is the additional phenomenon of *sequential onset of collapse at different scales*. This process was described in Section 7.5.3, and has no parallel in TS. For convenience, we briefly describe it here again, in the context of the evolution of MCs.

As discussed in the previous section, according to GHC, molecular clouds are in a regime of "collapses within collapses", but small-scale, low-mass fluctuations can only begin to collapse locally at later times. This is because GHC, similarly to TS, also considers MCs to be turbulent, but considers the *actual* turbulence (that is, motions unrelated to gravitational contraction) to be only moderately (not strongly) supersonic. Such turbulence does generate nonlinear density fluctuations, but does not drive them directly into collapse. Instead, they can only become locally Jeans unstable after the entire cloud has contracted sufficiently for the typical Jeans mass in the fluctuations, Equation (7.88), to be comparable to the mass of the fluctuations themselves. That is, at the onset of the large-scale collapse, the Jeans mass at the density fluctuations may be still significantly larger than the masses of the fluctuations themselves. However, as the cloud contracts, the density everywhere within it increases, and so the Jeans mass in the fluctuations decreases, until it becomes of the order of the clumps' masses, at which time they can begin to collapse locally. Indeed, simulations of cloud formation by compressions in the WNM do show the presence of density fluctuations from the start of the simulations, but they are initially transient, and only begin to collapse until after significant contraction of the whole cloud (Figure 11.4). However, they may still be able to terminate their own collapses earlier than the cloud at large, since they have a much shorter free-fall time.

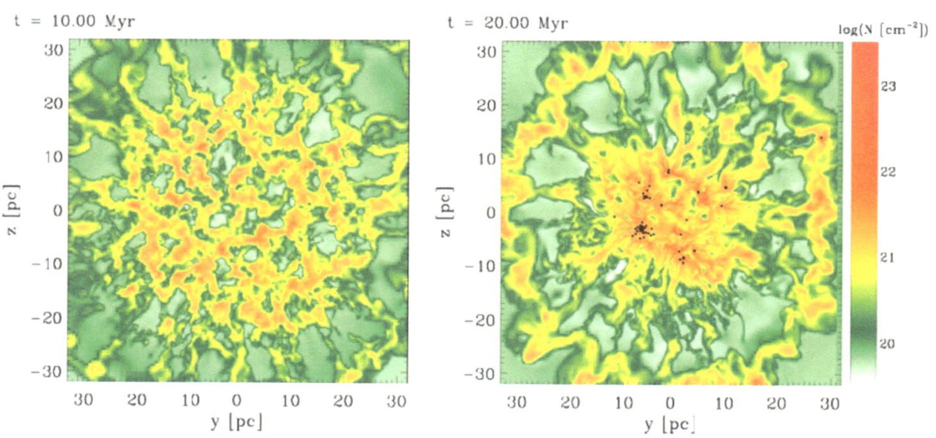

Figure 11.4. Column density images of a simulation of the formation of a giant molecular cloud the collision of WNM streams, and its subsequent gravitational contraction, at times $t = 10$ Myr (*left panel*) and $t = 20$ Myr (*right panel*) since the start of the simulation. The black dots represent *sink particles*, which in this simulation correspond to stellar groups of few tens of stars. It is seen that the left panel already contains numerous clumps, but no star formation, and that by the time there is significant star formation (*right panel*), the cloud at large has contracted to about half of its initial size (from initially spanning from ~ -20 to $+20$ pc in the left panel to ~ -10 to $+10$ pc in both axes). Reproduced from Vázquez-Semadeni et al., (2011) with permission from Oxford University Press on behalf the Royal Astronomical Society.

References

Bergin, E. A., Hartmann, L. W., Raymond, J. C., & Ballesteros-Paredes, J. 2004, ApJ, 612, 921
Elmegreen, B. G., & Scalo, J. 2004, ARA&A, 42, 211
Gómez, G. C., & Vázquez-Semadeni, E. 2014, ApJ, 791, 124
Hartmann, L., Ballesteros-Paredes, J., & Bergin, E. A. 2001, ApJ, 562, 852
Heitsch, F., & Hartmann, L. 2008, ApJ, 689, 290
Heitsch, F., Slyz, A. D., Devriendt, J. E. G., Hartmann, L. W., & Burkert, A. 2006, ApJ, 648, 1052
Ibáñez-Mejía, J. C., Mac Low, M.-M., Klessen, R. S., & Baczynski, C. 2016, ApJ, 824, 41
Ibáñez-Mejía, J. C., Mac Low, M.-M., Klessen, R. S., & Baczynski, C. 2017, ApJ, 850, 62
Kirk, H., Myers, P. C., Bourke, T. L., et al. 2013, ApJ, 766, 115
Klessen, R. S., & Glover, S. C. O. 2016, Saas-Fee Advanced Course, 43, 85
Kumar, M. S. N., Palmeirim, P., Arzoumanian, D., & Inutsuka, S. I. 2020, A&A, 642, A87
Longmore, S. N., Kruijssen, J. M. D., Bastian, N., et al. 2014, Protostars and Planets VI, ed. H. Beuther, R. S. Klessen, C. P. Dullemond, & T. Henning (Tucson, AZ: Univ. Arizona Press) 291
Mac Low, M.-M., & Glover, S. C. O. 2012, ApJ, 746, 135
Mac Low, M.-M., & Klessen, R. S. 2004, RvMP, 76, 125
Mac Low, M.-M., Burkert, A., & Ibáñez-Mejía, J. C. 2017, ApL, 847, L10
McKee, C. F., & Ostriker, E. C. 2007, ARA&A, 45, 565
Myers, P. C. 2009, ApJ, 700, 1609
Naranjo-Romero, R., Vázquez-Semadeni, E., & Loughnane, R. M. 2022, MNRAS, 512, 4715
Padoan, P., Pan, L., Juvela, M., Haugbølle, T., & Nordlund, Å. 2020, ApJ, 900, 82
Vazquez-Semadeni, E., Ostriker, E. C., Passot, T., Gammie, C. F., & Stone, J. M. 2000, Protostars and Planets IV, ed. V. Mannings, A. P. Boss, & S. S. Russell (Tucson, AZ: Univ. Arizona Press) 3
Vázquez-Semadeni, E., Ballesteros-Paredes, J., & Klessen, R. S. 2003, ApJL, 585, L131
Vázquez-Semadeni, E., Gómez, G. C., Jappsen, A. K., et al. 2007, ApJ, 657, 870
Vázquez-Semadeni, E., Gómez, G. C., Jappsen, A. K., Ballesteros-Paredes, J., & Klessen, R. S. 2009, ApJ, 707, 1023
Vázquez-Semadeni, E., Banerjee, R., Gómez, G. C., et al. 2011, MNRAS, 414, 2511
Vázquez-Semadeni, E., Palau, A., Ballesteros-Paredes, J., Gómez, G. C., & Zamora-Avilés, M. 2019, MNRAS, 490, 3061

Enrique Vázquez-Semadeni

Chapter 12

Star-Forming Regions and the Star Formation Rate

The last stage of the gravitational contraction of an interstellar cloud is the formation of stars, after the cloud has become molecular, and has developed multiple centers of collapse. During this stage, the flow in the cloud transitions from being dominated mostly by collapse to being dominated mostly by feedback, whereby *feedback* here we refer to the momentum and thermal and kinetic energy injection by the stars it forms. In this chapter, we discuss the properties and the theoretical description of star-forming regions during this stage.

12.1 Low-Mass and High-Mass Star-Forming Regions

Star-forming regions within molecular clouds can contain either only low-mass stars, like in the Pipe, Taurus, or Perseus clouds (left panel of Figure 12.1), or both low- and high-mass stars, like in the Orion or Monoceros R2 molecular clouds (right panel of Figure 12.1). These are known respectively as low-mass and high-mass star-forming regions. Traditionally, they have been considered as two distinct classes of objects. For example, in the magnetic-support theory (discussed in Section 9.2.1), low-mass regions were thought to arise in magnetically supported (or "subcritical") clouds, while high-mass regions were thought to arise in clouds lacking sufficient magnetic support (known as "magnetically supercritical clouds") to prevent collapse. This view had to be given up, however, when it was observationally determined that most molecular clouds are magnetically supercritical anyway (Crutcher 2012), and that the majority of low-mass stars form in regions that also form high-mass stars (e.g., Lada & Lada 2003).

Under the *turbulent support* (TS) scenario, the distinction between low- and high-mass star-forming regions has not been discussed specifically, although it can be argued that the proposed explanation is that turbulence is a stochastic phenomenon, and therefore it produces density fluctuations of both low and high masses, which

doi:10.1088/2514-3433/adcb28ch12

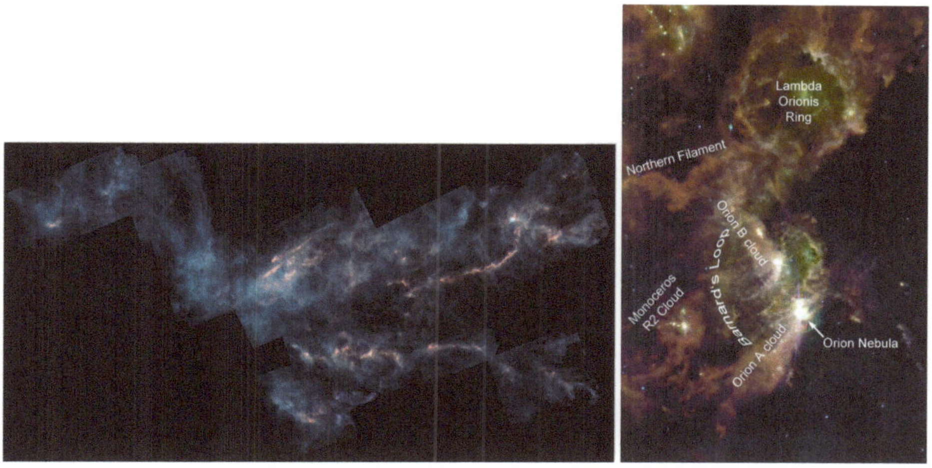

Figure 12.1. *Left:* The Taurus molecular cloud, a low-mass star-forming region (by ESA/Herschel/NASA/JPL-Caltech; acknowledgement: R. Hurt (JPL-Caltech), CC BY-SA IGO 3.0). *Right*: The Orion molecular cloud, a high-mass star-forming region. "This [Orion molecular cloud ver2 1] image has been obtained by the author(s) from the Wikimedia website where it was made available by [Meli thev] under a CC BY-SA 4.0 licence. It is included within this article on that basis. It is attributed to [IRIS + Dame et al. 2001]."

then give rise to the corresponding star-forming regions. Indeed, various theoretical works (Padoan & Nordlund 2002; Hennebelle & Chabrier 2008; Hopkins 2012) have proposed that this distribution of density fluctuation masses is responsible for the distribution of stellar masses at birth (the stellar initial mass function (IMF); see Section 12.5 below), an approach that implicitly assumes that the masses of the stars are determined by the masses of the dense cores in which they form.

Under the GHC scenario, on the other hand, star-forming regions *evolve* by accretion from their parent structures, so that their masses and sizes increase, together with their star formation rate (hereafter, SFR) and with the range of stellar masses they form per unit time, which grows toward larger masses. This growth continues until the feedback from the newly-formed stars is enough to disrupt the accretion flow onto the region, interrupting the local SF episode. The time at which this happens, and thus the maximum gaseous and stellar mass reached by the region, depend on the initial mass available for collapse. In this chapter we describe in detail the evolution of molecular clouds and their star formation (SF) activity according to GHC, since it is the only model that describes this evolution.

12.1.1 Multi-Stage Accretion and Cloud Growth

In Section 7.4.2 we discussed an analytical model describing a cloud-core-stars system undergoing multi-stage accretion, so that the core accretes from the cloud and the stars accrete from the core (see also Vázquez-Semadeni et al. 2024). There we showed that the gravity-driven accretion flow from the cloud onto the core can cause an increase of the latter's mass and mean density, first by a steepening of its density profile's slope (during the prestellar stage), and next by an increase of its radius while maintaining a constant profile slope (during the protostellar stage).

The gas mass increase in the core during the protostellar stage must happen because the mass transfer from the core to the stars in its center is somehow partially inhibited by some mechanism. This can either be that the accretion onto the stars is obstructed by various forms of feedback, or simply by the mechanism of *gravitational choking* described in Section 7.4.2.2, which amounts to an accumulation of the infalling material in the gaseous core if its density profile's logarithmic slope is $p < 2$. In this case, the radial mass distribution causes the gravity-driven accretion rate to decrease inwards, and so the gas tends to stagnate in the core.

Moreover, more realistic numerical simulations of the formation and evolution of turbulent clouds undergoing generalized gravitational contraction (e.g., González-Samaniego & Vazquez-Semadeni 2020; Camacho et al. 2020) also show that cores and clumps, defined as connected regions above a certain density threshold, $n_{\rm th}$, in general grow in size, mass, and mean density, as illustrated in Figure 12.2. Finally, observations of various classes of objects also indicate that they also evolve by increasing their masses. For example, Rivera-Ingraham et al. (2016) have shown that filamentary structures identified in the *Herschel* Galactic Cold Cores program evolve by increasing their masses and linear mass densities; multiple observations

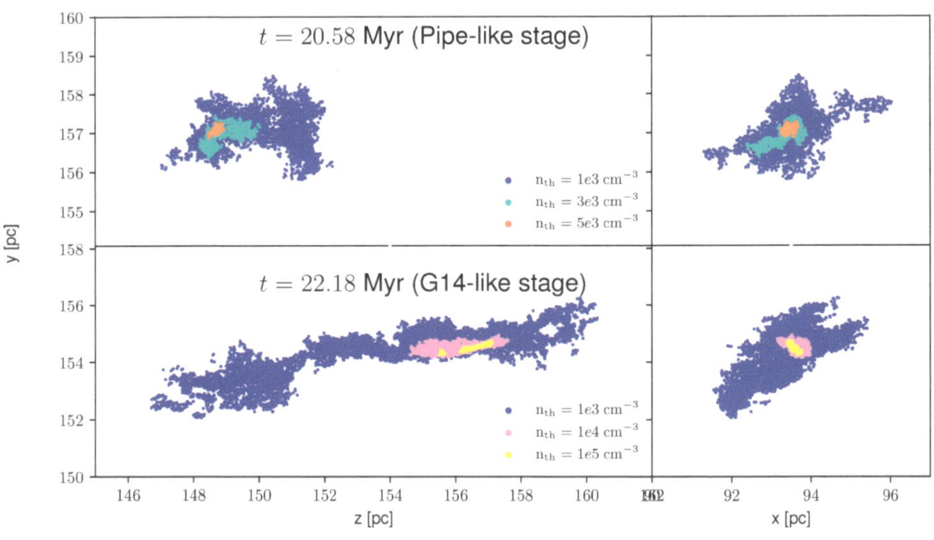

Figure 12.2. Evolution of a clump-core system in a numerical simulation of the formation and evolution of clouds in the turbulent WNM. The left panels show the column density projected onto the (z, y) plane of the simulation, and the right panels show the projection onto the (x, y) plane, at an early stage (20.58 Myr since the start of the simulation, top panels) and a later one (22.18 Myr, bottom panels). Hierarchically nested structures defined at various density thresholds are shown by different colors, as indicated in the insets. The lowest threshold ($n_{\rm th} = 10^3\ {\rm cm}^{-3}$) can be said to define "clumps", while the higher thresholds ($5 \times 10^3\ {\rm cm}^{-3}$ at the early stage and $10^5\ {\rm cm}^{-3}$ at the late stage) can be considered "cores". The clump has mass $M \approx 200 M_\odot$ at the early stage, and $M \approx 700 M_\odot$ at the late stage. Also, the density contrast between the cores and the cloud increases over time: the cores have densities of only $\sim 5\times$ the cloud's threshold density at the early stage, while at the late stage they have densities up to two orders of magnitude larger than the cloud's. Reproduced from Camacho et al. (2020).

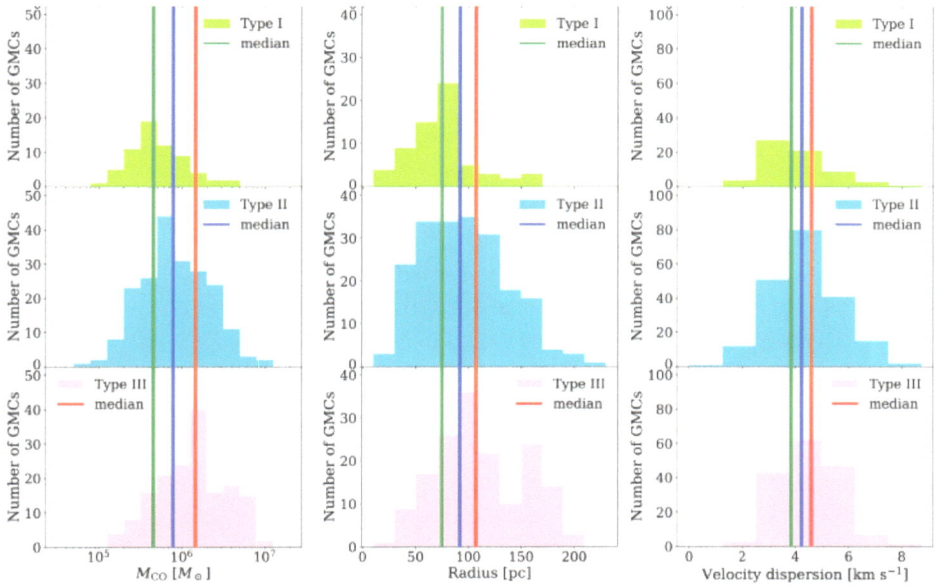

Figure 12.3. Histograms of mass (left column), radius (middle column), and velocity dispersion (right column) of a sample of 432 giant molecular clouds (GMCs) in the galaxy M78 (or NGC 628). The GMCs are classified as either Type I (without massive star formation), Type II (with Hα luminosity $L_{\mathrm{H}\alpha} < 10^{37.5}$ ergs s^{-1}), or Type III (with $L_{\mathrm{H}\alpha} > 10^{37.5}$ ergs s^{-1}). The vertical lines represent the median values of each of the cloud variables. It is seen that all three variables increase with increasing cloud type (i.e., with SFR and age), suggesting evolutionary growth. Reproduced from Demachi et al. (2024). CC BY 4.0.

suggest that filaments at a variety of size scales accrete *from* their parent structures while simultaneously accreting *onto* their central hubs (e.g., Kirk et al. 2013; Peretto et al. 2014; Juárez et al. 2017; Chen et al. 2019); observations of giant molecular clouds in external galaxies show that, as they evolve, their SFR and mass increase systematically (e.g., Kawamura et al. 2009; Demachi et al. 2024), as illustrated in Figure 12.3, and observations of the massive star-forming region W43 suggest it may have formed by accretion of mass from its environment (Motte et al. 2014). Therefore, analytical, numerical, as well as observational studies all suggest that clouds and their substructures undergo a period of growth in mass, mean density, and size by accretion from their parent structures.

12.2 The Star Formation Rate and Its Efficiency

One of the main puzzles about star formation in molecular clouds is that it is *highly inefficient*. As described in Section 9.1.1, Zuckerman & Palmer (1974) pointed out that the SFR in the Milky Way is much smaller than what it would be if all the mass in molecular clouds were converted into stars in a free-fall time. At the time of this writing, the precise cause of this large discrepancy remains one of the main open problems in star formation, although there is general agreement that the solution somehow involves stellar feedback.

12.2.1 The Slow and Fast Scenarios

In terms of the average SFR (hereafter denoted as $\langle \dot{M}_* \rangle$) and the duration of a local SF episode in a cloud ($\tau_{\rm SF}$; cf. Equation (8.10)), the inefficiency of star formation means that

$$\langle \dot{M}_* \rangle \ll \frac{M_{\rm mol}}{\langle \tau_{\rm ff} \rangle}, \tag{12.1}$$

where $M_{\rm mol}$ and $\langle \tau_{\rm ff} \rangle$ are respectively the molecular gas mass and the mean free-fall time in the cloud. This can be accomplished either if the SFR is continuously small over the entire star-forming stage of a cloud—i.e., if

$$\dot{M}_*(t) \approx \langle \dot{M}_* \rangle \ll \frac{M_{\rm mol}}{\langle \tau_{\rm ff} \rangle} \tag{12.2}$$

at all times during the SF episode, with $\tau_{\rm SF} \gtrsim \langle \tau_{\rm ff} \rangle$ (i.e., with the duration of the SF episode being significantly larger than the free-fall time), or, alternatively, if the SFR can reach high values

$$\dot{M}_* \sim \frac{M_{\rm mol}}{\langle \tau_{\rm ff} \rangle}, \tag{12.3}$$

but only over brief time intervals $\tau_{\rm SF} \ll \langle \tau_{\rm ff} \rangle$, and also only in a few locations within the clouds—that is, if star formation is allowed to be *bursty* or *intermittent*, in the sense of nonlinear dynamics, occurring in short but strong episodes at scattered locations in the clouds. Therefore, as also discussed in Chapter 9, the main proposed explanations for the inefficiency of star formation in molecular clouds are that star formation in molecular clouds is either

- *Slow and steady* (SS), because the clouds are supported globally by either magnetic fields (see, e.g., the review by Shu et al. 1987) or by turbulence (e.g., Zuckerman & Evans 1974; Mac Low & Klessen 2004; Krumholz & McKee 2005; Krumholz & Tan 2007; McKee & Ostriker 2007; Krumholz et al. 2019), or
- *Fast but brief* (FB), because, although the clouds are unsupported and collapsing freely, they are rapidly destroyed by the massive stars they produce, which interrupt the local star formation event long before most of the collapsing mass has been converted to stars (e.g., Field 1970; Franco et al. 1994; Zamora-Avilés et al. 2012; Colín et al. 2013; Mac Low et al. 2017; Vázquez-Semadeni et al. 2019; González-Samaniego & Vazquez-Semadeni 2020; Chevance et al. 2020, 2022).

The SS case is the situation assumed in the TS scenario, while the FB case is assumed in the GHC scenario.

12.2.2 When Is an Averaged Description Sufficient? An "Uncertainty Principle" for SF

From the above definitions, it is readily apparent that the two cases considered above are not necessarily mutually exclusive. Rather, the SS scenario can apply to regions large enough that a meaningful average can be taken over a multitude of clouds, each of which is locally evolving according to the FB scenario, or to long stretches of time, so that many FB cycles of one or a few clouds occur. The two scenarios are only contradictory if the SS scenario is assumed to hold everywhere and at all times throughout the entire volume and lifetime of individual clouds. Since the SS description involves averaging over a sufficiently large cloud ensemble, or over sufficiently large time spans, it is important to determine what those spatial and temporal threshold scales are, on the basis of observations.

The minimum spatial scales over which it is necessary to average in order to obtain a stationary SFR can be estimated through the result by Schruba et al. (2010) that the *gas depletion time* (cf. Section 8.5.2) measured in regions of external galaxy disks centered on CO peaks (signaling the presence of abundant molecular gas) begins to differ significantly from that measured in regions centered on Hα peaks (signaling the presence of abundant star formation) when the size of the regions is smaller than $\sim$ 300 pc, as shown in Figure 12.4. As stated by Chevance et al. (2020), the small-scale discrepancy of the depletion times in these two types of regions is caused by evolutionary processes taking place at the scale of molecular clouds, so that they are either undergoing a gas-accumulation stage (thus having a long depletion time) or a gas-consumption stage (and therefore have a short depletion time). This effect has been interpreted by Kruijssen & Longmore (2014) as evidence of an "uncertainty principle" for SF, caused by the fact that we cannot observe a star

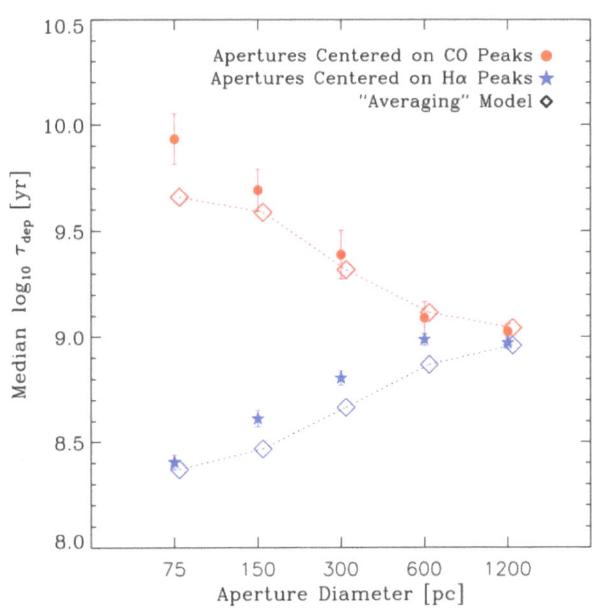

Figure 12.4. Depletion times, computed as $\tau_{\rm dep} = \Sigma_{\rm g}/\Sigma_{\rm SFR}$, determined from observations of regions of the indicated diameters when the regions are centered on either CO peaks (red symbols) or on Hα peaks (blue symbols). The measured $\tau_{\rm dep}$ is already seen to differ significantly between the two types of locations when the region size is $\lesssim$ 300 pc. Reproduced from Schruba et al. (2010).

cluster and its progenitor cloud at the same time. A single, global depletion time can only be obtained when it is averaged over regions or time spans large enough to contain a large number of star-forming regions at various evolutionary stages. The results from Schruba et al. (2010) and later similar studies suggest that this scale of sufficient statistical sampling is of the order of a few to several hundred parsecs.

12.3 Theoretical Descriptions of the SFR

12.3.1 SFR in the TS Model

Under the TS scenario, which assumes a stationary, slow and steady star formation process, molecular clouds are assumed to be in approximate virial equilibrium between their self-gravity and the turbulent pressure inside them. The magnetic field also contributes to the support of the cloud, although most likely only by slowing down the collapse rather than by preventing it, since, as discussed in Section 8.4.2, it is now commonly agreed that most MCs are magnetically supercritical (Crutcher 2012).

In this scenario, the SFR is mainly controlled by the properties of the turbulence and the magnetic field, such as the sonic and Alfvénic Mach numbers $\mathcal{M}_s$ and $\mathcal{M}_A$ (or, alternatively, one of these Mach numbers and the *plasma beta*, the ratio of the thermal to the magnetic pressure, $\beta \equiv P_{th}/P_{mag}$), the virial parameter α_{vir} (Section 5.2.5), and the fraction of compressible modes in the turbulence, given by the *forcing parameter b* (see Equation (3.74)). The basic framework is a quasistationary regime in which the cloud is globally supported by turbulence, but small regions within it are rendered gravitationally unstable by the supersonic turbulent compressions.

Both the Kolmogorov and the Burgers forms of the turbulent energy spectrum (cf. Sections 3.5 and 3.9.4.1, respectively) imply that the amplitude of the velocity differences in a turbulent flow decreases with decreasing scale. Therefore, there is a certain size scale, known as the *sonic scale*, below which the velocity dispersion is subsonic. Thus, in the TS/SS scenario, the interiors of these subsonic regions are assumed to be incapable of further turbulent fragmentation, and to be subject to the thermal Jeans criterion for collapse (Vázquez-Semadeni et al. 2003). Observationally, the sonic scale for molecular clouds, at least in the Solar neighborhood, is $\lambda_s \sim 0.1$ pc (e.g., Larson 1981; Heyer & Brunt 2004; see the top and bottom right panels of Figure 8.5).

In view of this, (Krumholz & McKee 2005; hereafter KM05) proposed that the star formation efficiency per free-fall time, ϵ_{ff} (cf. Section 8.5.1), is given by the mass fraction in fluctuations of size smaller than the sonic scale and with densities high enough that the local Jeans length L_J satisfies $L_J \approx \lambda_s$. Since molecular clouds are nearly isothermal, the density scales essentially as L_J^{-2} (cf. Equation (6.14)), and therefore the critical overdensity for local Jeans instability is given by

$$x_{crit} \equiv \frac{\rho_{crit}}{\rho_0} \approx \left(\frac{L_{J0}}{\lambda_s}\right)^2, \tag{12.4}$$

where ρ_0 is the background density and L_{J0} is the Jeans length at ρ_0. The mass fraction above this critical overdensity is then given by

$$f_{m,x_{\rm crit}} = \int_{x_{\rm crit}}^{\infty} xP(x)\,dx, \tag{12.5}$$

where $P(x)$ is the density PDF defined by Equation (3.68), albeit written in terms of $x \equiv \rho/\rho_0$, while that equation is written in terms of $s \equiv \ln x$.

If the mass in every density interval is collapsing on its own free-fall time $\tau_{\rm ff}(x)$, then the rate at which this fraction turns into stars is

$$\dot{f}_{m,x_{\rm crit}} = \int_{x_{\rm crit}}^{\infty} \frac{xP(x)}{\tau_{\rm ff}(x)}\,dx \approx \frac{f_{m,x_{\rm crit}}}{t_{\rm ff,e}}, \tag{12.6}$$

where $t_{\rm ff,e}$ is some effective timescale for the collapse of the gas with $\rho \geqslant \rho_{\rm crit}$. In turn, the total rate of gas mass conversion into stars (the SFR) is

$$\dot{M}_* = \dot{f}_{m,x_{\rm crit}} M_{\rm mol} \approx M_{\rm mol}\frac{f_{m,x_{\rm crit}}}{t_{\rm ff,e}}, \tag{12.7}$$

The *efficiency over a free-fall time* (Equation (8.7)) can then be written as

$$\epsilon_{\rm ff} \equiv \dot{f}_{m,x_{\rm crit}}\, t_{\rm ff,e} = f_{m,x_{\rm crit}}\,. \tag{12.8}$$

It is important to note that $\epsilon_{\rm ff}$ is a mass fraction, not a rate, as it is sometimes denoted. The fractional *rate* of mass conversion into stars is given by Equation (12.6), and the total mass rate is given by Equation (12.7).

A number of variations of the KM05 model have been proposed in the literature (Padoan & Nordlund 2011; Hennebelle & Chabrier 2011; Federrath & Klessen 2012), with the differences among them arising basically from different prescriptions for estimating the critical overdensity $x_{\rm crit}$ and the characteristic collapse timescale $t_{\rm ff,e}$. These include *multi-free-fall* prescriptions, obtained by keeping the free-fall time inside the integral in Equation (12.6). A thorough compilation of the models, and numerical testing of the various models, comparing their predictions on the dependence of the SFR with the turbulent parameters with the results of numerical simulations of driven turbulence in isothermal boxes, has been provided by Federrath & Klessen (2012).

12.3.2 SFR in the GHC Model

12.3.2.1 Cloud Evolution and Its Determining Processes

In the GHC/FB scenario, the SFR is an *evolving* quantity, which is primarily controlled by the temporal evolution of a collapsing mass of gas, with turbulence playing mostly the role of generating seed density fluctuations, which may then proceed to collapse once the mean density in the cloud has increased due to the global contraction and thus the Jeans mass has decreased sufficiently to be comparable to the mass in the fluctuations (cf. Sections 7.5.3 and 11.5). That is, *the collapse starts at the large scales* and propagates inward as the density increases and the Jeans mass decreases. During this early stage, the cloud's mass increases by accretion, and its density increases by contraction, causing an increase of the SFR. Later, once a sufficient number of massive stars have appeared (Section 12.1), the energy and momentum injection by stellar feedback rapidly disperse the cloud and the accretion flow, causing a reduction of the SFR. Although some turbulence is generated by the collapse itself (cf. Section 3.11), numerical simulations of the process suggest that it dissipates so rapidly that it cannot significantly delay the gravitational contraction (Guerrero-Gamboa & Vázquez-Semadeni 2020).

Under this evolutionary prescription, the fundamental temporal variable is the cloud's mass, which increases due to accretion from the environment, and decreases due to gas consumption by star formation and by dense gas destruction due to stellar feedback. Therefore, the cloud mass as a function of time can be generically written as

$$\begin{aligned} M_c(t) &= M_0 + M_{\rm acc}(t) - M_*(t) - M_{\rm ion}(t), \\ &= \int_0^t \dot{M}_{\rm acc}(t')\,dt' - \int_0^t \dot{M}_*(t')\,dt' - \int_0^t \dot{M}_{\rm ion}(t')\,dt', \end{aligned} \tag{12.9}$$

where M_0 is the initial mass of the region, $M_{\rm acc}(t)$ is the total mass that has been accreted by the cloud over its evolution, $M_*(t)$ is the mass that has been converted to stars, $M_{ion}(t)$ is the mass that has been photoionized by massive stars, and $\dot{M}_{\rm acc}(t')$, $\dot{M}_*$, and $\dot{M}_{\rm ion}(t')$ are the corresponding rates at time t'. In particular, $\dot{M}_*$ is precisely the SFR. This general prescription can be adapted to a variety of specific scenarios of cloud evolution. An early semi-empirical model providing a first approximation to the evolution of the SFR during the gravitational contraction of a molecular cloud was provided by Zamora-Avilés et al. (2012, hereafter, the ZV12 model; see also Zamora-Avilés & Vázquez-Semadeni 2014). Several other models were discussed by Krumholz & McKee (2020).[1] In the remainder of this section we follow the discussion of ZV12's, since it takes into account all three processes in a simple manner.

12.3.2.2 Accretion

The accretion rate is perhaps the most uncertain ingredient in this class of models, since it depends on larger-scale processes, and therefore it is essentially a boundary condition for the evolution of the molecular cloud. ZV12 made the simple assumption of a fixed accretion rate during a finite time interval, whose duration can range from $\lesssim$1 Myr, for gas swept-up and compressed by expanding SN and HII-region shells (e.g., Elmegreen & Lada 1977), to several tens of megayears, for compressions exerted, for example, by passages of the gas through the gravitational potential well of stellar spiral arms. Furthermore, ZV12 assumed that, being the result of a compression, the cloud is a thin circular layer of radius $R_c(t)$, whose initial

[1] Note, however, that Krumholz & McKee (2020) incorrectly describe the GHC scenario as lacking accretion, while in fact GHC does include accretion. The GHC scenario is best described by the "Conveyor Belt" model considered by those authors.

value is a free parameter, and afterwards it is determined by the gravitational contraction (Section 12.3.2.3).

Due to the accretion, the cloud will in general grow in mass, mean density, and even radius (when defined through a fixed density threshold), as discussed in Section 12.1.1. However, for computational simplicity, the model uses a mixed description that lagrangianly follows the gravitational contraction of the initial gas mass, and separately adds to it the accreted mass through the instantaneous surface area of the cloud. Therefore, in this model, the mass increases, but the radius decreases. In this sense, this model is not fully self-consistent, and it must be considered only as a first-order semi-analytical approximation to the complex problem of modeling an open system, which is itself a part of an also evolving, larger-scale one, and can only be solved in full through numerical simulation.

Under these assumptions, the accretion rate is given by

$$\dot{M}_{acc} = 2\pi \rho_{acc} v_{acc} R_c^2(t), \tag{12.10}$$

where ρ_{acc} and v_{acc} are, respectively, the density and inflow speed of the background medium that accretes onto the cloud. The factor of 2 arises from the assumption that the cloud accretes from both sides.

It is important to note that a crucial consequence of the accretion is that it can cause the instantaneous SFE (Equation (8.6)) to remain low at all times, because both the gaseous and stellar masses grow together. Measurements of the SFE in numerical simulations by González-Samaniego & Vazquez-Semadeni (2020) showed that it remained $\sim 6.5\%$ even in the absence of stellar feedback.

12.3.2.3 Hierarchical Cloud Contraction and Early Evolution of the SFR

Similarly to the strategy of the TS/SS models, ZV12's model also computes the SFR as the mass fraction above a certain threshold density ρ_{crit}, divided by the free-fall time corresponding to this threshold, as described by Equations (12.4), (12.6), and (12.7). The threshold ρ_{crit} is taken as a density whose free-fall time is much smaller than that of the whole cloud, so that Equation (12.7) represents the "instantaneous" SFR in the cloud.[2] However, the main difference with the TS/SS prescription is that ZV12's model takes into account the global gravitational contraction of the gas parcel considered, which leads to a *temporal evolution of the density PDF* itself, such that the mean density of the cloud increases over time. In turn, this leads to a systematic increase of the mass fraction above ρ_{crit} and therefore to an increase of the SFR. ZV12's model made the simple assumption that the density PDF remains as a lognormal function and just shifts to a higher mean density as the cloud contracts, assuming a constant turbulent Mach number.[3] This model then includes the crucial

[2] Specifically, ZV12 used the value $n_{crit} = 10^6\,cm^{-3}$, which allowed the model to match the evolution of the SFR measured in the numerical simulations of Vázquez-Semadeni et al. (2010).

[3] In reality, it is now known that, as collapse centers arise in a cloud, its density PDF evolves by developing a power-law high-density tail Ballesteros-Paredes et al. (2011; Kritsuk et al., 2011). This was taken into account, for example, in the SFR model by Burkhart (2018), although accretion and global contraction were not, and so here we follow the prescription by ZV12, which does take them into account.

hierarchical nature of gravitational collapse: on the one hand there is a large-scale collapse, which causes a systematic increase of the mean density, while on the other hand there is a small-scale collapse that forms stars, accounted for by modeling the SFR as the time-dependent mass fraction above $\rho_{\rm crit}$ divided by its corresponding free-fall time, both of which vary as dictated by the large-scale contraction.

Specifically, ZV12 followed the large-scale contraction by assuming a sheet-like cloud and numerically integrating the gravitational acceleration exerted on its edge to follow the evolution of its radius $R_c(t)$. The fact that the cloud is assumed to be sheet-like is important, since it implies a slower contraction rate than that for a spherical cloud of the same density and radius (cf. Section 7.3.2.3).

12.3.2.4 Early Near-Exponential Acceleration of the SFR

As mentioned in the previous section, the large-scale collapse of the cloud implies that its SFR increases systematically during early stages of evolution, before feedback becomes important. The rate of increase of the SFR remains somewhat uncertain. Some numerical simulations suggest a near-exponential growth of the SFR. In particular, Zamora-Avilés et al. (2024) have found an explicit exponential temporal increase of the SFR in simulations of global hierarchical collapse of a molecular cloud, during the early epochs where feedback should not be dynamically important (Figure 12.5). Over longer time periods, the feedback begins to oppose the accretion, thus reducing the SFR (Sections 12.3.2.5 and 15.3).

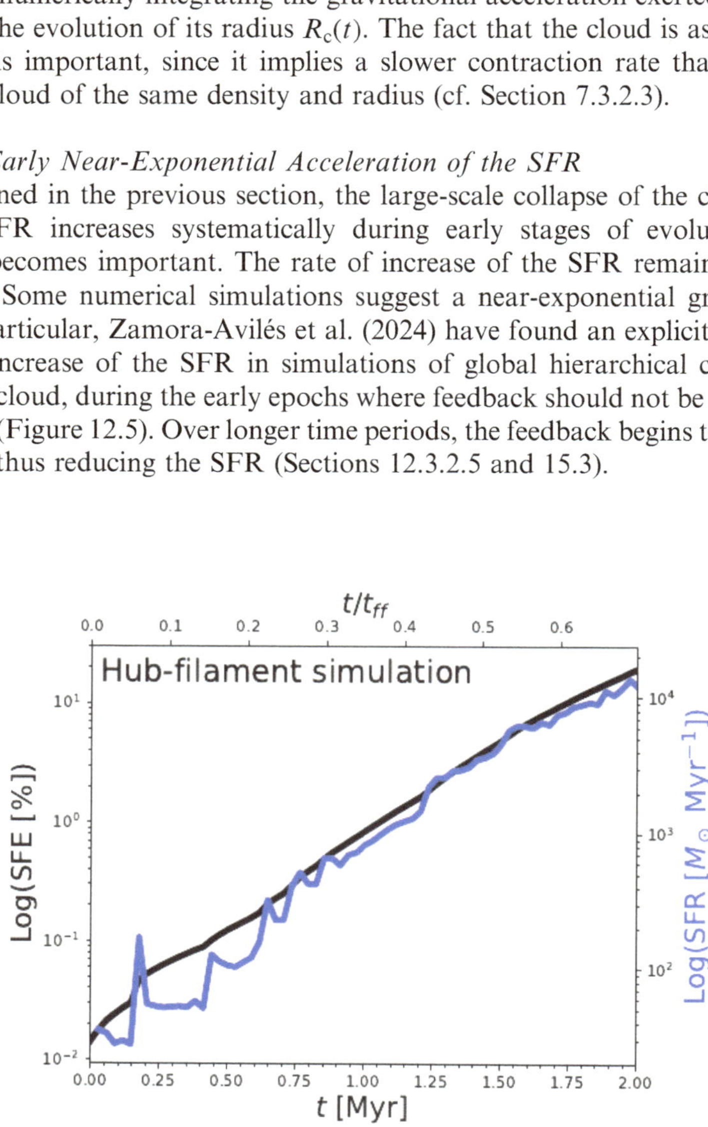

Figure 12.5. Early evolution of the instantaneous star formation efficiency, SFE=$\mathcal{M}_*/(M_g + M_*)$ (Equation (8.6)) and the star formation rate, SFR, in a numerical simulation of the formation and gravitational contraction of a hub-filament system, showing a nearly exponential growth of both quantities, while stellar feedback is not important. Reproduced from Zamora-Avilés et al. (2024).

To date, we know of no theoretical explanation for the nearly exponential growth of the SFR, but it may possibly be related to the exponential growth of the density fluctuations during the early stages of the Jeans gravitational instability (Section 6.1). In fact, the growth of the peak density is super-exponential during the nonlinear stage of gravitational instability, since the density diverges in a *finite* time (the free-fall time; Section 7.2.1), while, although very fast, exponential growth still requires an infinite time to reach a diverging density.

12.3.2.5 Cloud Destruction or Dispersal by Internal Stellar Feedback

The model can be numerically integrated to evolve the mass, radius, and SFR of the cloud. The total stellar mass M_* is then given by Equation (8.9), from which the fraction corresponding to massive stars that can photoionize the gas is obtained by assuming a standard IMF. ZV12 used the simple prescription of assuming a characteristic massive-star mass, and then computing its corresponding ionizing flux. Multiplying this by the total instantaneous number of massive stars, the cloud mass destruction rate by photoionization can be computed, thus completing the necessary ingredients to insert in Equation (12.9). Note that, as the cloud begins to be eroded by feedback, its mass and mean density begin to decrease again, and so does its SFR. The resulting evolutionary plots for the masses, radii, SFRs and SFEs of clouds shown in Figure 12.6.

Several points are important to note about this model for the evolution of clouds and their SFR under GHC:

- The SFR of an individual star-forming region first increases, due to the increase in mass and mean density of the cloud, and then decreases, due to the local erosion of the cloud performed by the stellar feedback and to the termination of the accretion flow.
- In the GHC model, the role of turbulence is only to provide the width of the density PDF, but the evolution of the cloud is determined by the development of the collapse. In turn, the latter is essentially determined by the initial amount of mass available for collapse and its spatial extent. These are the two main parameters determining the evolution of a star-forming molecular cloud.
- Since GHC assumes that the molecular line widths have a significant (and often dominant) contribution from the cloud-scale infall motions (Section 9.3.2), the *virial parameter* $\alpha_{\rm vir}$ of the clouds (Equation (5.24)) is *not* an external parameter for the clouds, but rather a *cloud variable*, whose value is determined by the collapse and the interaction with feedback. This is exemplified by the left panels of Figure 12.7, which show the evolution of $\alpha_{\rm vir}$ in a numerical simulation of the formation and evolution of a cloud by Colín et al. (2013). It is seen that $\alpha_{\rm vir} \sim 1$ during the gravitational contraction of the cloud, while $\alpha_{\rm vir} \gg 1$ after feedback begins to disperse the cloud.
- Under the GHC scenario, the fundamental reason why the *final* SFE (given by evaluating Equation (8.6) at the end of the SF episode) is low is that the collapse of the cloud is interrupted early in its evolution due to the onset of strong feedback from massive stars, so that only a small fraction of the total

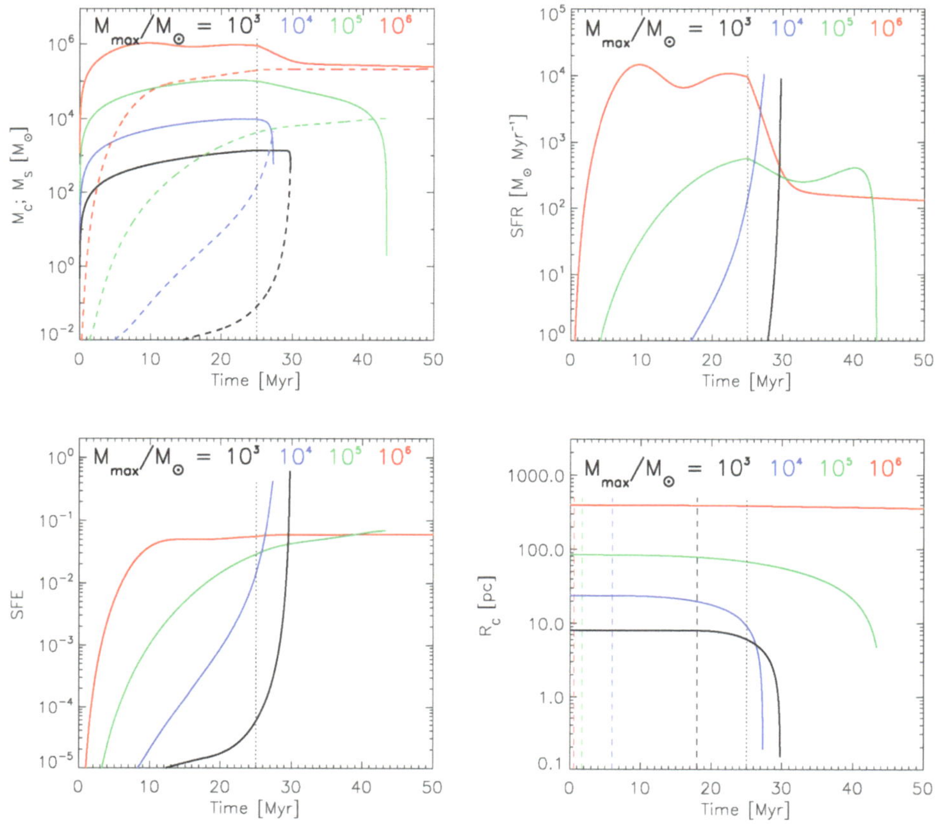

Figure 12.6. *Top left:* Evolution of the gaseous (*solid lines*) and stellar (*dashed lines*) masses of four clouds, according to the evolutionary model for GHC. The clouds are characterized by the maximum gas mass they reach during their evolution, indicated by the line colors, as labeled in each panel. In turn, this maximum mass is determined by the accretion rate given by Equation (12.10). *Top right:* Evolution of the SFR in each model cloud. *Bottom left:* Evolution of the SFE, as given by Equation (8.6). *Bottom right:* Evolution of the cloud's radius. The black vertical dotted line indicates the time when accretion is stopped (25 Myr). The colored vertical dashed lines indicate the times at which each cloud begins forming stars. Reproduced from Zamora-Avilés & Vázquez-Semadeni (2014).

mass in the cloud is ever deposited into stars. The majority of the cloud's mass is dispersed or returned to the diffuse ISM phase by the stellar feedback.

- On the other hand, under GHC, the *instantaneous* SFE (Equation (8.6) at arbitrary t) is low at all times because the accretion onto the cloud causes the cloud mass to grow, so that both the gaseous and the stellar masses grow together.

12.3.2.6 Results from Numerical Simulations

As mentioned above, the ZV12 model we have discussed here is a strong idealization of the real process, since it assumes specific cloud shapes and accretion flows, and it neglects magnetic fields and other forms of feedback besides photoionizing radiation. Nevertheless, it provides a useful first approximation to the evolution of a

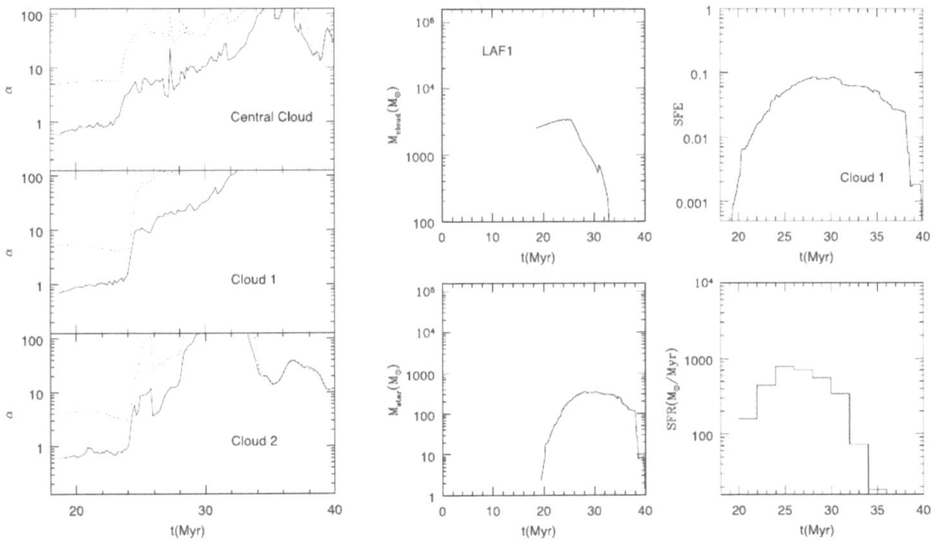

Figure 12.7. *a) (Upper, middle, and lower left panels)*: Evolution of the virial parameter $\alpha_{\rm vir}$ (Equation (5.24)) weighted by mass (thus corresponding roughly to the dense gas; *solid lines*) and weighted by volume (corresponding roughly to the diffuse gas; *dotted lines*) in three cylindrical regions of radius 10 pc and length 20 pc centered at the three main gravitational potential minima in a numerical simulation of cloud formation and evolution by Colín et al. (2013). Note that, *before* stellar feedback becomes dominant, $\alpha_{\rm vir} \sim 1$ for the dense gas, due to the gravitational contraction, while, during the stage of feedback domination, $\alpha_{\rm vir} \gg 1$, indicating the destruction of the cloud by stellar feedback. *(b)* Evolution of the dense gas mass (with density $n > 100{\rm cm}^{-3}$; *upper middle panel*), SFE (*upper right panel*), stellar mass (*lower middle panel*), and SFR (*lower right panel*) in a cylindrical region of radius 10 pc and length 20 pc centered at the gravitational potential minimum corresponding to the so-called "Cloud 1" in this simulation. Note the qualitative consistency with the plots in Figure 12.6. Reproduced from Colin et al., (2013) with permission from Oxford University Press on behalf the Royal Astronomical Society.

molecular cloud and its SFR when accretion and global gravitational contraction are taken into account.

Moreover, numerical simulations of the full cloud evolution process including photoionizing stellar feedback show good qualitative agreement with the model predictions, as shown in the middle and right panels of Figure 12.7. These show the evolution of a cloud formed by the collision of WNM streams that triggers a phase transition to the CNM, forming a cloud that then begins to undergo gravitational contraction. The cloud here ("Cloud 1") is defined as the dense gas mass ($n \geqslant 100{\rm cm}^{-3}$) within a fixed cylindrical region of radius 10 pc and length 20 pc centered at the gravitational potential minimum in that part of the simulation. The cloud mass (*top middle panel*) is seen to reach a maximum value of $\sim 4 \times 10^3\ M_\odot$ at $t \approx 25.5$ Myr after the collision of the streams. The "stellar" mass (*bottom middle panel*) is the mass in stellar particles in the same region. Star formation is seen to begin at $t \approx 19.5$ Myr. The stellar mass in the region later decreases because the stellar particles begin to leave the volume considered. The SFR (*bottom right panel*) is seen to peak at $t \approx 25$ Myr, roughly at the same time when the dense gas mass is maximum. The star formation efficiency here is defined as SFE $=M_*/M_{max}(t)$, where $M_{max}(t)$ is the maximum total mass (dense gas + stars) ever reached up to time t. This definition gives a closer approximation to the real efficiency, since at later times part of the dense gas mass has been either pushed away from the region or else returned to the diffuse phase, and so the SFE computed with the instantaneous dense gas mass may appear artificially high. The SFE thus defined is seen to peak at $\sim$10%, and to decrease at later times again due to the stellar particles leaving the region.

This simulation exhibits a qualitative behavior consistent with that of model clouds labeled 10^4 and $10^5 M_\odot$ in Figure 12.7 (blue and green lines). Both of these curves are seen to have final SFEs hovering around 10% and an SFR that peaks between 25 and 30 Myr after the stream collision. Finally, these numbers are also consistent with the cloud timescales inferred observationally, discussed in Section 14.3.6. This suggests that the essential evolutionary aspects of the clouds are captured to first order by a simple model that exclusively considers accretion, gravitational contraction, and cloud destruction by photoionizing radiation.

12.4 The Nearly Stationary Gravity-Driven Gas Flow from Low- to High-Density Regions and the Intermittent Nature of Star Formation

A crucial feature of the SF process in molecular clouds is that it occurs exclusively in the densest parts of the clouds. That is, it is a highly spatially-localized (or *spatially intermittent*, in the sense of nonlinear dynamics) phenomenon. This is actually a natural consequence of the fact that gravitational collapse generates a continuous, and approximately stationary *flow* from low to high densities. Indeed, as discussed in Section 7.3.2.2 and also pointed out by Murray & Chang (2015), the *cruising* asymptotic solution for spherical collapse flow of Whitworth & Summers (1985) implies that the density profile in a dense core becomes stationary (i.e., time-independent; Equation (7.56)) as long as the central protostellar mass does not become comparable to the gaseous mass. The "cruising" name of the solution itself describes the fact that the gas *travels* along a stationary density profile from low to high densities, generating a gravity-driven *collapse flow*, analogous to the flow of water down a funnel, except that the gas is compressed as it falls down the "density funnel". It is important to note, however, that the establishment of a stationary density profile requires the existence of an unlimited reservoir of gas at large scales and low densities, which supplies fresh gas to the molecular clouds at their boundaries, replenishing the material transferred to stars. At the scale of entire molecular complexes, a possible such reservoir is the continuous flow of the warm diffuse gas across the stellar spiral arms.

Also, note that the stationary nature of the density profile implies that the clouds, defined through *feature-tracking* criteria (such as through a density-threshold criterion; cf. Section 1.2.1) *do not* contract as they evolve, but rather remain at an approximately constant size, increasing only their stellar content. Moreover, if a *gravitational-choking* process occurs (Section 7.4.2) that causes the gaseous mass to increase with a roughly constant average slope of the density profile shallower than

-2, then the clouds may even grow over time. That is, the clouds can be considered just as the funnels themselves (i.e., the regions in space through which the gas is flowing), which are different from the Lagrangian gas parcels passing through them, and getting compressed as they do so. This, plus the amplification of anisotropies during nearly pressureless collapse, may be the origin of the *conveyor belt* flow observed along filaments toward the dense hubs (Section 11.4). The filaments remain, while the gas flows along them. Indeed, numerical simulations show that the clouds, defined as connected regions above a certain threshold density, do not contract, and in fact may expand as they evolve and collect gas from their surroundings (see, e.g., Camacho et al. 2020, and Figure 12.2 here).

In the context of this stationary collapse flow, it becomes clear that star formation can only occur at a few localized sites in the clouds, rather than throughout their entire volume, as it is sometimes assumed. Specifically, the gas must travel "down" the density profile funnel until it reaches stellar densities, which are many orders of magnitude higher than the typical densities of molecular clouds and their dense cores, and this implies traveling distances of the order of the cloud size itself.

As an illustration, consider a collapse flow with an average power-law density profile given by $\rho(r) = \rho_0(r/r_0)^{-p}$ (Equation (7.70)), with $p = 2$. For such a profile, $r \propto \rho^{-1/2}$, and so, if the gas is traveling down this profile, it needs to contract by, say, 5 orders of magnitude in size, in order to increase its density by 10 orders of magnitude. Alternatively, it can be stated that regions with densities n orders of magnitude larger than the density at the cloud boundary have sizes $\sim n/2$ orders of magnitude smaller, and therefore volumes $\sim 3n/2$ orders of magnitude smaller. This shows the extremely spatially intermittent nature of the star formation process.

12.5 The Delayed Formation of Massive Stars and the Stellar IMF

An important consequence of the growth of the cloud and core masses and mean densities is that the probability of forming massive stars increases with time. This is due to two main effects:

1. The SFR increases systematically, implying that, if the IMF is interpreted as the fraction of the total number of newly-formed stars with masses in a certain mass interval, then, as the total number of stars forming per unit time increases, so does the number of new massive stars formed per unit time.
2. As the core becomes more massive and denser, the mass contained within a given radius increases, and thus the accretion rate at that radius increases too (see Equation (7.80) for the accretion rate in the idealized spherical case), thus creating a larger ram pressure that the feedback from the stars must overcome.

The net result is that massive stars appear with a certain delay from the onset of SF in numerical simulations of cloud formation and evolution. This is illustrated in Figure 12.8, which shows the evolution of the distribution of stellar masses in a numerical simulation of the formation and evolution of a molecular cloud by Colín et al. (2013). The appearance of "massive" stars (i.e., with masses $\gtrsim 10 M_\odot$) is seen to

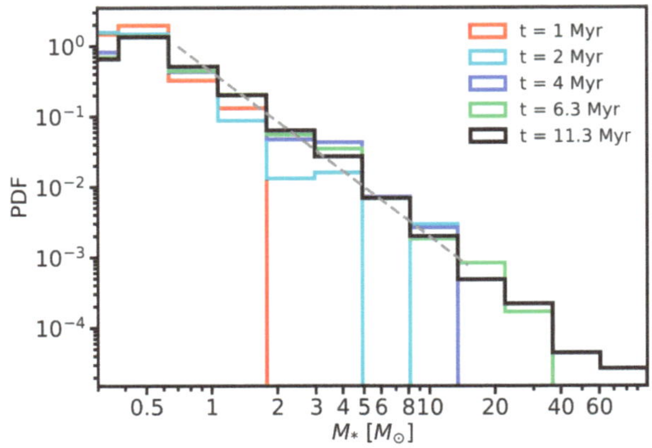

Figure 12.8. Evolution of the stellar mass distribution (IMF) in a cloud in a numerical simulation of cloud formation and evolution. The times indicated in the inset give the elapsed time since the onset of SF in the cloud. The high-mass end of the IMF is seen to increase from $\sim$1 $M_\odot$ at 1 Myr to $\sim$30 $M_\odot$ at $t \sim 6$ Myr. Reproduced from Vázquez-Semadeni et al (2024). CC BY 4.0.

occur a few megayears after the onset of SF in that cloud. This property implies that the clouds can grow for some time until the stellar feedback becomes capable of destroying them. This will be the subject of Chapter 14.

12.6 Highlights

- The mean density of a cloud with a power-law density profile with logarithmic slope $p < 3$ truncated at radius R differs only by a factor of order unity from the density at the truncation radius, $\rho(R)$ (cf. Equation (6.33)). That is, the mean density is very close to the *minimum* density of the cloud, because most of the cloud's mass is at the lowest densities. In consequence, most of the cloud material contracts on a timescale close to the *maximum* free-fall time of the cloud.
- Both the TS and the GHC models involve turbulence, but TS assumes *strongly supersonic* turbulence ($\mathcal{M}_s \gtrsim 10$) that can provide support against self-gravity for all but the smallest scales ($\gtrsim$1 pc) in the clouds, while GHC assumes *moderately supersonic* turbulence ($\mathcal{M}_s \sim 3$), unable to provide support, so that gravitational contraction motions can occur at up to the largest scales in molecular clouds. The infall motions in general constitute the dominant contribution to the nonthermal motions in the clouds.
- Under the TS scenario, the width of the various molecular lines corresponds *only* to the turbulent motions. Under GHC, on the other hand, the linewidth contains both a (perhaps dominant) contribution from large-scale infall motions and a (perhaps subdominant) turbulent contribution.
- In consequence, under TS/SS, the virial parameter $\alpha_{\rm vir}$ is an *external parameter* of the system, determined by the strength of the turbulence (itself determined by the rate of turbulent energy injection, most often assumed to be due to stellar feedback). Under GHC/FB, on the other hand, $\alpha_{\rm vir}$ is an

internal variable of the problem, which evolves together with the system, since the infall motions contributing to it are determined by the evolution of the gravitationally-driven flow.

- In TS, the clouds are considered as *objects* in isolation and equilibrium with a relatively well defined mass. In GHC, they are considered as quasistationary *flow features* (patterns, or *loci* in the flow) through which the gas *passes* in its journey along the gravitationally-driven mass cascade from low to high densities. This can be considered the mass analogue of the kinetic energy cascade of turbulence theory. This generates accretion at all scales.
- The interpretation of the SFE, $\epsilon_{\rm ff}$ and $\tau_{\rm dep}$ is radically different for isolated objects and for quasistationary flow features. For the former, the mass reservoir is fixed, and star formation necessarily consumes it, and so these indicators represent this consumption. In the presence of accretion, however, these indicators are incomplete, as they lack information as to the mass accretion rate onto the cloud. In particular, *the SFE and* $\epsilon_{\rm ff}$ (Equations (8.6) and (8.7), respectively) *are determined by both the cloud growth and SF rates, and not only by the SFR.*

References

Ballesteros-Paredes, J., Vázquez-Semadeni, E., Gazol, A., et al. 2011, MNRAS, 416, 1436

Burkhart, B. 2018, ApJ, 863, 118

Camacho, V., Vázquez-Semadeni, E., Palau, A., Busquet, G., & Zamora-Avilés, M. 2020, ApJ, 903, 46

Chen, H.-R. V., Zhang, Q., Wright, M. C. H., et al. 2019, ApJ, 875, 24

Chevance, M., Kruijssen, J. M. D., Hygate, A. P. S., et al. 2020, MNRAS, 493, 2872

Chevance, M., Kruijssen, J. M. D., Krumholz, M. R., et al. 2022, MNRAS, 509, 272

Colín, P., Vázquez-Semadeni, E., & Gómez, G. C. 2013, MNRAS, 435, 1701

Crutcher, R. M. 2012, ARA&A, 50, 29

Dame, T. M., Hartmann, D., & Thaddeus, P. 2001, ApJ, 547, 792

Demachi, F., Fukui, Y., Yamada, R. I., et al. 2024, PASJ, 76, 1059

Elmegreen, B. G., & Lada, C. J. 1977, ApJ, 214, 725

Federrath, C., & Klessen, R. S. 2012, ApJ, 761, 156

Field, G. B. 1970, MSRSL, 19, 29

Franco, J., Shore, S. N., & Tenorio-Tagle, G. 1994, ApJ, 436, 795

González-Samaniego, A., & Vazquez-Semadeni, E. 2020, MNRAS, 499, 668

Guerrero-Gamboa, R., & Vázquez-Semadeni, E. 2020, ApJ, 903, 136

Hennebelle, P., & Chabrier, G. 2008, ApJ, 684, 395

Hennebelle, P., & Chabrier, G. 2011, ApJL, 743, L29

Heyer, M. H., & Brunt, C. M. 2004, ApJL, 615, L45

Hopkins, P. F. 2012, MNRAS, 423, 2037

Juárez, C., Girart, J. M., Zamora-Avilés, M., et al. 2017, ApJ, 844, 44

Kawamura, A., Mizuno, Y., Minamidani, T., et al. 2009, ApJS, 184, 1

Kirk, H., Myers, P. C., Bourke, T. L., et al. 2013, ApJ, 766, 115

Kritsuk, A. G., Norman, M. L., & Wagner, R. 2011, ApL, 727, L20

Kruijssen, J. M. D., & Longmore, S. N. 2014, MNRAS, 439, 3239

Krumholz, M. R., & McKee, C. F. 2005, ApJ, 630, 250
Krumholz, M. R., & McKee, C. F. 2020, MNRAS, 494, 624
Krumholz, M. R., & Tan, J. C. 2007, ApJ, 654, 304
Krumholz, M. R., McKee, C. F., & Bland-Hawthorn, J. 2019, ARA&A, 57, 227
Lada, C. J., & Lada, E. A. 2003, ARA&A, 41, 57
Larson, R. B. 1981, MNRAS, 194, 809
Mac Low, M.-M., & Klessen, R. S. 2004, RvMP, 76, 125
Mac Low, M.-M., Burkert, A., & Ibáñez-Mejía, J. C. 2017, ApL, 847, L10
McKee, C. F., & Ostriker, E. C. 2007, ARA&A, 45, 565
Motte, F., Nguyên Luong, Q., Schneider, N., et al. 2014, A&A, 571, A32
Murray, N., & Chang, P. 2015, ApJ, 804, 44
Padoan, P., & Nordlund, Å. 2002, ApJ, 576, 870
Padoan, P., & Nordlund, Å. 2011, ApJ, 730, 40
Peretto, N., Fuller, G. A., André, P., et al. 2014, A&A, 561, A83
Rivera-Ingraham, A., Ristorcelli, I., Juvela, M., et al. 2016, A&A, 591, A90
Schruba, A., Leroy, A. K., Walter, F., Sandstrom, K., & Rosolowsky, E. 2010, ApJ, 722, 1699
Shu, F. H., Adams, F. C., & Lizano, S. 1987, ARA&A, 25, 23
Vázquez-Semadeni, E., Ballesteros-Paredes, J., & Klessen, R. S. 2003, ApL, 585, L131
Vázquez-Semadeni, E., Colín, P., Gómez, G. C., Ballesteros-Paredes, J., & Watson, A. W. 2010, ApJ, 715, 1302
Vázquez-Semadeni, E., Palau, A., Ballesteros-Paredes, J., Gómez, G. C., & Zamora-Avilés, M. 2019, MNRAS, 490, 3061
Vázquez-Semadeni, E., Gómez, G. C., & González-Samaniego, A. 2024, MNRAS, 530, 3445
Whitworth, A., & Summers, D. 1985, MNRAS, 214, 1
Zamora-Avilés, M., & Vázquez-Semadeni, E. 2014, ApJ, 793, 84
Zamora-Avilés, M., Vázquez-Semadeni, E., & Colín, P. 2012, ApJ, 751, 77
Zamora-Avilés, M., Camacho, V., Ballesteros-Paredes, J., et al. 2024, arXiv e-prints, arXiv:2409.11588
Zuckerman, B., & Evans, N. J. I. 1974, ApJL, 192, L149
Zuckerman, B., & Palmer, P. 1974, ARA&A, 12, 279

AAS | IOP Astronomy

Interstellar Flow and Star Formation

Enrique Vázquez-Semadeni

Chapter 13

Evolution of the Energy Balance and Scaling Relations in Molecular Clouds

13.1 Interpretation of the Scaling Relations

In Section 8.4, we discussed the observed scaling relations between the velocity dispersion (measured by the widths of molecular lines), mass, magnetic field strength, and size of molecular clouds. In this chapter we now discuss how these relations reflect the energy balance of the clouds, both in the stationary framework of the TS scenario, as well as in the evolutionary description of the GHC scenario, both of which were described in Section 11.2.

13.1.1 Quasistatic Interpretation under the TS Model

Under the TS scenario, the interpretation of the observed scalings in molecular clouds is that the linewidth-size relation, $\sigma \propto L^{1/2}$ (cf. Equation (8.3)), indicates that molecular clouds are supersonically turbulent,[1] Equation (8.1) indicates that the mass surface density does not vary significantly among molecular clouds, and Equation (8.4) indicates that, to first order, there is near equipartition between the gravitational and kinetic energies in molecular clouds (Larson 1981; Heyer et al. 2009).

Under this interpretation, the velocity dispersion measured by molecular linewidths corresponds to the turbulent motions in the clouds. Thus, the near equipartition property indicates that, within the scatter in the data, molecular clouds and their substructures are in near virial equilibrium between turbulence and gravity. Indeed, in this case, $2E_k = |E_g|$, with E_k and E_g, respectively, corresponding to the terms K and W in the virial theorem, as discussed in Section 5.2.2. Therefore,

[1] Larson (1981) interpreted his original scaling exponent of $\sim$0.38 (cf. Equation (8.2)) as consistent with incompressible Kolmogorov turbulence (cf. Equation (3.31)). However, the presently accepted value of $\sim$0.5 is more consistent with the expectation for strongly compressible, Burgers-like turbulence (cf. Equation (3.61)).

doi:10.1088/2514-3433/adcb28ch13

in this interpretation, gravitational contraction of the clouds is prevented by the turbulence.

Moreover, under the TS interpretation, the additional variation around virial equilibrium manifest in the dependence of the virial parameter on mass (cf. Figures 8.6 and 8.7) has been explained by Krumholz et al. (2019) also as a consequence of Larson's linewidth-size relation. In this case, $\sigma_v \propto R^{1/2}$, and so the kinetic energy per unit mass of a cloud scales as $e_k \approx \sigma_v^2 \propto R$. On the other hand, its gravitational energy per unit mass is $e_g \approx -GM/R$. If "one chooses a region close to the mean density", so that $M \approx \langle\rho\rangle R^3$, with $\langle\rho\rangle \approx$ cst., then $e_g \propto R^2$. In consequence, $\alpha_{vir} \equiv 2e_k/|e_g| \propto R^{-1}$. Furthermore, this same set of assumptions implies that $\alpha_{vir} \propto M^{-1/3}$, which is *qualitatively* consistent with the observed trend of decreasing α_{vir} with M, although it deviates considerably from the typical observed scaling $\alpha_{vir} \propto M^{-0.55}$ (see, e.g., the *left* panel of Figure 8.7). Note, in addition, that the assumption $M \propto R^3$ violates Larson's density-size relation $\langle\rho\rangle \propto R^{-1}$, or its equivalent scaling, $M \propto R^2$. Thus, it is not possible to simultaneously satisfy the α_{vir}-M scaling and the Larson relations.

13.1.2 Evolutionary Interpretation under the GHC Model

13.1.2.1 Preliminary Considerations

Contrary to the TS interpretation, under GHC the velocity dispersion derived from molecular linewidths is interpreted primarily as representative of the infall motions in the clouds, perhaps complemented by a truly turbulent (disordered) component. In this case, the kinetic energy is derived from the gravitational energy released during the collapse. In particular, if the motions correspond exclusively to infall, then they must satisfy the condition $E_k + E_g = E_{tot}$ (Ballesteros-Paredes et al. 2011). Moreover, if the total energy is defined to be zero at the time when collapse starts, then the free-fall condition becomes $E_k = -E_g$, which corresponds to $\alpha_{vir} = 2$. Equivalently, the free-fall scaling for the Larson ratio becomes

$$\mathcal{L}_{ff} \equiv \frac{v_{ff}}{R^{1/2}} = \left(\frac{2\pi G\Sigma}{5}\right)^{1/2} \tag{13.1}$$

(cf. Section 5.2.5.1, Equation (5.26)). In turn, this implies that Larson's original density-size and linewidth-size scalings are interpreted as a special case of Equation (13.1), occurring when the definition criteria for the objects in the sample causes them to have approximately the same column density. This is in fact the case for clouds with approximately power-law density profiles with slopes $p < 3$ and defined by means of density thresholds, since, in general, the mean density of such a cloud is within a factor of order unity of the density threshold used to define the cloud (cf. Equation (6.33)).[2] A similar situation occurs for clouds defined in terms of column density thresholds. Thus, samples of clouds defined in terms of specific CO intensity levels, for example, tend to have the same column density. In this case, the Larson

[2] This means that for such clouds, most of their mass is at the lowest densities.

linewidth-size relation follows directly from Equation (13.1) with $\Sigma = \text{cst}$. But, as soon as a larger dynamic range in Σ is allowed, as in the sample of Heyer et al. (2009), the linewidth-size relation is generalized to Equation (13.1) with $\Sigma \neq \text{cst}$.

In addition to the approximate equipartition, the variation of $\alpha_{\rm vir}$ and $\mathcal{L}$ with M and Σ, respectively, is interpreted under GHC as a consequence of the *evolution* of a gas parcel as it is first compressed by an external agent and then becomes dominated by its self-gravity and begins to collapse, generating a *collapse flow*. The initial compression itself can be either *inertial*, such as, for example, a compression by a passing SN shell, or *gravitational*, but due to an *external* large-scale potential, such as, for example, if the parcel is falling into the gravitational potential of a stellar spiral arm, or if the parcel is embedded in a larger-scale gravitationally unstable and collapsing gaseous region. In any case, it is important to note that the *initial* compression triggering the collapse of a gas parcel *cannot* be due to its own self-gravity, since we are assuming that the fluctuation is initially stable. We therefore refer to the initial compression as *external*.

13.1.2.2 Evolution of the Larson Ratio for the Induced Collapse of a Cloud

We now discuss the evolution of a gas parcel (generically, a "cloud") in a *Lagrangian* description in the $\mathcal{L}$-Σ diagram (which we will also refer to as the *Keto-Heyer*, or KH, diagram), following the treatment by Vázquez-Semadeni et al. (2019).

Consider a spherical cloud of fixed-mass M and uniform density, which suffers an initial external compression $v_{\rm comp}$ at some time t_0, at which it has a radius R_0, and therefore, an initial column density $\Sigma_0 = M/\pi R_0^2$. As the cloud is compressed, it increases its gravitational energy, and eventually becomes gravitationally unstable and begins to collapse due to its self-gravity. We refer to this process as the *induced collapse* of the cloud. Σ_0 can be taken as the column density of the cloud's background, from which the cloud decouples at the onset of its local collapse. From this time on, we neglect any form of support[3] and assume the cloud collapses freely.

On the other hand, let us also consider the gravitational infall velocity of this cloud, $v_{\rm inf}$, and its associated kinetic energy, $E_{\rm k,inf} = 1/2Mv_{\rm inf}^2$. At $t = 0$, at the moment it begins contracting, the cloud's total energy equals its potential energy, $E_{\rm tot} = E_{\rm g,0} = -3GM^2/5R_0$, where we have used the assumption that the cloud is spherical and of uniform density to write $E_{\rm g,0}$ (cf. Equation (5.13)). Afterwards, during the compression and subsequent collapse flow, the infall kinetic energy satisfies

$$E_{\rm k,inf}(t) = E_{\rm tot} - E_{\rm g}(t), \tag{13.2}$$

where $E_{\rm g}(t) = -3GM^2/5R(t)$ is the instantaneous gravitational energy at time t. Therefore, the infall speed satisfies

[3] If the compression is acting on isothermal gas, such as the interior of a molecular cloud, this assumption clearly overestimates the infall speed. However, if the compression acts on a WNM gas parcel, inducing a transition to the CNM, then the cooling causes a large sudden drop on the Jeans mass, as described in Section 11.1, and the assumption is more realistic.

$$v_{\rm inf}^2 = \frac{6GM}{5}\left(\frac{1}{R} - \frac{1}{R_0}\right). \tag{13.3}$$

This can be written in terms of the Larson ratio associated to $v_{\rm inf}$ and R, $\mathcal{L}_{\rm inf}^2 = v_{\rm inf}^2/R$, using the fact that M remains constant, to obtain

$$\mathcal{L}_{\rm inf}^2 = \frac{6\pi G\Sigma}{5}\left[1 - \left(\frac{\Sigma_0}{\Sigma}\right)^{1/2}\right]. \tag{13.4}$$

We refer to $\mathcal{L}_{\rm inf}$ as the Larson ratio corresponding to *pure collapse* (i.e., non-induced) from $R = R_0$.

Finally, to the above result we can add in quadrature the Larson ratio corresponding to the external compression, $v_{\rm comp}$, $\mathcal{L}_{ext} = v_{\rm comp}^2/R$, to obtain the total Larson ratio as

$$\begin{aligned}\mathcal{L}_{\rm tot} &= \sqrt{\mathcal{L}_{\rm inf}^2 + \mathcal{L}_{\rm ext}^2} \\ &= \sqrt{\frac{v_{\rm inf}^2}{R} + \frac{v_{\rm comp}^2}{R}} \\ &= \sqrt{\frac{6\pi G\Sigma}{5}\left[1 - \left(\frac{\Sigma_0}{\Sigma}\right)^{1/2}\right] + \frac{v_{\rm comp}^2}{R}}.\end{aligned} \tag{13.5}$$

We refer to $\mathcal{L}_{\rm tot}$ as the Larson ratio corresponding to induced collapse.

The evolution of $\mathcal{L}_{\rm tot}$ as the cloud is first externally compressed and later collapsing under its own self-gravity is given by Equation (13.5), and constitutes an evolutionary track in the KH diagram. This evolutionary track is shown in Figure 13.1 for two different cases: that of a "cloud" (denoted $\mathcal{L}_{\rm tot,cld}$ and the dashed lines) of mass $M = 10^4 M_\odot$, initial radius $R_0 = 20$ pc, initial column density $\Sigma_0 = 8 M_\odot$ pc^{-2}, induced by a compression velocity $v_{\rm comp} = 4$ km s^{-1}, and that of a "massive core" (denoted by $\mathcal{L}_{\rm tot,core}$ and the dash-triple-dot lines), with $M = 100 M_\odot$, $R_0 = 0.4$ pc, $\Sigma_0 = 200 M_\odot$ pc^{-2} and $v_{\rm comp} = 0.6$ km s^{-1}. For each case, we plot two subcases, corresponding to different assumptions for the evolution of the initial compression velocity as the cloud contracts: one assuming that $v_{\rm comp} \propto R^{1/2}$ (upper curves), as would be the case for a compressible turbulent spectrum (Section 3.9.4.1) and another assuming $v_{\rm comp} \propto R^{-2}$ (lower curves), loosely representing strong dissipation due, for example, to shocks. Also plotted in Figure 13.1 are the evolutionary tracks corresponding to pure collapse from $R = R_0$ for the two cases, respectively denoted $\mathcal{L}_{\rm inf,cld}$ and $\mathcal{L}_{\rm inf,core}$, and represented by the solid lines, and the evolutionary track for free-fall starting from $R \to \infty$, denoted $\mathcal{L}_{\rm ff}$, given by Equation (13.1) and represented by the dotted line. Note that $\mathcal{L}_{\rm ff}$ depends only on the column density Σ, and thus is the same for both the cloud and the core.

The two curves for pure collapse from $R = R_0$, $\mathcal{L}_{\rm inf,cld}$ and $\mathcal{L}_{\rm inf,core}$, are seen to start out at zero at their initial column densities, and to approach the curve for

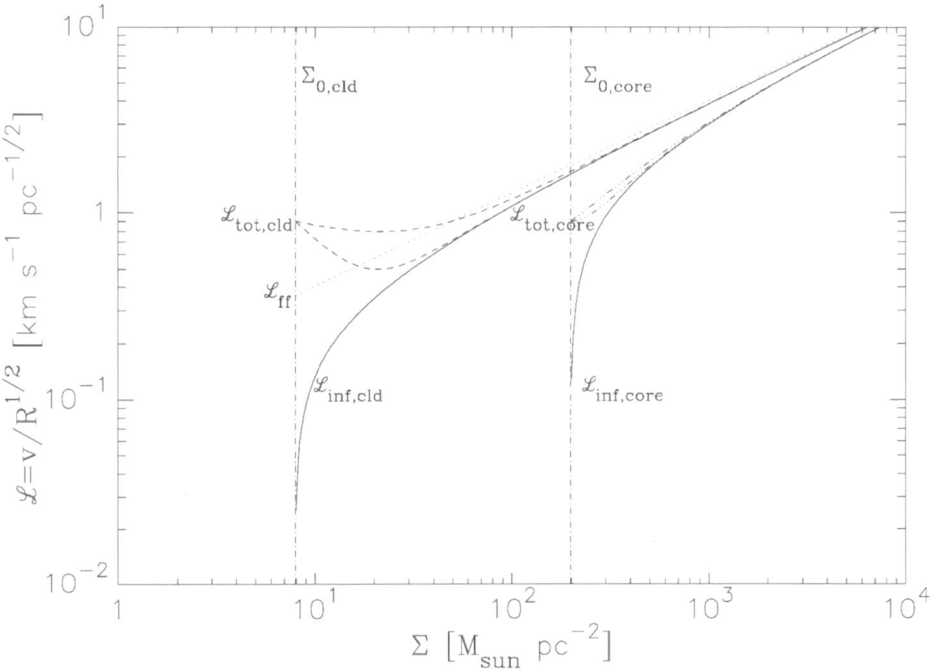

Figure 13.1. Evolution of a GMC (subindex "cld") and a dense core (subindex "core"), each of fixed mass, in the $\mathcal{L}$-Σ diagram as they are induced into collapse by an external compression, starting from rest. Both are assumed to be spherical, and the cloud is assumed to have a mass $M_{\rm cld} = 10^4 M_\odot$, an initial radius $R_{0,\rm cld} = 20$ pc, and compressed at an initial velocity of $v_{\rm comp,cld} = 4$ km s^{-1}, while the core is assumed to have mass $M_{\rm core} = 100 M_\odot$, initial radius $R_{0,\rm core} = 0.4$ pc, and to be compressed at an initial velocity $v_{\rm comp,core} = 0.6$ km s^{-1}. The lines labeled "tot" represent the evolution according to Equation (13.5), assuming two different dependences of $v_{\rm comp}$ on R: $v_{\rm comp} \propto R^{1/2}$ (upper curves) and $v_{\rm comp} \propto R^2$ (lower curves). The lines labeled "inf" represent the purely gravitational component of $\mathcal{L}$, given by Equation (13.4). Finally, the curve labeled $\mathcal{L}_{\rm ff}$ represents the Larson ratio for a cloud infalling from infinity, and is the same for both the cloud and the core, as it only depends on Σ (see Equation (13.1)).

collapse from infinity, $\mathcal{L}_{\rm ff}$, as the radius decreases and the column density increases. On the other hand, the curve for the cloud starting from initial velocity $v_{\rm comp}$, $\mathcal{L}_{\rm tot,cld}$, starts out supervirial, because it is initially dominated by the contribution from $v_{\rm comp}$, which is larger than $v_{\rm ff}$. However, as the cloud contracts, the initial compression dissipates while the infall velocity increases, and so the cloud transitions to appearing subvirial for a while, especially in the case of $v_{\rm comp} \propto R^{-2}$. Finally, late during the collapse, at large Σ, $\mathcal{L}_{\rm tot,cld}$ approaches the locus of collapse from infinity, $\mathcal{L}_{\rm ff}$, given by Equation (13.1).

The situation for the induced collapse of the core is different, however. In this case, because the core is initially already quite dense, its reference free-fall speed is larger than the initial compression,[4] and therefore it starts out and remains subvirial

[4] Note that this does not imply that the core was already collapsing, because it is assumed to initially have a mass smaller than its Jeans mass; i.e., it is assumed to be initially pressure-supported.

during its entire evolution. Note, however, that this does not need to the case for all cores. A sufficiently strong initial compression will make the core appear supervirial as well.

13.1.3 Insight

Equation (13.5) determines the evolution of a fixed-mass gas parcel as it is induced to collapse by an initial external compression $v_{\rm comp}$. Several points are worth noting:

- First, Equation (13.5) illustrates the fact mentioned in Section 13.1.2.1, that the initial stages of the contraction *cannot* be due to the *self*-gravity of the parcel. This can be seen from the fact that, at the initial condition, when $\Sigma = \Sigma_0$, the factor within the square brackets vanishes, and so the contribution of self gravity to $\mathcal{L}$ is null. Therefore, the only contribution to the Larson ratio of the parcel at the onset of the contraction is that due to the external compression, $v_{\rm comp}$.
- The external compression *can* nevertheless still be driven by gravity, although not the parcel's self gravity, but rather the infall of its parent structure. This is the case, for example, of the central, uniform-density region of a prestellar core (cf. Section 7.3.2), whose diameter is of the order of its own Jeans length (the region with $|\xi| < 1$ in Figure 7.2). Therefore, this region is not bound (or, at most, it is marginally bound) by its self-gravity, in spite of being in a state of fully dynamical contraction. This contraction is due to the compression exerted on this central region by the ram pressure of the infalling material from the gravitationally-collapsing envelope, not by the self-gravity of the central region. This is an example of *hierarchical collapse*, in which the small-scale structures become unstable themselves *at later times* and *within* larger-scale structures.
- On the other hand, the external compression can also be due to a purely *inertial* agent, not directly related to gravity, such as the passage of a shock front from a nearby HII region or SN explosion, or simply a generic turbulent compression.
- For mathematical tractability, the discussion in this chapter has assumed that the object has a fixed mass. This description is appropriate from a Lagrangian point of view, but neglects the essential property of hierarchical collapse flow that all scales accrete from their parent structures. A description in an Eulerian reference frame would be required, for example, for describing the evolution in the virial parameter $\alpha_{\rm vir}$ as a function of mass (cf. Section 8.4.1). Nevertheless, the treatment here provides a proof-of-concept for the also essential property that the Larson ratio $\mathcal{L}$ of a cloud (and therefore its virial parameter $\alpha_{\rm vir}$ also) *evolve* during the collapse.

References

Ballesteros-Paredes, J., Hartmann, L. W., Vázquez-Semadeni, E., Heitsch, F., & Zamora-Avilés, M. A. 2011, MNRAS, 411, 65

Heyer, M., Krawczyk, C., Duval, J., & Jackson, J. M. 2009, ApJ, 699, 1092

Krumholz, M. R., McKee, C. F., & Bland-Hawthorn, J. 2019, ARA&A, 57, 227
Larson, R. B. 1981, MNRAS, 194, 809
Vázquez-Semadeni, E., Palau, A., Ballesteros-Paredes, J., Gómez, G. C., & Zamora-Avilés, M. 2019, MNRAS, 490, 3061

AAS | IOP Astronomy

Interstellar Flow and Star Formation

Enrique Vázquez-Semadeni

Chapter 14

Formation of Stellar Clusters and Associations, and Cloud Dispersal

14.1 Introduction: The Relationship between the Evolution of a Cloud and of Its Stellar Products

The final products of the gravitational contraction of a gas cloud are stars, which form in groups that are themselves grouped into larger structures known as *stellar clusters*, following the hierarchical structure of the clouds and of the fragmentation process. Having formed from the gaseous material in the cloud, the stellar clusters acquire several properties that are derived from the processes in the clouds. We discuss these in the present chapter.

According to the discussion in the previous chapters, in primarily atomic environments such as that at the solar radius in the Milky Way, molecular clouds first come into existence through compressions in the WNM that induce a phase transition to the CNM (Section 4.5), to form flattened cold atomic clouds, which are highly prone to rapidly become Jeans unstable because of the drastic cooling and condensation. Furthermore, as they continue to accrete material from the compressive flow, they become magnetically supercritical, and therefore the gas begins to undergo gravitationally driven contraction. This contraction, however, proceeds in a highly anisotropic (Section 7.2.2) and hierarchical (Section 7.5) manner, amplifying anisotropies, and thus forming filamentary structures. Because the flow is turbulent, there exist density fluctuations within the filaments, which contract faster than the bulk of the filament, because of the different timescales for contraction of roundish and filamentary structures (Section 7.2.2). The densest among these fluctuations (known as *hubs*) exert a gravitational pull on the neighboring filament material, generating a filamentary flow onto them (see Section 11.4 and Figure 11.1). This constitutes a *hierarchical collapse flow*, which consists of

doi:10.1088/2514-3433/adcb28ch14

1. large, cloud-scale contraction that feeds the filaments;
2. intermediate, filament-scale collapse that feeds the hubs, and
3. small, core-scale collapse that eventually feeds protostellar disks through filamentary structures known as *streamers.*

Moreover, secondary collapses may occur at other locations in the filaments, if eventually they locally exceed the critical linear mass density for radial collapse (Equation 6.39). These secondary collapses within the filaments produce isolated *cores.*

Because the SFR in the cores and hubs increases with time (Section 12.3.2), the number of stars in a cluster increases in an accelerated manner. Moreover, the mass of the most massive stars forming in the hubs and cores also increases with time (Section 12.5). Therefore, the massive-star content of a cluster increases with time, up to a moment when the energy injection (*feedback*) from the massive stars may become capable of destroying the cloud and/or the accretion flow onto it, thus halting the local SF episode. In this case, the dense gas may be either "evaporated" (i.e., returned to the hot or warm diffuse phases) or just torn into small molecular shreds that are pushed away from the local site, possibly initiating a new SF episode elsewhere in the cloud, or forming new clouds.

In this chapter we will first discuss the imprint of the molecular cloud structure on the stellar clusters formed within them, and next we discuss the destruction of the clouds by the most massive cluster member stars. However, before that, it is important to clearly define our terms.

14.2 Definitions and Terminology

14.2.1 What Is a Cluster?

In spite of being widely used, at the time of this writing there is no consensus in the literature regarding the precise meaning of the term "stellar cluster", and this has caused significant confusion and probably unnecessary disputes. Most commonly, the term specifically denotes a large, *bound*, group of stars. Less massive, unbound groups in the process of dispersal are generally referred to as *associations* (e.g., Lada & Lada 2003; Portegies Zwart et al. 2010; Longmore et al. 2014). By "bound" it is specifically meant that the group's age significantly exceeds its dynamical crossing time (the orbital time of a typical star; Portegies Zwart et al. 2010). Note that this definition does not specifically require the binding to be due to the self-gravity of the cluster members, and that a stellar group may remain together merely because they share a common velocity pattern, derived from a previous epoch when they were part of a stars+gas bound region. This is often the case of the groups known as *open*, or *galactic* clusters, as opposed to the case of *globular clusters*, which are extremely massive and dense, and therefore have remained gravitationally bound over time-scales comparable to the age of the Universe.

In this chapter we will not be concerned about whether a stellar group ends up being bound or not after losing its parental gas, and so, for conciseness, we will use the term cluster in a rather loose sense, simply denoting a group of stars that share a common origin within a molecular cloud. In this sense, our use of the term "cluster" should be interpreted as meaning "clusters and/or associations", when their bound nature is considered.

Also, in the remainder of this chapter we shall not distinguish between a "star" (already in the main sequence) and a "young stellar object" (YSO, a protostar or a pre-main sequence object), since the numerical simulations that will be discussed do not keep track of the stellar evolution, and, in any case, it is not very relevant for the cluster dynamics that we will discuss.

14.2.2 What Is a Clump?

As discussed in Section 1.2.1, the definition of any density structure (a cloud, clump, filament, core, etc., which we will collectively refer to as "clumps") within the ISM is also an elusive task, since the ISM is a *continuous* medium, in which most boundaries are purely a matter of convention or definition, and not necessarily of physical differentiation.[1] The most common way to define a generic clump is as a connected region above some threshold in the medium, such as a number density threshold in numerical simulations, or intensity of column density thresholds in observational data. In observations, this can be only done at a particular moment in time. However, in numerical simulations, the evolution of the material belonging to the clump *can* be followed over time. This amounts to a *Lagrangian* description (following the flow), which numerically can be accomplished using a *passive scalar*[2] in grid-based codes, or directly tracing particles in SPH codes.

Consider, for example, the gravitational collapse of a clump defined in a Lagrangian way—that is, as all of the material within a certain density isocontour in the medium at some moment in time. In this case, the material in the clump will contract, and eventually some parts of it will end up in a singularity. That is, the gas parcels in the clump begin to move from their initial location toward the collapse center, and therefore a *collapse flow* is established. Moreover, for any realistic initial condition with local density inhomogeneities, the collapse will occur non-homologously, meaning that different parts of the clump will reach the singularity at different times. That is, different parts of the clump pass through a given spatial position at different times.

Now consider a clump defined by the same isodensity contour, but in an Eulerian frame, so that the initial boundary defined by the contour remains fixed in space. In this case, the onset of collapse of the material initially inside the contour will leave

[1] Observationally, apparent "boundaries" may arise from the use of specific tracers, such as emission lines of certain atoms or molecules, which may form or disappear depending on the physical conditions of the gas (density, temperature, etc.), and sometimes with delays, depending on the ratio of the timescale for the physical conditions to change and the chemical reaction speed. However, from the pure gas-dynamical viewpoint, these boundaries are in general of little or no relevance. More physically relevant boundaries may arise from phase transitions, such as that between the WNM and the CNM in the atomic gas (cf. Section 4.5), but even in this case, the gas can flow across the phase transition front, and so the dense medium is dynamically interacting with the diffuse phase.

[2] A passive scalar is a fluid quantity that follows the continuity equation, Equation (2.74) but does not affect the flow in any form—i.e., it does not appear in the momentum nor the energy equations—, similar to ink in water.

behind a low-pressure region, causing the gas in the surroundings to begin moving into the volume defined by the Eulerian boundary. That is, a *mass flux* (with units of mass per unit time per unit area) will be established across the Eulerian boundary.

A third case of boundary was discussed in Section 1.2.1, termed *feature based*, such as a density isocontour, which is redefined at the same density at each moment in time, and therefore changing in time. In the case of a collapsing region in a simulation, for example, the contour in general *does not shrink*—in fact, it often becomes larger (cf. Figure 12.2), since the collapse flow can accumulate more material within the contour, causing the contour to *expand* (cf. Section 7.4). Similarly to the case of the Eulerian boundary, a mass flux is established across the contour, but in this case the contour is free to move in space.

So, what is a clump? No unique definition exists, but it may be argued that our most common clump-defining procedure involves feature-based boundaries, and therefore, in what follows, we will define a clump as the connected region within a closed isodensity contour, which in general will vary over time.

14.3 Imprint of the Parent Cloud's Structure on the Nascent Cluster

14.3.1 Spatial Structure of the Parent Filamentary Cloud

The hierarchical and filamentary structure produced by the combined action of cloud-scale gravitational contraction and turbulent fluctuations in the clouds becomes imprinted on the spatial distribution of the stars formed within them. As discussed in Section 11.4, the cloud-scale anisotropic contraction forms filaments, which constitute linearly extended (rather than point-like) centers of collapse, and which can grow by accretion. Moreover, the filaments necessarily contain random internal turbulent fluctuations, which grow together with the filament by the global accretion onto it. It is important to note that the filaments and the more roundish fluctuations they contain *are not initially gravitationally unstable themselves*, but are rather being compressed and fed by the infall of material from the globally collapsing cloud, similarly to the inner, flat-density region of a prestellar collapsing core (Section 7.3.2.3).

However, as they grow by this accretion, the most massive of these fluctuations can become locally Jeans unstable themselves, and therefore begin their own local collapse. At this point, the local collapsing regions within the filaments (hubs and cores) begin to grow faster than the rest of the filament because of their approximately spherical shape, for which the collapse timescale is shorter than for thin sheets and filaments of the same density (Section 7.2.3). The gravity of the hubs then creates an acceleration region at their interface with the filament, as observed both in observations and simulations (Figure 14.1). Further away from the hub, its mass becomes subdominant for accelerating the gas, and thus the filament material moves at nearly constant speed, as also seen in the figure. Near the secondary cores, the pattern is repeated, while the cores themselves follow the longitudinal flow imposed by the dominant hub. This constitutes a "conveyor-belt" (Longmore et al. 2014) type of flow (Section 11.4), characterized by a filamentary large-scale collapse flow toward a remote hub, on top of which local collapses toward the smaller, local

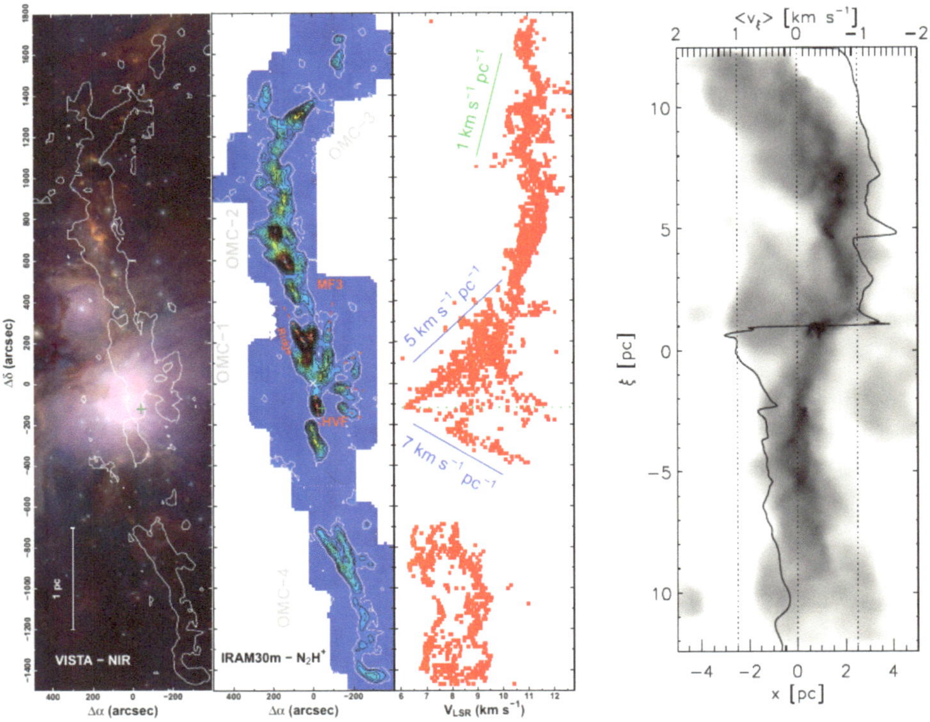

Figure 14.1. Longitudinal flow in filaments. *Left:* Three views of the *integral shaped filament* in the Orion A cloud in the near infrared, N_2H^+ (1 – 0) emission, and N_2H^+ (1 – 0) line velocity centroid. The latter image shows two filamentary streams of gas at LSR velocities $\sim$ 8 km s^{-1} in the lower part and $\sim$11 km s^{-1} in the upper part, and a strong velocity gradient near the central hub (Hacar et al., (2017), reproduced with permission © ESO. *Right:* An image and a longitudinal velocity plot of a filament formed by large-scale gravitational contraction in a numerical simulation of cloud formation and evolution. The grey scale shows the column density in a plane approximately containing the filament, with ξ being the longitudinal coordinate along it, and x being one of the axes in the simulation. The solid line shows the longitudinal velocity, with values indicated by the upper horizontal ruler. A sharp velocity jump is observed at the central hub located at $\xi \approx 0$, where the incoming gas from the opposite sides of the filament collides. Similarly, lower-amplitude secondary jumps are observed at the secondary cores located at $\xi \approx -3$ pc and $\xi \approx 5$ pc, indicating the hierarchical nature of the collapse process in the filaments and the “conveyor belt” nature of the flow, as these local collapse centers “ride” on top of the large-scale filamentary collapse flow. Reproduced from Gomez and Vazquez-Semadeni (2014).

cores occur (Figures 11.1 and 11.3). Indeed, note in the rightmost panel of Figure 14.1 that a major velocity jump is observed at the central hub located at $\xi \approx 0$, and similar, lower-amplitude secondary jumps are observed at the secondary cores located at $\xi \approx -3$ pc and $\xi \approx 5$ pc, indicating the self-similar, hierarchical nature of the collapse process in the filaments.

14.3.2 Spatial Structure of the Nascent Cluster

Stars form in both the primary and secondary collapse centers of the conveyor-belt flow (respectively, the hubs and the cores), and observations suggest that they inherit the spatial and kinematic structure of the gas, as shown in Figure 14.2.

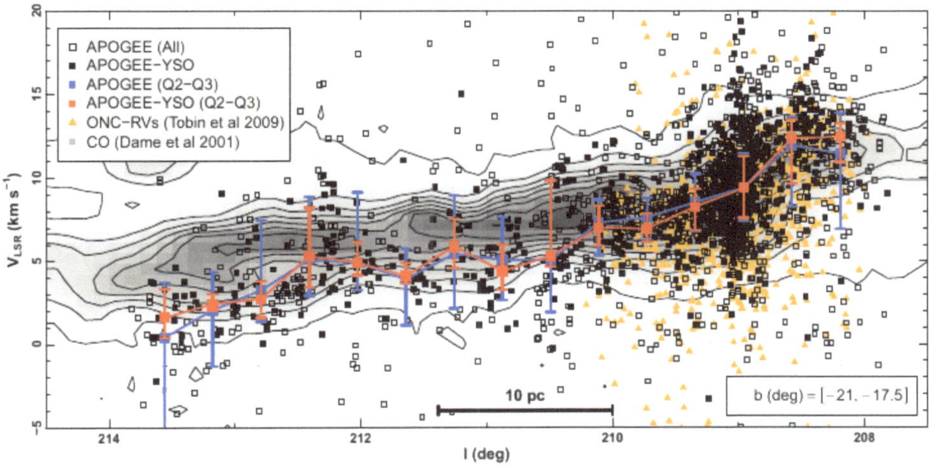

Figure 14.2. Position–velocity diagram of stellar (dots) and gas (contours and gray scale, from Dame et al. 2001) emission along the Orion A cloud. The stellar data include all stellar sources in the field detected by APOGEE (black empty squares), APOGEE sources also identified as YSOs by Megeath et al. (2012, black solid squares), and stellar sources from Tobin et al. (2009, orange triangles). The red and blue solid squares represent the median of the full APOGEE sample and of the APOGEE-YSO sample, respectively, while the colored bars represent the corresponding 25%–75% interquartile range. The YSOs are seen to qualitatively share both the spatial and the kinematic distribution of the gas. From Hacar et al. (2016), reproduced with permission © ESO.

The mechanism likely responsible for this to occur is illustrated in Figure 14.3, both of whose panels are available as an animation in the electronic version of this book. The figure shows, in the top panel, a bent, flattened, and filamentary cloud formed in a numerical simulation of a collision of moderately-turbulent warm atomic gas (WNM) streams in a numerical box of 256 pc per side, including model heating and cooling functions, and simple prescriptions for emulating photoionizing heating and for forming stellar-mass particles, originally presented by Colín et al. (2013). The stream collision constitutes a nonlinear perturbation in the WNM, which is strong enough to trigger a transition to the cold atomic phase (the CNM; see Section 4.5), and to produce a sheet-like cloud that fragments into filaments due to the continuing nonlinear growth of the thermal instability (Vázquez-Semadeni et al. 2006).

The cloud eventually becomes gravitationally unstable, and continues to accrete onto the filaments, which later also become locally gravitationally unstable ("supercritical", in the sense of exceeding the critical linear mass density for thermal support given by Equation (6.39)), and begin forming stars along their full length. In this way, three clusters of different masses and sizes are seen to form: the most massive one is an extremely long filament, with a length of over 50 pc and a width of only a few pc. The other two are less massive and less extended, although a zoomed-in view of the intermediate-mass cluster (bottom panel) shows that it is filamentary as well.

The bottom panel of Figure 14.3, and the animation associated with it, show the development and evolution of a hub-filament system (HFS) arising by the process described above in the region marked by the square in the top panel, and the intermediate-size cluster forming in this HFS. Similarly to the case in the upper

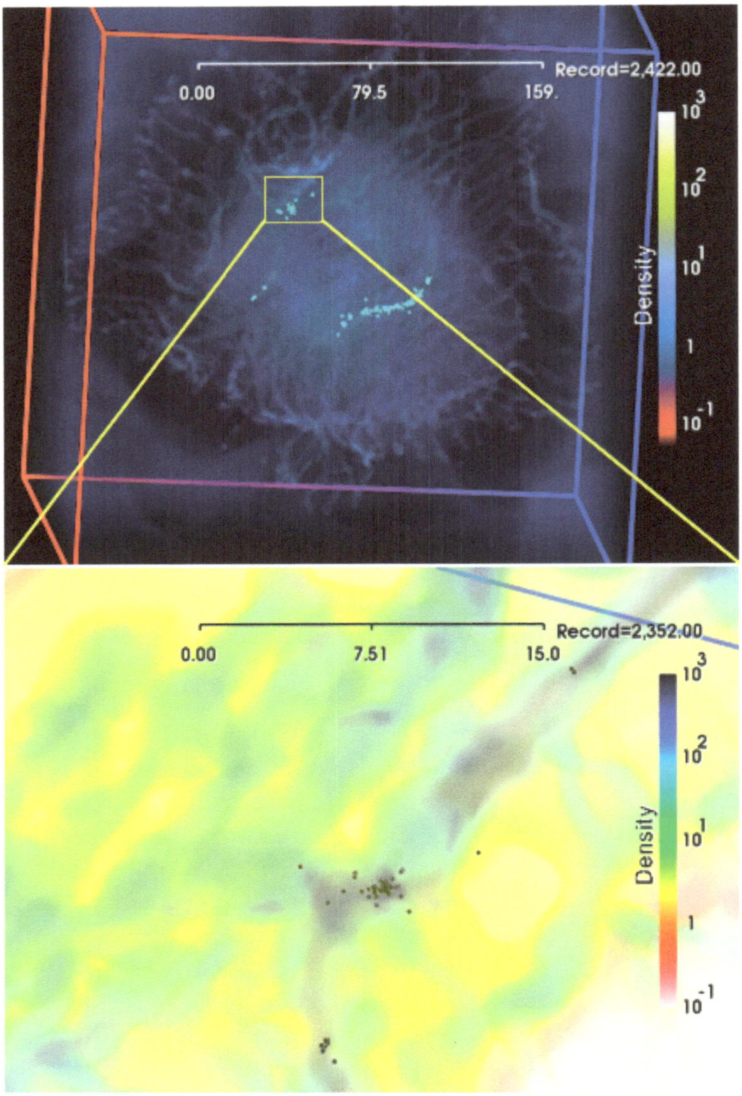

Figure 14.3. *Top panel:* Large-scale view of a numerical simulation of cloud formation and evolution by a compression in the turbulent warm neutral medium (WNM), which nonlinearly triggers a phase transition to the CNM, producing a curved sheet-like cloud that subsequently fragments into filaments due to the nonlinear evolution of the thermal instability. The cloud eventually becomes gravitationally unstable, and continues to accrete onto the filaments, which later also become locally gravitationally unstable, and begin to form stars, represented by the white dots. These correspond to individual stars, and are seen to form along the filamentary structures in the cloud, arranged in a fractal-like hierarchy of stellar associations (compare the top and bottom panels). *Bottom panel:* A zoomed-in view of the gas and stars within the yellow rectangle in the top panel, showing the accretion flow along the filament toward the main hub, while forming stars also at secondary sites (cores). The rulers show the scale in parsecs. The "Record" shows time in units of 10^4 yr. The color bar indicates the volume density of the semi-transparent voxels in units of cm^{-3}. Reproduced from González-Samaniego & Vazquez-Semadeni (2020) with permission from Oxford University Press on behalf the Royal Astronomical Society. Both images are available as animations in the electronic edition of this book (https://doi.org/10.1088/2514-3433/adcb28).

panel, this 30-pc long filament is seen to fragment into several substructures: a main *hub* (a main condensation, toward which the gas in the filament flows, as seen in the animation) in which the largest group of stars is forming, and two secondary, smaller *cores* in the upper right and the bottom-center parts of the image. In the animation, the secondary groups are seen to share the accretion flow toward the hub along the filament, although it is important to note that this filamentary flow is rather slow ($\lesssim 1$ km s^{-1}); this is discussed at more length in Section 14.3.4. At later times, more secondary star-forming cores appear along the filament.

Figure 14.3 thus illustrates the *hierarchical* nature of the gravitational fragmentation process, since the structure in the bottom panel is essentially a scaled-down version of the structure in the upper panel, with the clusters or associations consisting of distinct, smaller stellar subgroups at each stage. That is, the process of fragmentation is self-similar, as discussed in Section 7.5.4: as the density increases and the Jeans length decreases, a given scale passes from being supported by thermal pressure to being in near pressureless collapse. Therefore, filament-generating anisotropic gravitational contraction can occur at at any scale sufficiently larger than the local, instantaneous Jeans length, and so it is observed from scales of hundreds of parsecs down to the filamentary accretion flow onto protostellar accretion disks known as *streamers*.

14.3.3 Stellar and Gaseous Mass Growth, Increase of the SFR, and the Stellar Age Distribution

The combined effect of the increasing SFR discussed in Section 12.3.2 and the development of the filamentary flow discussed in Sections 7.2.2 and 14.3.2 causes a number of specific patterns in the spatial and age distributions of the member stars of the cluster.

The distribution of ages of the cluster member stars is directly affected by the rate at which stars were forming in the cluster at a certain time in the past. As an illustration, let us again consider the simulation of Figure 14.3, whose evolving SFR is shown in the *bottom-right* panel of Figure 12.7. It is seen that star formation starts in this simulation at $t \approx 19$ Myr.[3] At that time, the SFR was $\sim 150\ M_\odot$ Myr^{-1}, and increases to $\sim 1000\ M_\odot$ Myr^{-1} by $t \approx 26$ Myr. Simultaneously with the increase of the stellar mass, the gas mass of the cloud is also increasing, as shown in Figure 14.4, as well as the mass of the most massive star contained in the cluster (given by the extent of the high-mass tail of the stellar mass distribution, shown in Figure 12.8). The formation of increasingly more massive stars over the course of the evolution of the star-forming region can be understood, from a probabilistic point of view, as a consequence of the higher SFR, which increases the probability per unit time of forming a massive star. However, it can also be understood from a physical point of view, as a consequence of the increased accretion ram pressure onto the stars due to

[3] It is worth recalling that this simulation follows the collision of oppositely-directed gas streams in the WNM, so that the prior evolution of the cloud includes the entire duration since the cloud first appeared as a thin CNM cloud in the WNM until the time when it begins forming stars. In this simulation, this process takes $\sim$19 Myr.

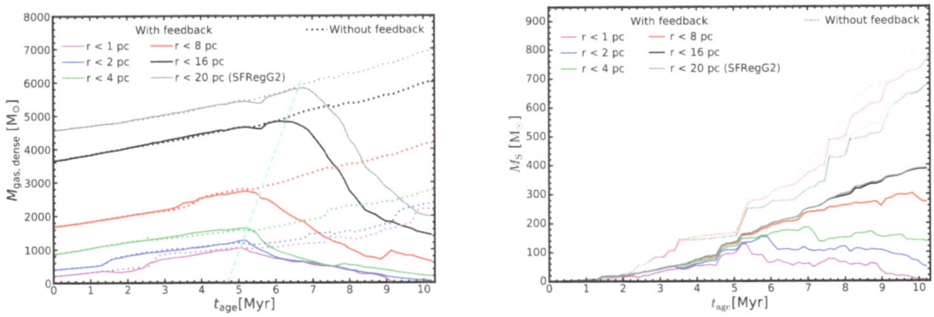

Figure 14.4. Evolution of the dense gas mass (*left panel*) and the stellar mass (*right panel*) within spheres of the radius indicated by the line color and the inset. The spheres are centered at the instantaneous center of mass of the cluster shown in Figure 14.3 for two numerical simulations, one without photoionizing feedback from massive stars (*dotted lines*) and one including it (*solid lines*). The times in this figure are given since the time when the first star formed. The *cyan dashed line* indicates the locus of the expanding shell from a massive HII region formed by the appearance of a 30 $M_\odot$ star at $t \approx 4.8$ Myr. Reproduced from González-Samaniego & Vazquez-Semadeni (2020) with permission from Oxford University Press on behalf the Royal Astronomical Society.

the increasing depth of the region's gravitational potential well (Vázquez-Semadeni et al. 2024). Finally, the ionizing feedback from the most massive stars eventually becomes capable of eroding or disrupting the cloud and/or the accretion flow onto it, and so both its gas mass and its SFR begin decreasing at $t \approx 27$ Myr, as seen in the *middle upper* and *bottom-right* panels of Figure 12.7.

As a consequence of the evolution of the SFR, the observed age distribution of the stars in the cluster therefore evolves as well. Note that the cluster age is essentially equal to the age of its oldest stars. Then, if a nascent cluster is observed at a time *before* the SFR has reached its peak, the observed stellar age distribution will peak at zero age, because the present SFR is larger than it has ever been so far in the region, and so the most numerous stars are the ones just formed in the cluster. However, if the cluster is observed *after* the SFR peaked, then the age of the most numerous stars will correspond to the lookback time at which the SFR peaked. For example, if the cluster of Figure 14.3 is observed at $t = 25$ Myr, and the SFR peaked at $t \approx 23$ Myr, the most numerous stars will be those $\sim$2 Myr old. The evolution of the age distribution in the same numerical simulation is shown in Figure 14.5, and clearly illustrates that the distribution peaks at the present time (zero age) for times earlier than $\sim$23 Myr, but peaks at an age corresponding to the lookback time to the maximum of the SFR at times later than that.

14.3.4 Primordial Cluster Kinematics

14.3.4.1 Primordial Kinematics and Effects of the Feedback

The conveyor-belt kinematics undergone by the gas during its gravitational contraction is imprinted in the stellar products. We refer to this as the "primordial" kinematics of the cluster, in the sense that it is imprinted on the cluster members at birth, as opposed to being "acquired", through, for example, N-body interactions or gas removal by the stellar feedback. This is illustrated in Figure 14.6, which shows

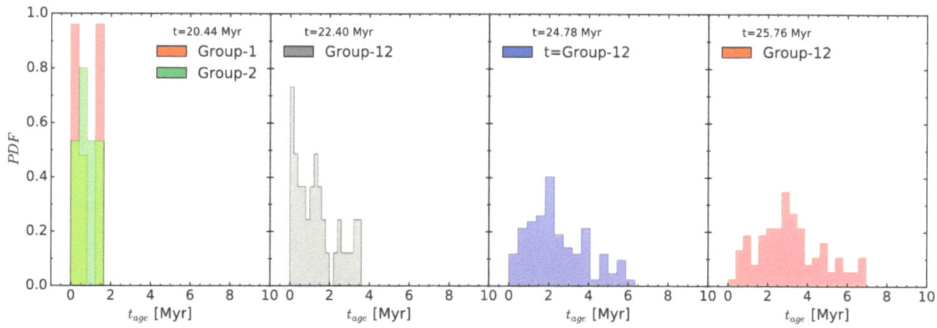

Figure 14.5. Evolution of the age histogram for the cluster shown in Figure 14.3. The time since the start of the simulation is indicated in each panel. In the first panel, the cluster consists of two subgroups, represented with separate histograms. At later times, the subgroups have merged, and thus only one histogram is shown. Reproduced from Vázquez-Semadeni et al., (2017) with permission from Oxford University Press on behalf the Royal Astronomical Society.

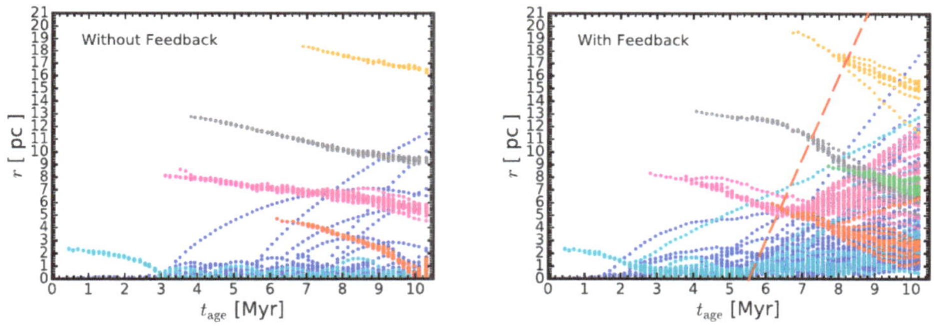

Figure 14.6. Distance from each stellar particle (colored dots) to the global center of mass of the main hub (located at $r = 0$, with its stars colored dark blue) as a function of time since the formation of the first star, t_{age}, for the cluster shown in the bottom panel of Figure 14.3. *Left panel:* Simulation with no feedback. *Right panel:* Simulation with feedback. The dot colors indicate membership to the various stellar subgroups that appear along the filament. The slope of each dot sequence gives its infall speed toward the center of mass, which for this simulation is $\sim$ 1 km s^{-1}. The long-dashed maroon line on the right panel indicates the average expansion rate (with speed $\sim$ 7 km s^{-1}) of the HII-region front from the most massive star forming at the central hub at $t_{age} \approx 5.5$ Myr. The blue dots moving away at high speed from the central hub are stars that have been ejected from it by N-body interactions. Reproduced from González-Samaniego & Vazquez-Semadeni (2020) with permission from Oxford University Press on behalf the Royal Astronomical Society.

the distance of each one of the cluster members (colored dots) to the center of mass (CM) of the main hub (at $r = 0$) as a function of the time since the formation of the first star, t_{age}. The *right panel* of Figure 14.6 shows the distances for the cluster appearing in the *bottom panel* of Figure 14.3, while the *left panel* shows the same, but for a simulation without stellar feedback. The blue dots denote the stars formed in the main hub, while the other colors indicate the various subgroups formed along the filament. The long-dashed maroon line indicates the evolution of the average size of the HII-region around the most massive star, which forms in the central hub at $t_{age} \approx 5.5$ Myr. Several features are worth noting in this figure:

1. The group indicated by the cyan dots, which forms very nearby to the main hub, merges with it within 2-3 Myr. None of the other groups reaches the main hub before gas dispersal in the simulation with feedback.
2. The slopes of the dot sequences indicate their infall speeds toward the CM, which for this simulation are in the range $0.5 \lesssim v_{\rm inf} \lesssim 2$ km s^{-1}.
3. The infall speeds are of the order of the gravitational speed $v_{\rm g} \approx (2GM/r)^{1/2}$. For example, consider the set of yellow dots, which move from $r \sim 19.5$ to $r \sim 17.5$ pc from $t_{\rm age} \sim 7$ Myr to $t_{\rm age} \sim 8$ Myr, implying an infall speed of 2 km s^{-1}. On the other hand, $v_{\rm g}$ can be estimated from Figure 14.4, which shows that the dense gas mass contained within 20 pc of the center of mass at $t_{\rm age} \sim 7$ Myr is $\sim$5800 $M_\odot$, implying $v_{\rm g} \sim 1.6$ km s^{-1}.
4. The groups are seen to be more scattered in general in the simulation with feedback than in the one without it, even from early times, when only low-mass feedback is present. This is because the feedback stirs the gas, and thus the stars form with a larger velocity dispersion in the simulation with feedback.
5. A few stars are ejected from the main hub (dark blue dots) by pure N-body interactions. The ejections occur more frequently and with larger velocities in the case without feedback because the groups are more compact and therefore the N-body interactions are stronger in this case.
6. In the case with feedback, the appearance of a very massive ($65M_\odot$) star in the central hub at $t_{\rm age} \approx 5.5$ Myr creates an HII region whose shell expands with an angle-averaged speed of 7 km s^{-1}, depicted by the long-dashed maroon line. It is seen that, as the shell passes through each group, it causes the group to begin dispersing, because it removes the remaining gas around each group, and leaves the stellar group gravitationally unbound.
7. In the case of this particular simulation, the final kinematics consists of a set of individually-expanding groups, whose centers of mass nevertheless maintain the initial infall motion toward the hub. This need not be the case in general, however, because it depends on whether a peripheral core begins to form stars before or after it has received a "kick" from the stellar feedback. The general conclusion is that the motions in the cluster consist of a combination of the pre-feedback motions inherited from the cloud kinematics, and the dispersal that begins when the feedback destroys the star-forming hub and cores.

14.3.5 Cloud Dispersal and the Most Massive Star a Core Can Harbor

Numerical simulations (e.g., Haid et al. 2019) as well as observations (e.g., Chevance et al. 2022) suggest that the main mechanism responsible for molecular cloud destruction is pre-supernova stellar feedback (particularly photoionizing radiation and stellar winds). Chevance et al. (2022) conclude, comparing observations of gas (CO) and star formation (Hα) tracers, that GMCs are typically dispersed within $\sim$ 3 Myr after the emergence of unembedded high-mass stars. Numerical simulations in general take times of the same order of magnitude, as seen in the accompanying video to Figure 14.3.

An important feature of the stellar feedback is that the feedback does *not* precisely balance the self-gravity of the cloud to maintain it near virial equilibrium. This can be appreciated both from the evolution of the SFR in the simulation (lower-right panel of Figure 12.7) and from the accompanying animation to Figure 14.3. At early stages, when the feedback is dominated by low-mass stars, it is incapable of preventing the infall and accretion from large scales onto the star-forming region. Therefore, the cloud continues to increase its mass and its SFR. Later, once sufficiently massive stars begin to form, the feedback suddenly becomes capable of overwhelming gravity, and thus of dispersing the cloud. The dispersal occurs on a short timescale compared to that of the previous evolution of the cloud, due to the extremely nonlinear dependence of the ionizing feedback strength with the stellar mass, and because the gas is dispersed at a speed comparable to the sound speed in the ionized warm gas.

A simple, idealized toy model of this process that considers a two-stage accretion flow (onto a dense core and onto the core's central region where stars are forming) can be constructed (Vázquez-Semadeni et al. 2024) using the core growth model from Section 7.4.2 in the spherical case for a power-law density profile of the form

$$\rho(r) \propto r^{-p}, \tag{14.1}$$

where p is the *logarithmic slope* (cf. Equation (7.70)). In this case, the accretion rate $\mathcal{F}$ as a function of radius r is given by Equation (12.10), which we repeat here for convenience:

$$\dot{M}_{\rm acc} = 2\pi \rho_{\rm acc} v_{\rm acc} R_{\rm c}^2(t), \tag{14.2}$$

where $v_{\rm acc} = \sqrt{2GM(r)/r}$ is the accretion velocity and $M(r)$ is the total mass (gas + stars) contained within radius r. The *left panel* of Figure 14.7 shows the evolution of both the mass in the core's inner region, $M_{\rm i}$ (which can be considered as the total mass in "stars"), and the core's mass, $M_{\rm c}$, resulting from the numerical integration of the accretion rate at the inner and outer boundaries of the core, for various values of the density profile's logarithmic slope, p. In addition, the mass gained by the "stars" is lost by the core, and therefore, in some cases, the core can be completely "deflated" for $p > 3/2$, as is the case of the orange and blue lines.

The accreting mass carries a kinetic energy density $e_{\rm k} = \rho(r) v_{\rm ff}^2(r)/2$, where $\rho(r)$ is the density at radius r. Therefore, the power density (energy per unit volume per unit time) associated to the accretion at radius r is

$$\begin{aligned} P_K &= 4\pi r^2 e_{\rm k} v_{\rm ff}(r) \\ &= \frac{\left[2\left(\pi \rho_0 r_0^p\right)^5 G^3\right]^{1/2} r^{\frac{5}{2}(2-p)}}{(3-p)^{3/2}} \\ &\sim L_\odot \left(\frac{r_{\rm c}}{1\ {\rm pc}}\right)^5 \left(\frac{n_0}{10^3\ {\rm cm}^{-3}}\right)^{5/2}, \end{aligned} \tag{14.3}$$

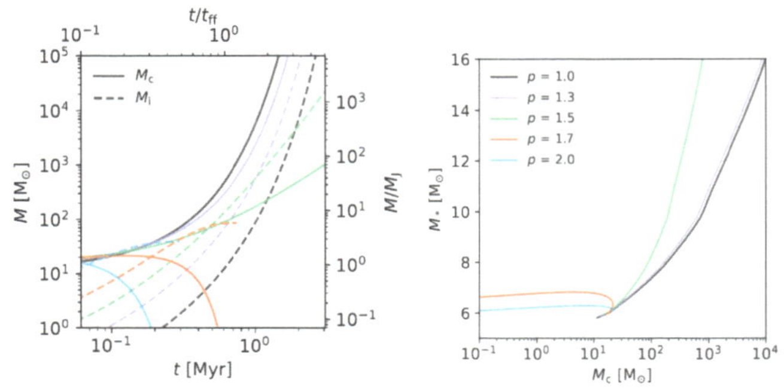

Figure 14.7. *Left:* Evolution of the mass M_c of a core with density profile with slope p, immersed in a uniform background medium (assumed to be flowing uniformly toward a distant, global collapse center), and of the mass M_i within a certain inner radius where the mass is assumed to be exclusively in stars. The core's radius is taken as 0.22 pc, corresponding to half the Jeans length at the background density of $3 \times 10^3\,\mathrm{cm}^{-3}$ and a temperature of 10 K. The inner radius is taken as 3000 au, representing the size of a large accretion disk, where the spherical symmetry is broken. The evolution of both masses is obtained by numerical integration of Equation (14.2) evaluated at the outer and inner radii. The mass entering the inner region is subtracted from the core's mass, and so the core may ultimately be deflated, as is the case of the blue and orange curves. Reproduced from Vázquez-Semadeni et al (2024). CC BY 4.0. *Right:* Mass M_* of the star whose ionizing radiation power equals the accretion power onto the core (Equation (14.3)) and can thus terminate the SF episode, plotted against the core's mass M_c, as the latter evolves due to the mass flux through its outer and inner boundaries. Reproduced from Vázquez-Semadeni et al (2024). CC BY 4.0.

where ρ_0 and r_0 are the normalization constants of the power-law density profile, and can be taken as the density and Jeans length of the background medium in which the core is embedded. For reference, the last line of this equation expresses this power at the boundary of a clump of radius $r_0 = r_c$ in terms of typical sizes and densities of molecular cloud clumps. $L_\odot \approx 3.8 \times 10^{33}$ erg s^{-1} is the solar luminosity.

In order to stop the clump growth and the SF episode, this accretion power needs to be overcome by some form of feedback from the stars forming in the core. For example, the *right panel* of Figure 14.7 shows the mass of the star whose photoionizing radiation (according to the ionizing photon fluxes for a range of stellar masses calculated by Diaz-Miller et al. 1998) can overcome the accretion power onto the core, plotted against the core mass, as the latter evolves due to the mass flux through its outer and inner boundaries.

Clearly, this calculation is highly simplified, as it assumes spherical geometry, as well as a 100% efficiency for the photoionizing feedback to stop the accretion flow. Also, it assumes that all stellar products are located precisely at the core's center, strongly deepening the gravitational potential there. These idealizations are probably the cause that the model predicts *deflation* of the core for density profile slopes steeper than -1.5. Nevertheless, the model can be considered as a proof-of-concept for the property that stars of progressively larger masses can form as the core evolves and gains mass itself, because both its SFR and the accretion it generates onto the stars increase over time. Also, stars of ever larger masses are progressively needed to stop the core's growth as it becomes more massive, showing that cores of larger masses can harbor larger-mass stars. Finally, an important corollary is that, by the time massive stars can begin to form in a core, it must have formed already a significant number of low-mass stars, which were the only ones it could form while it had a low mass.

14.3.6 Timescale for Cloud Clearance

The simple two-stage accretion model outlined above, as well as numerical simulations and observations, indicate that the typical time for disruption of a local SF site and episode ranges from a few to $\sim$ 10 Myr, typically increasing with the mass of the cloud. For example, for the simulation discussed in this chapter and Chapter 12, both the cloud mass and the SFR are seen to drop to $\sim$10% of their peak values in $\sim$8 Myr (from $t \approx 24$ to $t \approx 32$ Myr; *top-center* and *bottom-right* panels of Figure 12.7).

The timescale for cloud clearance can also be estimated statistically from observations taking advantage of the "uncertainty principle for star formation" (Kruijssen & Longmore 2014). As discussed in Section 12.2.2, those authors postulated that the Galactic SF laws relating the gas content and the SFR, such as the Schmidt–Kennicutt (SK) relation, can only be recovered over large enough regions that all the evolutionary stages of a cloud and its star formation activity are sampled within the observed region, so that a statistically valid average is performed. This sets a minimum scale over which the Galaxy-averaged SF laws are recovered. On scales smaller than that, which sample only a single or a few clouds, the gas content and the SFR will depart from the average values, since both of these quantities evolve in a cloud (cf. Section 12.3.2), and thus their ratio is not in a unique relationship. In other words, clouds that have the same mass can have different SFRs depending on their evolutionary stage, as shown by Vázquez-Semadeni et al. (2018).

In practice, this size scale has been estimated by observing regions of different sizes in the disks of nearby galaxies, using both a gas tracer (such as CO emission) and a massive-star-formation tracer (such as Hα emission, produced by the recombination of protons and electrons in the ionized gas surrounding massive stars). From these, the gas-depletion timescale ($\tau_{\rm dep} \equiv M_{\rm g}/{\rm SFR}$, where $M_{\rm g}$ is the gas mass) can be derived. Pointing the observations at either CO or Hα peaks on the disks, the desired size scale is the minimum region size over which $\tau_{\rm dep}$ is equal for both types of pointings (Figure 12.4). In addition, assuming that the relative abundance of gas- and SF-dominated regions is proportional to their relative lifetimes, the relative durations of the two stages can be estimated. Finally, the timescale over which the cloud is destroyed can be estimated through the relative abundance of regions in which both the gas and the SF tracers are visible.

Using this technique, Chevance et al. (2020) have estimated the cloud dispersal timescale to be $\sim$1–5 Myr once massive stars have appeared in the clouds (Figure 14.8). Adding the time for massive stars to appear since the formation of the first stars ($\sim$2 Myr in the two-stage accretion model and $\sim$5 Myr in the simulation by Colín et al. 2013), we again recover a timescale from the onset of SF to the dispersal of the cloud of order $\sim$10 Myr.

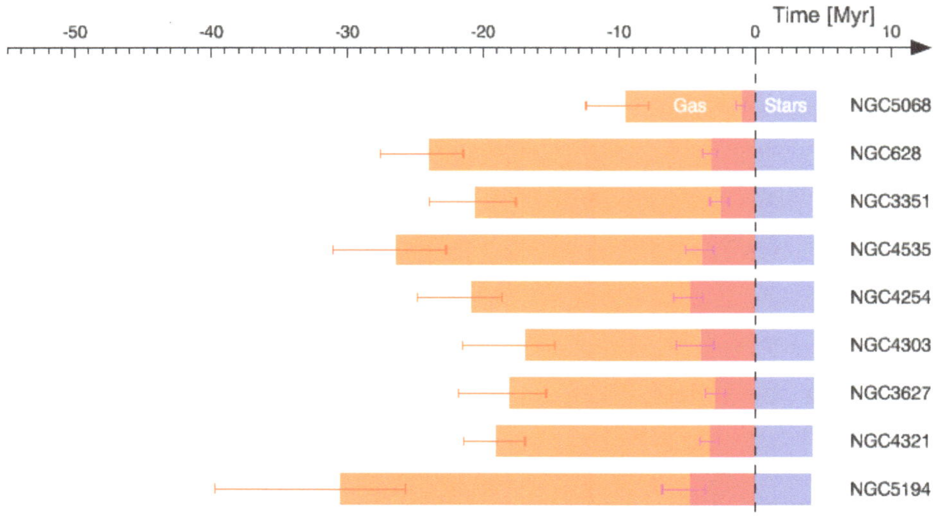

Figure 14.8. Evolutionary timeline of molecular clouds, including the gas (CO emission)-dominated stage (orange), the massive-star ($H\alpha$ emission)-dominated stage (purple), and the overlap (maroon) stage, during which stars and gas coexist. A median duration of the overlap (clearance) stage of ~3 Myr is observed. Reproduced from Chevance et al., (2020) with permission from Oxford University Press on behalf the Royal Astronomical Society.

14.3.7 The Size Scale of the Cloud Dispersal and the Net Reduction of the SFR by Feedback

Although in the previous sections we have discussed the cloud "dispersal", it is important to note that this is a relative term. As we have mentioned repeatedly in this book (Sections 1.2.1 and 14.2.2), the notion of a "cloud" is poorly constrained. Ultimately, most molecular clouds in a galaxy belong to large molecular cloud complexes, which can extend over several kiloparsecs. In grand-design spiral galaxies, the density-wave theory of stellar spiral arms proposes that the stellar spiral arms are density waves that sweep the Galaxy's disk, creating a potential well which the gas falls into, is compressed and shocked, and finally exits (Lin & Shu 1964; Roberts 1969). As the gas passes through a spiral arm, it forms large molecular complexes, in which stars form. Molecular clouds within these complexes are not isolated entities, but rather they are just somewhat denser regions along these huge kpc-long molecular complexes. Such large complexes cannot be fully destroyed by a local star formation episode, even if the latter is very strong (such as the "mini-starbursts" occurring in various regions of the Galaxy's spiral arms; e.g., Nguyen-Luong et al. 2016). Instead, the SF bursts simply carve holes and rearrange the gas within the complexes (Figure 14.9). From a Lagrangian point of view, a gas parcel passing through an arm most likely just cycles through a sequence of collapse and dispersal events until the parcel leaves the spiral arm. A particular star formation episode disperses the molecular gas *locally*, on scales of up to tens or hundreds of parsecs, but star formation continues elsewhere in the complex, either spontaneously due to the gas continuing to fall into the potential well, or by *triggering* induced by

Figure 14.9. Combined optical and mid-infrared image of the M74 galaxy, also known as the Phantom Galaxy, by the James Webb Space Telescope, showing the multiple cavities carved in the gaseous spiral arms by the stellar feedback. Nevertheless, the arms themselves are not destroyed by the feedback. Credit: ESA/Webb, NASA & CSA, J. Lee and the PHANGS-JWST Team.

the neighboring star formation episode. Indeed, recent observations suggest that several of the star-forming regions in the Solar neighborhood lie at the intersections of superbubbles produced by supernova explosions (Zucker et al. 2022).

On the other hand, outside the main spiral arms, or in more flocculent galaxies, molecular clouds may be truly relatively isolated, and surrounded by lower density atomic gas. In this case, a local SF event may be capable of entirely destroying or dispersing the cloud and completely halting itself.

In any case, the net effect of the feedback is still to strongly *reduce* the large-scale averaged SFR, although not by maintaining the clouds locally in near virial equilibrium, but rather by maintaining a *cycle* of gas assembly, collapse and dispersal, in which only the collapse phase forms stars, and with low efficiency, because it only manages to convert a small fraction (~10% as indicated by numerical simulations in general) of the collapsing gas mass into stars before the SF episode is interrupted by feedback. This solves the Zuckerman & Palmer (1974) SF conundrum (Section 9.1.1).

It is also important to note that pure pre-supernova feedback may be incapable of destroying the most massive clouds, even if these are isolated (e.g., Dale et al. 2012). In this case, the SF continues to accelerate, and the cloud continues to produce massive stars. The process continues until a *super star cluster* (Melnick et al. 1985) forms, in which the final disruption of the cloud is accomplished by *clustered supernovae*. A comprehensive review on the subject has been provided by Dale (2015). Nevertheless, photoionizing radiation and stellar winds appear to be indispensable for producing the observed kinematics of the star-forming ISM (e.g., Rathjen et al. 2021).

14.3.8 The Spatially Intermittent Nature of Star Formation and the Density of Clusters

An important feature of star clusters is that they appear to be significantly denser than gas clumps out of which they form (e.g., Section 3.4.2 and Figure 14 of Krumholz et al. 2019). As we will discuss now, this is a natural consequence of

1. the spatially intermittent nature of SF (Section 12.4),
2. a density profile near r^{-2} and a continuous accretion inflow onto the star-forming region, and
3. a near-exponentially increasing pace of the SFR.

Item (1) refers to the fact that the mathematical counterpart to the formation of a star (or, more properly, a protostar) corresponds to the appearance of a *singularity* in the solution of the collapse equations (Section 7.3.2). As discussed in that section, the appearance of a singularity marks the transition from the *prestellar* to the *protostellar* stage of collapse, therefore marking a specific moment in time. Moreover, it occurs at an isolated point in space. This is known as a *spatially intermittent* process, meaning that it occurs on a very small (infinitesimal) subset of space.

In practice, of course, stars have finite dimensions, and the regions in which they appear ("star-forming regions") also have finite spatial extents, although they still occupy a very small fraction of a cloud's volume. It is convenient, then, as a practical representation of a singularity, to define the "instantaneously-star-forming gas" within a cloud as a small subregion of it, in which the density is so high that its local free-fall time is much shorter than the evolutionary timescale of the clump. Specifically, we require this gas to have a free-fall time at least one order of magnitude smaller than that of the parent clump. For example, in reference to entire molecular clouds, whose mean densities are of the order of a few 10^2 cm^{-3}, and their free-fall times are a few to several 10^6 yr, the "instantaneously-star-forming gas" could be considered as that whose free-fall time is a few times 10^5 yr; i.e., with densities $\sim 10^4$ cm^{-3}. On the other hand, within cores of densities $\sim 10^4$ cm^{-3}, the instantaneously-star-forming gas would have densities $\sim 10^6$ cm^{-3}.

Item (2) above refers to the property that a nearly-stationary infall regime develops during the protostellar stage (i.e., after the formation of a singularity, or protostar) of core collapse (Section 7.3.2). This is due to the establishment of a nearly-stationary density profile, and therefore a nearly-stationary gravitational potential well, akin to a funnel, along which the gas is continuously infalling from the low- to the high-density regions, therefore causing an also nearly-stationary infall flow (Section 7.3.2.2; see also Murray & Chang 2015). As mentioned in that section, the strictly stationary density profile in the spherical solution has a logarithmic slope of -2 (i.e., $\rho \propto r^{-2}$), although in practice real cores tend to have somewhat shallower slopes (see the compilation by Gómez et al. 2021), and thus the profile is only *approximately* stationary.

Now, for a power-law density profile with logarithmic slope p, the mass increases with radius as r^{3-p}, and the free-fall (or the e-folding) time scales as $r^{p/2}$. Moreover, in Section 7.4.2.1 we mentioned that $p \to 2$ during the protostellar stage in the outer region of cores. The slope $p \sim 2$ of the density profile has the remarkable property that both M and $\tau_{\rm ff}$ scale linearly with r. Therefore, the condition that the instantaneously-star-forming gas must have a free-fall time much shorter (say N times) than that of the whole clump, implies that this gas has a mass N times smaller than the clump, and a density N^2 times larger. If the cluster inherits the density of the gas from which it directly formed, it must be typically much denser than the full extent of the clumps containing the star-forming regions. Again, this is a consequence of the strongly concentrated (spatially intermittent) nature of the star-forming gas process within molecular clouds and clumps.

Concerning item (3), in Section 12.3.2.4 we pointed out that, in numerical simulations, the local SFR (denoted $\dot{M}_*$) in a star-forming region accelerates at a nearly exponential rate before feedback interrupts the process. Therefore, we write, approximately,

$$\dot{M}_*(t) \equiv \frac{dM_*}{dt} = \frac{M_*(t)}{\tau} \approx \frac{M_0}{\tau} e^{t/\tau}, \tag{14.4}$$

where $M_*(t)$ is the instantaneous total mass in stars, t is the time measured since the onset of star formation, τ is the e-folding time, and M_0 is a reference mass, of the order of the stellar mass present at time $t = \tau$. For the Jeans instability (Section 6.1), the e-folding time is given by Equation (6.15), which for convenience we reproduce here:

$$\tau = \sqrt{\frac{\pi}{G\rho_0}}, \tag{14.5}$$

where ρ_0 is the mean density of the cloud or clump being considered. It is worth recalling that this timescale is longer than the free-fall time by a factor $\sqrt{32/3} \approx 3.27$ (cf. Equation (7.8)). We assume that this is also the relevant timescale for the growth of the SFR.

An important implication of the near-exponential time dependence of the SFR, Equation (14.4), is that the total stellar mass, M_*, also grows exponentially. In turn, this implies that *most of the total stellar mass at any given time after the onset of SF has been produced over the last e-folding time of evolution*, since for an exponential, the increment in the variable is proportional to its present value. Furthermore, we can assume that the mass converted to stars over the last e-folding time is the gas mass that was contained in the instantaneously-star-forming region one e-folding time ago. Therefore, for any generic clump, the *stellar* mass within it is of the order of the *gaseous* mass in the "instantaneously-star-forming" region. Finally, due to the scaling of the mass with radius, this mass is then $\sim N$ times smaller than the mass of the whole clump, and is confined to a region $\sim N$ times smaller than the clump.

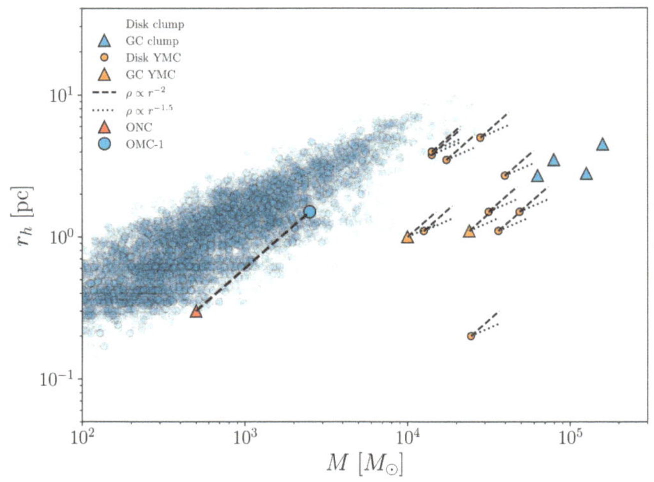

Figure 14.10. Size *versus* mass of molecular clumps in the Galactic disk from ATLASGAL (light blue circles), clumps near the Galactic center (blue triangles), young massive clusters in the disk (YMCs; orange circles), Galactic center YMCs (orange triangles), the Orion Nebula Cluster (ONC; red triangle) and its parent clump OMC-1 (blue circle). The black dashed and dotted lines represent the radius as a function of mass for density profiles $\rho \propto r^{-2}$ ($r \propto M$) and $\rho \propto r^{-3/2}$ ($r \propto M^{2/3}$), respectively. The black dashed line joining the ONC and OMC-1 has a slope of 1. Reproduced from Zamora-Aviles et al. (2024).

The spatially intermittent nature of the SF process explains the observation that clusters are systematically denser and more compact than the clumps in which they form. This would constitute a problem if stars formed throughout the entire volume of their parent clumps, since in this case one would expect to find clumps as dense and massive as the clusters (Krumholz et al. 2019). This is illustrated in Figure 14.10, which shows a size *versus* mass plot for a collection of young massive clusters and massive molecular clumps in the Milky Way. The dashed and dotted lines coming out from the orange circles and triangles indicate $M \propto r$ ($\rho \propto r^{-2}$) and $M \propto r^{3/2}$ ($\rho \propto r^{-3/2}$) scalings, respectively, and extend over factors ~1.5 in mass and radius, as drawn by Krumholz et al. (2019), under the assumption that SF occurs throughout the volume and mass of a clump, and implying that clumps are in general larger but less dense that the clusters.

However, the points corresponding to the Orion Nebula Cluster (ONC, red triangle) and its parent clump OMC-1 (blue circle) illustrate that, in fact, the cluster is much more compact than its parent clump, because it is forming only in the densest part of the latter. The dashed line shows that the ONC almost precisely connects with the OMC-1 clump along the $M \propto r$ line, but is a factor ~5 smaller and less massive than the cloud, and therefore a factor ~25 times denser. This suggests that, in fact, the dashed lines coming out from the YMCs should be extended to greater lengths (~10×) in order to correctly connect the massive clusters with their parent clumps. This is also consistent with the typical final SFE observed in star-forming clouds, ~10%, since the typical cluster mass is therefore ~10% of its parent clump's mass.

14.4 Triggered versus Primordial Peripheral Star Formation

An important suggested mechanism for inducing the formation of new star-forming regions is that of *triggered*, or *sequential* star formation, in which the expansion of an H II region or supernova shell compresses the surrounding gas, eventually inducing instability in the compressed gas, thus producing new star-forming regions in the compressed layers. This process is known as the *collect and collapse* mechanism (Elmegreen & Lada 1977), and has become accepted as one of the main mechanisms for producing new star-forming regions.

However, an important consequence of the hierarchical nature of the cloud formation and collapse model is that multiple local star-forming sites arise in the hub-filament systems generated by the anisotropic, hierarchical gravitational contraction in molecular clouds. Therefore, an important question is what fraction of the regions formed in the periphery of the main hubs were pre-existing, formed just by hierarchical collapse, and what fraction is actually induced to collapse by the expansion of the shell.

In the particular case of the simulations by González-Samaniego & Vazquez-Semadeni (2020) discussed in this chapter, it was found that, out of a total of six local groups appearing in the simulation with feedback (right panel of Figure 14.6), five also appear in the simulation without feedback (left panel). Indeed, the group denoted by green dots in the right panel is missing in the left panel. This implies that, for the setup of these simulations, $\sim$80% of the star forming sites were pre-existing, and only $\sim$20% was really produced by the compression caused by the passing shock. To the time of this writing, a systematic statistical study remains to be performed, to quantify the relative effectiveness of the triggered star-formation scenario, in comparison with that of the primordial SF.

References

Chevance, M., Kruijssen, J. M. D., Hygate, A. P. S., et al. 2020, MNRAS, 493, 2872
Chevance, M., Kruijssen, J. M. D., Krumholz, M. R., et al. 2022, MNRAS, 509, 272
Colín, P., Vázquez-Semadeni, E., & Gómez, G. C. 2013, MNRAS, 435, 1701
Dale, J. E. 2015, NewAR, 68, 1
Dale, J. E., Ercolano, B., & Bonnell, I. A. 2012, MNRAS, 424, 377
Dame, T. M., Hartmann, D., & Thaddeus, P. 2001, ApJ, 547, 792
Diaz-Miller, R. I., Franco, J., & Shore, S. N. 1998, ApJ, 501, 192
Elmegreen, B. G., & Lada, C. J. 1977, ApJ, 214, 725
Gómez, G. C., & Vázquez-Semadeni, E. 2014, ApJ, 791, 124
Gómez, G. C., Vázquez-Semadeni, E., & Palau, A. 2021, MNRAS, 502, 4963
González-Samaniego, A., & Vazquez-Semadeni, E. 2020, MNRAS, 499, 668
Hacar, A., Alves, J., Forbrich, J., et al. 2016, A&A, 589, A80
Hacar, A., Alves, J., Tafalla, M., & Goicoechea, J. R. 2017, A&A, 602, L2
Haid, S., Walch, S., Seifried, D., et al. 2019, MNRAS, 482, 4062
Kruijssen, J. M. D., & Longmore, S. N. 2014, MNRAS, 439, 3239
Krumholz, M. R., McKee, C. F., & Bland-Hawthorn, J. 2019, ARA&A, 57, 227
Lada, C. J., & Lada, E. A. 2003, ARA&A, 41, 57

Lin, C. C., & Shu, F. H. 1964, ApJ, 140, 646
Longmore, S. N., Kruijssen, J. M. D., Bastian, N., et al. 2014, Protostars and Planets VI, ed. H. Beuther, R. S. Klessen, C. P. Dullemond, & T. Henning (Tucson, AZ: Univ. Arizona Press) 291
Megeath, S. T., Gutermuth, R., Muzerolle, J., et al. 2012, AJ, 144, 192
Melnick, J., Moles, M., & Terlevich, R. 1985, A&A, 149, L24
Murray, N., & Chang, P. 2015, ApJ, 804, 44
Nguyen-Luong, Q., Nguyen, H. V. V., Motte, F., et al. 2016, ApJ, 833, 23
Portegies Zwart, S. F., McMillan, S. L. W., & Gieles, M. 2010, ARA&A, 48, 431
Rathjen, T.-E., Naab, T., Girichidis, P., et al. 2021, MNRAS, 504, 1039
Roberts, W. W. 1969, ApJ, 158, 123
Tobin, J. J., Hartmann, L., Furesz, G., Mateo, M., & Megeath, S. T. 2009, ApJ, 697, 1103
Vázquez-Semadeni, E., Ryu, D., Passot, T., González, R. F., & Gazol, A. 2006, ApJ, 643, 245
Vázquez-Semadeni, E., Gómez, G. C., Jappsen, A. K., Ballesteros-Paredes, J., & Klessen, R. S. 2009, ApJ, 707, 1023
Vázquez-Semadeni, E., González-Samaniego, A., & Colín, P. 2017, MNRAS, 467, 1313
Vázquez-Semadeni, E., Zamora-Avilés, M., Galván-Madrid, R., & Forbrich, J. 2018, MNRAS, 479, 3254
Vázquez-Semadeni, E., Gómez, G. C., & González-Samaniego, A. 2024, MNRAS, 530, 3445
Zamora-Avilés, M., Camacho, V., Ballesteros-Paredes, J., et al. 2024, arXiv e-prints, arXiv:2409.11588
Zucker, C., Goodman, A. A., Alves, J., et al. 2022, Natur, 601, 334
Zuckerman, B., & Palmer, P. 1974, ARA&A, 12, 279

Enrique Vázquez-Semadeni

Chapter 15

Conclusions: The Galactic Flow

15.1 A Change in Perspective

In this book we have attempted to cover some of the essential theoretical foundations underlying our present understanding of ISM hydro- and thermodynamics, and how they enter our present understanding of the SF process. Most of the theoretical foundations presented here are not new, but rather are part of the standard training of an astrophysicist. However, this standard training relies heavily in a "classical" perspective dating back to the mid-20th century, which was based on simplifying assumptions such as equilibrium, stationarity, similarity, and objects in isolation.

However, during the last quarter of the 20th century, a "numerical revolution" took place, which revealed a rich new world of complexity, nonlinearity, chaos, and, importantly, the truly *fluid* nature of the ISM. Therefore, in this book we have attempted to present the theoretical foundations from a different perspective which, while preserving (and some times recovering) the fundamental results from the early analytical studies, takes them as the starting point of the out-of-equilibrium, transient, and fluid-like behavior apparent in numerical simulation. This perspective involves the notions of continuous flows across scales and therefore across the "boundaries" of the various structures (clouds, clumps, filaments, etc., all the way down to stars), which, in the context of a continuum are necessarily defined quite arbitrarily, often based on instrument sensitivity or tracer excitation, rather than on actual structural boundaries.

Similarly, in regards to the theory of star formation, as derived from the physical properties and thermo- and magneto-hydrodynamics of the ISM, the theory has evolved from an often static, equilibrium perspective, to an evolutionary one. In this book we have attempted to provide a comprehensive view of the entire process of molecular cloud evolution, from their formation from the WNM through the onset of star formation, the erosion and destruction of the local star-forming site, and the

doi:10.1088/2514-3433/adcb28ch15

resulting stellar clusters and associations. Below we summarize the main highlights of this change of perspective.

15.1.1 The Theory Highlights

1. Concerning *turbulence*:
    - (a) *Turbulent pressure is **not** isotropic.* Thermal pressure is isotropic because the mean free path of the atoms and molecules is enormously smaller than the size of the clouds, and therefore all directions of molecular motion are present within any gas parcel in the flow, thus producing an isotropic pressure. Instead, turbulence contains most of its kinetic energy in motions at the largest scales, of sizes comparable to the size of the clouds themselves. Therefore, these motions occur in a few well-defined directions, and thus they cause a *distortion* of the clouds due to the ram pressure, rather than providing isotropic support that can keep the clouds in equilibrium.
    - (b) *Turbulence is not the only contribution to the linewidth.* The non-thermal component of molecular linewidths in general corresponds not only to the turbulent motions, but also to any other systematic motions, in particular gravitational contraction. Moreover, since the virial parameter is generally derived from the linewidths, this consideration is especially important when considering the evolution of the virial parameter during the gravitational contraction of clouds and cores.
2. Concerning *thermal instability (TI)*:
    - (a) *There is a constant exchange of material among the WNM and the CNM*. In particular, the equality of the thermal pressure in the WNM and the CNM only reflects the variation of the cooling rate with density and temperature of the gas as it flows, and does not imply any form of confinement. Instead, there is a constant mass flux between the phases, often mediated by *phase transition fronts*. In the presence of strong turbulence, such that the turbulent crossing time across a given spatial scale is shorter than the cooling time, cooling gas out of thermal equilibrium appears across such scale, constituting the so-called *lukewarm, unstable phase* (UNM), which is typically *flowing* between thermally-stable regions.
    - (b) *The density PDF of a thermally bistable medium* in the isobaric mode ranges from having a sharply bimodal shape in the quasistatic case (the modes corresponding to the WNM and CNM) to a single-mode PDF with a power-law tail in the case of strong turbulence. The transition is controlled by the ratio of the turbulent crossing time across the scale of interest (which decreases with increasing turbulent velocity dispersion) to the cooling time.

3. Concerning the *Virial Theorem*:
   (a) *Difference between the Virial Theorem and virial equilibrium.* Although the distinction should be obvious, the Virial Theorem (VT) is often confused with the *condition* of virial equilibrium. The scalar VT is a fundamental, time-dependent, fluid dynamics equation, which is satisfied at all times by a finite-volume fluid parcel, and which expresses the conservation of the parcel's kinetic energy. It is analogous to the other conservation equations of gas dynamics, such as the continuity, momentum, and internal energy equations (Section 2.5), except for the fact that it holds for a finite, rather than an infinitesimal, volume.

   Instead, virial *equilibrium* is a specific condition frequently *imposed* on the VT, which amounts to setting the time derivative term (d^2I/dt^2) in the Lagrangian version of the VT equal to zero.[1] This is a special condition, which is satisfied only in special cases. Moreover, the well-known condition $2E_{\rm k} = |E_{\rm g}|$ does *not* necessarily imply equilibrium, as it can arise also in fully dynamic (time-dependent) situations, in which other terms of the VT are non-zero.
   (b) The most widely used form of the VT is the *Lagrangian version*, which considers always the same set of fluid particles. However, in practice, this does *not* correspond to "clouds" as we generally define them, since this definition typically involves some form of boundary that does *not* move with the flow, but is rather based on some density, column density, or intensity threshold. Therefore, mass, momentum, and energy fluxes across these boundaries are expected, and must be taken into account.
   (c) *The mass-to-magnetic flux ratio (MFR) in a continuous medium is a boundary condition, not a fundamental property of a gas clump.* When considering a clump (a density fluctuation) immersed in a continuous background, the mass and the magnetic flux in general do not scale equally with radial distance from the center. Therefore, the value of the MFR depends on the choice of boundaries. Considering that the medium can extend to arbitrarily large distances, any clump is part of a magnetically supercritical medium. However, the relevant instantaneous MFR is given by the mass of the *dense gas* that dominates the gravitational potential. This mass may increase due to accretion, making the MFR a physical *variable* of the clump, rather than a fixed *parameter*.
4. Concerning *hydrostatic equilibrium*:
   (a) *The singular isothermal sphere (SIS)* ***cannot*** *be the initial condition for gravitational collapse.* The SIS is the most unstable hydrostatic spherical equilibrium configuration, with a singular (diverging)

[1] In the Eulerian version, it corresponds to setting *all* time derivatives to zero.

density at its center, and a wide range of unstable equilibrium solutions exists between the SIS and the first unstable BE sphere. Spherical collapse from hydrostatic conditions must start from the *first* unstable Bonnor–Ebert sphere, for which the central density is only $\sim$ 14 times the density at the boundary. Therefore, by the time a singularity develops at the center, the collapse has already become fully dynamic.

(b) *The inner, flat-density region of a Bonnor–Ebert sphere has a size of the order of the local Jeans length.* Since, in spherical geometry, the gravitational force at every radius is due only to the mass inside of it, self-gravity is negligible in the innermost parts of a hydrostatic sphere, which therefore behaves as a pressure-confined sphere. Since gravity becomes important at scales of the order of a Jeans length, the inner flat part of the BE sphere has a diameter of L_J.

5. Concerning *gravitational contraction*:

(a) *Spherical collapse flow from non-singular initial conditions occurs from the outside-in.* Since the Jeans length L_J is the size of the smallest region that can collapse in a given medium, when a marginally unstable region begins to collapse, it does so *as a whole*. Its internal parts, which are smaller than L_J, are dominated by the balance between the internal and external pressures, and thus cannot collapse on their own. Instead, these inner regions simply undergo *compression* by the outer parts, which are falling onto them. Therefore, non-singular initial conditions undergo collapse *from the outside-in*. The only hydrostatic configuration that can undergo inside-out collapse is the SIS, since in this case, self-gravity remains important all the way to the center, and so it can undergo collapse at all scales. However, the SIS is unrealizable in practice (item 4a above).

(b) *Spherical prestellar gravitational contraction develops a very similar density profile to that of non-singular hydrostatic equilibrium.* For nearly half a century, it has been thought that the Bonnor–Ebert-like density profile that develops during spherical collapse requires the establishment of detailed pressure balance in spherical geometry, and thus requires subsonic flow. In turn, this notion leads to the interpretation that Larson-Penston flow should be an exception, rather than the norm. However, in the last decade several analytical studies have shown that this type of flow is the natural result of gravitational contraction driven by the mass internal to each radius (always under spherical symmetry), because in this case the r^{-2} outer profile is an *attractor*, meaning that spherical collapse flow *approaches* this profile, regardless of the initial conditions.

(c) *Gravitational contraction in isothermal media increases the number of Jeans masses within a Lagrangian parcel.* As a polytropic

Lagrangian gas parcel with $\gamma_e < 4/3$ (including the isothermal case, for which $\gamma_e = 1$) contracts gravitationally, its Jeans length shrinks faster than its own radius, and therefore becomes ever more prone to undergo Hoyle's gravitational fragmentation. At any moment in time, the fragmentation ends at the scale of the Jeans length at that time. The sequence may then repeat as the Jeans length continues to decrease.

(d) *Gravitational contraction starts out extremely slowly.* This is due to mainly two reasons. First, even pressureless contraction proceeds in an extremely nonlinear manner. For example, after 50% of a free-fall time, the radius of a spherical collapsing gas parcel has decreased only by ~16%. Second, in the presence of thermal support, a gas parcel starting its collapse from just over the Jeans mass is very near equilibrium with thermal pressure support. Therefore, the net force driving the gravitational contraction is very small. However, as the parcel contracts and its Jeans mass decreases, the collapse begins to approach the free-fall rate.

(e) *Nothing collapses on one free-fall time.* As a consequence of the previous item, when an object becomes Jeans-unstable and starts to collapse, it is generally just above marginal stability, and so the net force compressing it is very small, making the contraction timescale significantly longer than the free-fall time, which applies to pressureless collapse. Furthermore, for filamentary structures formed by either turbulent compressions or anisotropic gravitational fragmentation, the longitudinal collapse timescale is larger than the standard spherical free-fall time by factors of the order of the aspect ratio or its square root, depending on whether the collapse proceeds edge-on or homologously. Therefore, the free-fall time should be considered a *lower limit* to the actual collapse timescale.

(f) *In Hoyle's gravitational fragmentation, small-scale collapses must start* ***later*** *than the large-scale collapse*, as they can only start when the global Jeans mass has decreased to a value of the order of the density fluctuations' masses.

(g) *Pressureless gravitational contraction tends to drift* ***away*** *from sphericity and to amplify anisotropies.* A pressureless collapsing triaxial ellipsoid contracts fastest along its shortest dimension. Therefore, the gravitational contraction sequentially produces sheets, filaments, and finally roundish cores.

(h) *Isothermal gravitational fragmentation produces a hierarchy of collapses within collapses.* Because small-scale collapses start later than the large-scale ones, they are part of a large-scale accretion flow oriented toward the large-scale collapse center. This generates a *conveyor-belt* type of flow.

15.1.2 The Cloud and SFR Evolution Highlights

Concerning the full evolutionary process of molecular clouds, from their first appearance from the warm diffuse medium to their destruction by stellar feedback, the main highlights are:

1. *The definition of a "cloud" is mostly a matter of convention.* Clouds and their substructures in the ISM are parts of a continuum, often highlighted by some practical, rather than physical, property, such as being above a density or chemical abundance threshold defined by sensitivity or simply convention. As such, clouds are not isolated and are continuously exchanging material with their surroundings.
2. *New molecular clouds start their lives as cold atomic clouds.* Generic compressions in the WNM can nonlinearly induce a phase transition to the CNM phase. These compressions can be due either to *inertial* motions, such as the passage of SN shells or generic turbulent motions, or to *gravitationally-driven* motions, such as those produced by the gravitational potential well of a stellar spiral arm.
3. *The Jeans mass drops precipitously upon the phase transition from the WNM to the CNM.* Being proportional to $\rho^{-1/2}T^{3/2}$, the upwards jump of the density by a factor $\sim 100\times$ and the corresponding downward jump in temperature imply a drop of the Jeans mass by a factor of $\sim 10^4$. This is a likely cause for the observation that molecular clouds contain hundreds or thousands of Jeans masses.
4. *Continuing accretion onto the CNM clouds causes them to roughly simultaneously* (at column densities $N \sim 10^{21}\,\mathrm{cm}^{-2}$ for solar neighborhood conditions):
   (a) become dominated by self-gravity;
   (b) become magnetically supercritical;
   (c) become molecular, and
   (d) acquire *moderately* ($\mathcal{M}_s \sim$ a few) supersonic turbulence.
5. *The large number of Jeans masses in the clouds makes them nearly pressureless,* and thus they can undergo anisotropic, hierarchical gravitational contraction, which forms filaments, hubs, and cores. The contraction generates a multi-stage accretion flow, from clouds onto filaments, and from filaments onto hubs and cores.
6. *The main role of turbulence may be just to create a population of moderate density fluctuations.* These fluctuations can eventually become local collapsing regions, after the large-scale contraction has reduced the global Jeans mass to the value of the fluctuations' masses.
7. *Two physical parameters* (such as mass-size, mass-density, or density-size pairs) *are needed to completely specify the physical state of a cloud or clump,* since the gravitational potential depends on both the amount of mass and the size of the region over which it is distributed.
8. *Observations indicate that the most massive objects in a given density range within molecular clouds are the most strongly gravitationally bound,* even

considering the kinetic energy in nonthermal motions. This applies in particular to full molecular clouds themselves, the most massive of which (which contain most of the molecular mass in the Galaxy) have $\alpha_{\rm vir} \sim 1$ and are thus bound at the largest scales.

9. *Stars begin forming in the cores and hubs within the filaments, and the system's SFR increases.* The continued accretion causes the hubs and cores to grow in mass, and therefore to increase their SFR. The stars inherit the spatial and kinematic distribution of the gas.
10. *As a region's mass and SFR increase, so does the maximum mass of the stars formed in the region.* Eventually, sufficiently massive stars begin to form that can disrupt the accretion flow, and stop the core/hub growth. The number of massive stars and the form of feedback (ionizing radiation, winds, SN explosions, etc.) required to stop the accretion flow and disperse the cloud depends on the total mass of collapsing gas. This terminates the local SF episode.
11. *The stars lose the gravitational potential of the gas and begin to locally expand.* The expansion, however, occurs from the multiple, hierarchical sites where the cores and hubs were located, generating a complex kinematic distribution consisting of expansion from multiple centers, mixing the stellar products from the various subgroups, and with only moderate overall expansion.
12. *The solution to the Zuckerman & Palmer (1974) "star formation conundrum" is that the local SF episodes are rapidly interrupted by the feedback from massive stars.* Low-to-moderate-mass molecular clouds ($M \lesssim 10^5 M_\odot$) may be dispersed by the photoionizing feedback from the first few massive stars formed in the clouds. More massive clouds ($M \gtrsim 10^5 M_\odot$) may require the feedback from clustered supernovae.

15.2 Useful Analogies

15.2.1 ISM Dynamics as the Galactic Weather

In this book we have presented, first, the basic concepts of thermo- and hydrodynamics of the ISM, as well as of hydrostatic equilibrium and gravitational collapse, that are relevant to the process of star formation (Part I). In Part II, we then presented a comprehensive view of the formation and evolution of molecular clouds and their star formation activity, from the formation of the clouds to their destruction by their own stellar products.

One particular goal of this monograph has been to emphasize the *fluid* nature of the ISM, in the sense that the processes discussed in Part I generate *flows* in the gas, akin to the flows of the ocean and atmosphere on Earth, which carry material from one location to another. In the ISM, these flows generate structures like clouds, clumps, and filaments. These should be regarded as *regions* (or *sections*) in the flow through which the gas *transits*, rather than as coherent *objects*, made up of the same gas parcels at all times, and moving as a physical unit.

In this sense, the dynamics of the ISM and the star formation resulting from it can be thought of as constituting the *Galactic weather*, in which the interstellar gas forms clouds, out which tiny condensations (the stars) form collectively, akin to water droplets forming in atmospheric clouds on Earth.

15.2.2 Filament Formation as the Galactic Hydrology

In turn, the intermediate step of the formation of hub-filament systems within molecular clouds can be likened to *hydrology*, the study of the motions and accumulation of water on the surface of the Earth. In this sense, rain water on Earth that lands on the mountain tops (regions of shallow gravitational potential, analogous to the lower-density regions in molecular clouds), slides down the gravitational potential to regions of somewhat deeper potential (the rivers, analogous to the filaments in the clouds), until it finally reaches the sites of deepest potential (the lakes and oceans, analogous to the cores and hubs in the clouds).

Both this and the previous analogies deeply involve the fluid nature of the motions, in contrast to the more ballistic, object-like approach frequently used in astronomy to describe interstellar objects, such as cloud-cloud collisions, clouds in equilibrium, wall-like pressure confinement, and *in situ* gravitational collapse.

15.3 The Galactic Mass Cascade

One fundamental aspect of the continuum and flow-like nature of the ISM and the circumgalactic medium (CGM) in general, is the possible existence of a continuous flow of material starting perhaps already at the CGM, and continuing all the way down to individual stars, as suggested by the *bathtub model* for the evolution of the SFR in galaxies across cosmic time (e.g., Bouché et al. 2010). In this model, the ultimate regulator of the SFR in galaxies is the gas accretion rate onto them. A scaled-down version of the bathtub model for molecular clouds has been presented by Burkert (2017), in which a cloud's SFR is regulated by the accretion rate onto it, rather than by its mass or collapse timescale. In this picture, all dense structures in the ISM (clouds, clumps, filaments, cores, fragments, protostellar disks, and stars) accrete from their parent structures. This is also the assumption of the Global Hierarchical Collapse model of molecular clouds (Section 9.3)

This continuous mass flow from large (diffuse) to small (dense) scales in a self-gravitating medium can be thought of as the mass analog of the turbulent energy cascade (Sections 3.3.1 and 3.5) in a turbulent medium, in which the energy is injected at large scales and dissipated at spatially intermittent locations (sites of large velocity gradient, where viscous dissipation is largest). Similarly, in the gravity-driven Galactic and ISM mass cascades, mass is accreted from the large scales into smaller-scale structures, and eventually forms spatially intermittent *singularities* (the stars).

Another similarity between the the turbulent energy cascade and the gravitational mass cascade is that, in the former, a stationary regime is established in which the energy injected at large scales is dissipated at the small scales by viscosity. Similarly, a near stationary regime may be established in the mass cascade, in which the mass accreted by the system at the large scales is *consumed* (rather than dissipated) by a collection of singularities (stars), onto which the mass is accreted at the final step of the cascade. Indeed, as discussed in Section 7.4.2.2, a stationary mass transfer rate across scales can be established for a density profile of the form $\rho(r) \propto r^{-2}$. Therefore, this density profile can be considered as the mass-cascade analogue of the stationary Kolmogorov $k^{-5/3}$ spectrum.

One important difference, however, is that, while in the turbulent energy cascade the energy is truly dissipated, and therefore removed from the system, the mass in the mass cascade accumulates in the collapsed objects, and continues to increase the gravitational pull toward the collapsed objects. This implies that the r^{-2} density profile can only remain stationary over timescales short compared to the growth timescale of the stellar mass at the centers of the collapse. Another feature exclusive of the mass cascade is the appearance of stars massive enough that their feedback can disrupt the accretion flow. Therefore, it appears that the gravitational mass cascade is less durable than the turbulent energy cascade. Nevertheless, there are clear similarities between the two that deserve further investigation.

15.4 The Road Ahead

Although, as described in this book, significant progress has been made in interpreting and applying the basic theoretical concepts, as well as in disentangling the evolutionary cycle of molecular clouds and their star formation activity, several questions remain open at the time of this writing. Some of them are:

1. How does the relative importance of turbulent and infall motions vary along the sequence of densities and scales from the CNM down to protostellar objects?
2. Can the gravity-driven accretion flow be extended up to kiloparsec scales, and how does it connect with galactic dynamics?
3. Is there a qualitative distinction between the formation mechanism of bound stellar clusters and unbound stellar associations, or is it just a matter of the total amount of mass involved at the largest scales of the collapsing cloud?
4. How does the multi-stage accretion flow drive the chemistry along the hierarchy of structures?
5. Is the magnetic field passively advected in general, or are there instances when it takes a dominant role?
6. Can a mass cascade theory analogous to the turbulent energy cascade be constructed?

As we continue to dig into the ISM dynamics and the star formation process, these questions may be resolved, while many other questions will undoubtedly appear.

References

Bouché, N., Dekel, A., Genzel, R., et al. 2010, ApJ, 718, 1001
Burkert, A. 2017, MmSAI, 88, 533
Zuckerman, B., & Palmer, P. 1974, ARA&A, 12, 279

AAS | IOP Astronomy

Interstellar Flow and Star Formation

Enrique Vázquez-Semadeni

Appendix A

Tensor Notation

In Chapter 2 extensive use is made of *tensor notation*. For reference, here we quickly summarize the conventions on which it is based:

- *Scalars* are written without subindices. For example, the density ρ.
- *Vectors* and *tensors* are written with one or more subindices; e.g., the velocity vector $u_i \equiv \boldsymbol{u}$ and the stress tensor $\sigma_{ij} \equiv \boldsymbol{\sigma}$. A *free* subindex (i.e., one that is not constrained by the summation convention) denotes any component of the vector or tensor.
- *Einstein's summation convention:* Repeated indices denote a summation over them. For example:

$$u_i u_i \equiv \sum_{i=1}^{3} u_i u_i = \boldsymbol{u} \cdot \boldsymbol{u} = u_x^2 + u_y^2 + u_z^2 = u^2.$$

- A vector equation must have a free subindex in each term, and represents one component of the vector (e.g., the momentum conservation equation, Equation (2.26)).
- In a vector equation, all indices other than the free index must be *dummy*: i.e., they must be summed over through the summation convention.
- The divergence of a vector is written as

$$\nabla \cdot \boldsymbol{u} = \frac{\partial u_i}{\partial x_i},$$

doi:10.1088/2514-3433/adcb28ch16

while the curl is written as

$$\nabla \times \boldsymbol{u} = \varepsilon_{ijk} \frac{\partial u_i}{\partial x_j},$$

where ε_{ijk} is the *totally antisymmetric Levi-Civita tensor*, defined as

$$\varepsilon_{ijk} \equiv \begin{cases} 0 & \text{if any of } i, j \text{ and } k \text{ are equal;} \\ 1 & \text{if } i, j \text{ and } k \text{ are a cyclic permutation of 1,2,3;} \\ -1 & \text{if } i, j \text{ and } k \text{ are not a cyclic permutation of 1,2,3.} \end{cases}$$

www.ingramcontent.com/pod-product-compliance
Ingram Content Group UK Ltd.
Pitfield, Milton Keynes, MK11 3LW, UK
UKHW061829100926
8556IPUK00011B/928
* 9 7 8 0 7 5 0 3 1 2 9 1 2 *